AF352800

Coatings for Harsh Environments

Coatings for Harsh Environments

Special Issue Editor

Shiladitya Paul

MDPI • Basel • Beijing • Wuhan • Barcelona • Belgrade • Manchester • Tokyo • Cluj • Tianjin

Special Issue Editor
Shiladitya Paul
TWI
UK
University of Leicester
UK.

Editorial Office
MDPI
St. Alban-Anlage 66
4052 Basel, Switzerland

This is a reprint of articles from the Special Issue published online in the open access journal *Coatings* (ISSN 2079-6412) (available at: https://www.mdpi.com/journal/coatings/special_issues/coat_harsh_environ).

For citation purposes, cite each article independently as indicated on the article page online and as indicated below:

LastName, A.A.; LastName, B.B.; LastName, C.C. Article Title. *Journal Name* **Year**, *Article Number*, Page Range.

ISBN 978-3-03936-174-8 (Hbk)
ISBN 978-3-03936-175-5 (PDF)

Contents

About the Special Issue Editor

Shiladitya Paul Director, Materials Innovation Centre, University of Leicester and Materials R&D Manager, TWI. He is an innovative technology specialist, with experience in the industrial development and application of specialised coatings, materials and corrosion mitigation methods. He specialises in coatings and corrosion. His chapters in the Encyclopaedia of Aerospace Engineering (Wiley, 2010) and the ASM Handbook (ASM International, 2013) are widely considered as authoritative reference works, providing critical scientific concepts and their application to engineering practice. He has authored over 80 papers/articles on materials, coatings and corrosion. He is active in the scientific community, as a peer-reviewer and member of the editorial team of several journals. He is also active in several international scientific and industrial committees (Vice Chair of the European Federation of Corrosion WP9, member of several NACE committees). He is a member (judge) of the International Thermal Spray Conference (ITSC) and the Journal of Thermal Spray Technology (JTST) best paper award committees. More information on Shiladitya's research can be found on his ORCID Profile: https://orcid.org/0000-0002-8423-313X.

Editorial

Special Issue: "Coatings for Harsh Environments"

Shiladitya Paul [1,2]

[1] School of Engineering, University of Leicester, University Road, Leicester LE1 7RH, UK

[2] TWI, Granta Park, Cambridge CB21 6AL, UK
    shiladitya.paul@twi.co.uk

Received: 14 April 2020; Accepted: 17 April 2020; Published: 20 April 2020

Abstract: The operation of numerous safety-critical components in industries around the world relies on protective coatings. These coatings often allow process equipment to be purposeful in environments well beyond the operational limit of the uncoated components. Durability, ease of application, repairability, reliability and long-term performance of such coatings are vital to their application. Therefore, this Special Issue of Coatings, "Coatings for Harsh Environments", is devoted to research and review articles on the metallic, non-metallic and composite coatings used in aggressive environments.

Keywords: corrosion; thermal spray; testing; geothermal; high temperature; hot corrosion; oxidation; crosion; wear

Preface

Humans have taken advantage of materials since prehistorical times, so much so that human civilisation has often been categorised into three archaeological periods: the Stone Age, the Bronze Age and the Iron Age [1]. Throughout much of human history, materials played a part in progressing human civilisation, and coatings developed slowly from artistry to technology. The initial developments in coatings were primarily decorative in nature. However, decorative values, which prevailed well past the Renaissance, gave rise to the protective function of coatings [2].

The technology associated with the development of coatings emerged after the Industrial Revolution. Further advancement was heralded by the need for technological developments during the World Wars, truly following the English saying, "necessity is the mother of invention". Not only did coatings fulfil a decorative function but they also allowed engineering to progress to a new level. Humans were able to use materials in extreme environments never before envisaged to be accessible by developing "engineered" coatings. These extreme environments are often called "harsh" environments.

Harsh environments cover areas such as high- and low-temperature, (bio)-chemical and mechanical disturbances (including extreme stresses and stress cycles), electromagnetic noise, pressure, radiation, or vacuum [3]. These environments by their very nature challenge the ability of materials to function. The complexity of components increasingly used for engineering applications mean that a wide variety of coating materials and methods are required for different coating applications. The purpose of these newly developed coatings is not only to protect the component exposed to the harsh environments but also to provide additional functionality where possible. These engineered coatings are able to decrease the production costs compared to other monolithic components. In addition, some of them are also able to reduce the running cost of plants in harsh environments by allowing component repair.

A high-temperature application offers harsh environments where conventional materials often reach the limit of their operability. In such environments, coatings offer affordability and functionality. A few examples of such environments include the gas turbines and the boilers for power generation. The environments in both systems include high temperatures, but the jet engines experience extreme temperatures, thermal cycles and stresses. The thermal barrier coatings offer thermal protection that allows the turbine blades to operate without excessive creep or melting. The coatings not only need to offer thermal protection but also operate under severe stress without cracking [4]. In addition to ceramic thermal barrier coatings, cermet coatings are also being developed to endure hot corrosion. These coatings when sealed by a high-energy laser offer better corrosion resistance than unsealed coatings [5].

In boiler services, high temperature corrosion and oxidation is a major issue and this is mitigated by using corrosion resistant coatings. Materials such as $MoSi_2$ allow alloys to be used in high-temperature oxidizing environments [6]. Novel coatings of Cr–N and Cr (N,O) are beginning to emerge that allow conventional steels such as 304 to be used in challenging environments [7]. These coatings are also being developed for applications where a high temperature exists in combination with wear and high stresses [8]. The combination of corrosion and wear at high temperatures makes the selection of coatings even more challenging. New concepts involving thermal diffusion coatings are being developed for boilers. Extensive corrosion testing of boronised, aluminised and chromised systems are being carried out to assess their applicability [9].

With the development of new materials such as high-entropy alloys (HEA), nanocrystalline and compositionally complex materials, new test methods need to be developed to assess these materials. If these materials need to be applied as coatings then coating methods need to be adopted to accommodate such materials' chemistries and properties. The phase evolution and deposition parameters need to be assessed. These will govern the performance of such coatings in service. However, prior to their adoption by the industry, testing needs to be carried out. The application of these materials requires corrosion testing, and the correlation between performance and microstructure needs to be established.

Recent developments in the above area have been encouraging. Arc-sprayed Fe-based amorphous/nanocrystalline materials have been developed and tested for cavitation erosion resistance [10]. Electro-spark deposition of CoCrFeNiMo HEA has been developed and the microstructures of the developed coatings have been studied [11]. Other compositionally complex alloys such as CoCrFeMo0.85Ni have been deposited by a high-velocity oxy-fuel (HVOF) spray. Their corrosion resistance in an aqueous chloride environment has also been studied [12]. These developments are in their early stages. Further improvements are required prior to the industrial uptake of these materials.

Harsh environments are also encountered by onshore and offshore structures. The combination of groundwater and vibrations for onshore, and seawater, sunlight, waves and tide for offshore have deleterious effects on structural steels. To mitigate corrosion of steel structures and pipelines, several coating systems are being developed. The application depends to some extent on the location of the part to be protected (or the environment). Broadly, two main coating concepts are adopted for corrosion mitigation. First, the use of dielectric barrier layers in conjunction with cathodic protection. Second, the use of sacrificial coatings which act as uniformly distributed anodes. Enamel coatings fall in the first category. These coatings offer protection in conjunction with cathodic protection. However, when damaged the efficacy might be compromised. Evaluation of these coatings when damaged offers new insights into the behaviour of these coatings [13].

Sacrificial coatings offer protection to steel structures by being preferentially consumed, or in other words by providing cathodic protection. In offshore structures and bridges, the choice of coating material depends on the location or the exposure environment. In the top side of the structure, thermally sprayed zinc (and zinc aluminium) alloy coatings in conjunction with a topcoat have a long history of adequate performance, particularly in the Norwegian bridge sector [14]. However, for the submerged zone and the splash zone, thermally sprayed aluminium (TSA) is often considered beneficial. The performance of TSA is dependent on several factors such as the alloy type, coating production method, surface preparation, coating thickness and sealant selection, among others. A review by Syrek-Gerstenkorn et al. [15] covers the area of TSA in detail.

Much of the work presented in this Special Issue focusses on thermal spray coatings. However, there are several other coating methods which are being explored by researchers around the world [16–23]. One such method is organic surface treatment. These treatments have been traditionally carried out using chromates. However, with the REACH regulation coming into force, there has been a concerted effort in finding alternatives to chrome treatments [24]. Treatment of copper by a non-chromate alternative has been found to offer an environmentally friendly technology for the electronics industry, where copper is indispensable [25].

Research work such as the above offers alternatives where the conventional technologies are unsustainable. Similar studies on coatings and surface modifications are ongoing where the preference to ensure the health and safety of fellow humans and the protection of the natural environment has overtaken the drive for quick financial gains. It is imperative that research continues to ensure improvement in the quality of human lives, but with a compassion for nature.

Conflicts of Interest: The author declares no conflict of interest.

References

1. Morse, M.A. Craniology and the adoption of the three-age system in Britain. *Proc. Prehist. Soc.* **1999**, *65*, 1–16. [CrossRef]
2. Myres, R.R. History of coatings science and technology. *J. Macromol. Sci. Part A Chem.* **1981**, *15*, 1133–1149. [CrossRef]
3. French, P.; Krijnen, G.; Roozeboom, F. Precision in harsh environments. *Microsyst. Nanoeng.* **2016**, *2*, 16048. [CrossRef] [PubMed]
4. Huang, J.; Wang, W.; Lu, X.; Liu, S.; Li, C. Influence of lamellar interface morphology on cracking resistance of plasma-sprayed YSZ coatings. *Coatings* **2018**, *8*, 187. [CrossRef]
5. Baiamonte, L.; Bartuli, C.; Marra, F.; Gisario, A.; Pulci, G. Hot corrosion resistance of laser-sealed thermal-sprayed cermet coatings. *Coatings* **2019**, *9*, 347. [CrossRef]
6. Choi, K.; Kim, Y.; Kim, M.; Lee, S.; Lee, S.; Park, J. Oxidation behavior of $MoSi_2$-coated TZM alloys during isothermal exposure at high temperatures. *Coatings* **2018**, *8*, 218. [CrossRef]
7. Dinu, M.; Mouele, E.; Parau, A.; Vladescu, A.; Petrik, L.; Braic, M. Enhancement of the corrosion resistance of 304 stainless steel by Cr–N and Cr(N,O) coatings. *Coatings* **2018**, *8*, 132. [CrossRef]
8. Zhang, C.; Gu, L.; Tang, G.; Mao, Y. Wear transition of CrN coated M50 steel under high temperature and heavy load. *Coatings* **2017**, *7*, 202. [CrossRef]
9. Mahdavi, A.; Medvedovski, E.; Mendoza, G.; McDonald, A. Corrosion resistance of boronized, aluminized, and chromized thermal diffusion-coated steels in simulated high-temperature recovery boiler conditions. *Coatings* **2018**, *8*, 257. [CrossRef]
10. Lin, J.; Wang, Z.; Cheng, J.; Kang, M.; Fu, X.; Hong, S. Effect of initial surface roughness on cavitation erosion resistance of arc-sprayed Fe-based amorphous/nanocrystalline coatings. *Coatings* **2017**, *7*, 200. [CrossRef]
11. Karlsdottir, S.; Geambazu, L.; Csaki, I.; Thorhallsson, A.; Stefanoiu, R.; Magnus, F.; Cotrut, C. Phase evolution and microstructure analysis of CoCrFeNiMo high-entropy alloy for electro-spark-deposited coatings for geothermal environment. *Coatings* **2019**, *9*, 406. [CrossRef]
12. Fanicchia, F.; Csaki, I.; Geambazu, L.; Begg, H.; Paul, S. Effect of microstructural modifications on the corrosion resistance of CoCrFeMo0.85Ni compositionally complex alloy coatings. *Coatings* **2019**, *9*, 695. [CrossRef]
13. Fan, L.; Reis, S.; Chen, G.; Koenigstein, M. Corrosion resistance of pipeline steel with damaged enamel coating and cathodic protection. *Coatings* **2018**, *8*, 185. [CrossRef]
14. Knudsen, O.; Matre, H.; Dørum, C.; Gagné, M. Experiences with thermal spray zinc duplex coatings on road bridges. *Coatings* **2019**, *9*, 371. [CrossRef]
15. Syrek-Gerstenkorn, B.; Paul, S.; Davenport, A.J. Sacrificial thermally sprayed aluminium coatings for marine environments: A review. *Coatings* **2020**, *10*, 267. [CrossRef]
16. Recent Developments on Functional Coatings for Industrial Applications. Available online: https://www.mdpi.com/journal/coatings/special_issues/funct_coat_ind_appl (accessed on 14 April 2020).
17. Current Research in Thin Film Deposition: Applications, Theory, Processing, and Characterisation. Available online: https://www.mdpi.com/journal/coatings/special_issues/film_depos_appl_theory_process_charact (accessed on 14 April 2020).

18. Current Research in Pulsed Laser Deposition. Available online: https://www.mdpi.com/journal/coatings/special_issues/pulse_laser_depos (accessed on 14 April 2020).

19. Advanced Thin Films Deposited by Magnetron Sputtering. Available online: https://www.mdpi.com/journal/coatings/special_issues/magnetron_sputter (accessed on 14 April 2020).

20. From Metallic Coatings to Additive Manufacturing. Available online: https://www.mdpi.com/journal/coatings/special_issues/met_coat_addit_manuf (accessed on 14 April 2020).

21. Advanced Thin Film Materials for Photovoltaic Applications. Available online: https://www.mdpi.com/journal/coatings/special_issues/adv_thin_film_mater_Photovolt_appl (accessed on 14 April 2020).

22. Active Organic and Organic-Inorganic Hybrid Coatings and Thin Films: Challenges, Developments, Perspectives. Available online: https://www.mdpi.com/journal/coatings/special_issues/org_hybrid (accessed on 14 April 2020).

23. Superhydrophobic Coatings. Available online: https://www.mdpi.com/journal/coatings/special_issues/superhydrophobic_coat (accessed on 14 April 2020).

24. ECHA. Substances Restricted under REACH. Available online: https://www.echa.europa.eu/substances-restricted-under-reach (accessed on 14 April 2020).

25. Shin, D.; Kim, Y.; Yoon, J.; Park, I. Discoloration resistance of electrolytic copper foil following 1,2,3-Benzotriazole surface treatment with sodium molybdate. *Coatings* **2018**, *8*, 427. [CrossRef]

Article

Influence of Lamellar Interface Morphology on Cracking Resistance of Plasma-Sprayed YSZ Coatings

Jibo Huang [1], Weize Wang [1,*], Xiang Lu [1], Shaowu Liu [1] and Chaoxiong Li [2]

[1] Key Laboratory of Pressure System and Safety, Ministry of Education, East China University of Science and Technology, Shanghai 200237, China; Y20150084@mail.ecust.edu.cn (J.H.); Y45160090@mail.ecust.edu.cn (X.L.); Y30150544@mail.ecust.edu.cn (S.L.)

[2] Shanghai Baosteel Industry Technological Service Co., Ltd., Shanghai 201900, China; lichaoxiong@baosteel.com

*   Correspondence: wangwz@ecust.edu.cn; Tel.: +86-21-6425-2819

Received: 21 March 2018; Accepted: 11 May 2018; Published: 15 May 2018

Abstract: Splat morphology is an important factor that influences the mechanical properties and durability of thermal barrier coatings (TBCs). In this study, yttria-stabilized zirconia (YSZ) coatings with different lamellar interface morphologies were deposited by atmospheric plasma spraying (APS) using feedstocks with different particle sizes. The influence of lamellar interface roughness on the cracking resistance of the coatings was investigated. Furthermore, the thermal shock and erosion resistance of coatings deposited by two different powders was evaluated. It was found that the particle size of the feedstock powder affects the stacking morphology of the splat that forms the coating. Coatings fabricated from coarse YSZ powders (45–60 μm) show a relatively rough inter-lamellar surface, with a roughness about 3 times greater than those faricated from fine powders (15–25 μm). Coatings prepared with fine powders tend to form large cracks parallel to the substrate direction under indentation, while no cracking phenomena were found in coatings prepared with coarse powders. Due to the higher cracking resistance, coatings prepared with coarse powders show better thermal shock and erosion resistances than those with fine powders. The results of this study provide a reference for the design and optimization of the microstructure of TBCs.

Keywords: YSZ; particle size; lamellar interface; cracking resistance; thermal shocks; erosion

1. Introduction

Plasma-sprayed (PS) ceramic coatings are widely used in engineering applications, such as the thermal barrier coatings (TBCs) applied to insulate the high-temperature components of gas turbines from hot gas [1–3]. The formation of the coatings in the plasma spraying process is characterized by the impingement of substantial molten and semi-molten particles, which are deposited on the previously deposited ones. Therefore, coatings fabricated by PS developed by successive build-up of solidified lamellae are known as splats [4]. The layering of the individual splats determines the microstructure and, ultimately, the quality of the coating [4,5]. Therefore, establishing the relationships between spraying parameters, the flattening behavior of splats, and ultimately the coating performance is of great significance, and has received considerable attention [6–8].

The primary factor determining splat formation and ultimately the coating properties has been identified as the melting degree of the individual particles at the impingement on the substrate [5]. The feedstock powder morphology is an important parameter affecting the properties of TBCs [9]. The feedstock size affects the deposition, solidification, and crystallization of molten droplets by influencing the particle state in the plasma, thus affecting the microstructure of the coatings [10].

Fine particles typically result in good melting during deposition, and promote the low porosity of the coatings due to the decreased volume of inter-lamellar gaps and voids [11].

The particle size has an effect on the mechanical properties of the coatings. Dwivedi et al. investigated the effect of particle size distribution on the fracture toughness of APS YSZ coatings. In their report, the fracture toughness of plasma-sprayed ceramics was significantly affected by the porosity of the coatings. Coatings sprayed with finer powders showed higher fracture toughnesses due to their denser microstructures. The fracture toughness values of the coatings deposited with particles in the range 10–45 μm and those in the range 79–117 μm showed significant differences, which ranged from 1.35 to 2.2 MPa·m$^{1/2}$ [12]. However, apart from the porosity, the mechanical properties of YSZ coatings also depend on the individual splat morphology, adhesion between the splat and the substrate, cohesive strength among individual splats, and the microstructure of the splats themselves [13–15]. Wang et al. studied the effects of pores and lamellar interfaces on the properties of plasma-sprayed zirconia coatings, and reported that the lamellar interfaces were equally as important as the porosity in defining the properties of the coatings [16]. Although the importance of the lamellar interface has been recognized, its exact impact on coating performance still remains unclear.

In plasma-sprayed TBCs, the YSZ coating is sprayed onto a rough bond coat (BC), guaranteeing good mechanical locking of the top coat (TC) to the bond coat [17]. Specifically, the interface roughness has been shown to play a major role in the development of the induced stresses and lifetime of TBCs [18,19]. The roughness of the bond coat and the previously deposited layer should be about 5–10 μm, as suggested by Vaßen et al., for good adhesion of the ceramic coating [17]. Li et al. also pointed out that asperities similar to the splat thickness might be beneficial to promoting mechanical bonding between the bond coat and the top coat [6]. In some cases, the spalling of the top coat occurs before the initiation and propagation of cracks at the BC/TC interface due to the cracking of the ceramic coating [15,20]. Therefore, the mechanical bonding at the ceramic–lamellae interface is also an important factor affecting the durability of the TBCs. Coating formation is comprised of the flattening and solidification of droplets when they impact the previously formed coating surface, which is undulating. The publications mentioned above showed that the interface roughness was able to affect the bonding strength between the bond coat and top coat. However, until now, the effect of the inter-lamellar surface morphology on the mechanical property of the plasma-sprayed ceramic coating has not been examined.

Previous work has found that the particle size of YSZ may significantly affect the microstructure and the thermal shock resistance of plasma-sprayed TBCs [21]. The surfaces of the coatings prepared by coarse particles are rougher than those prepared with fine ones [21]. The aim of this work is to obtain more information on the influence of the inter-lamellar surface morphology on the relevant coating properties. For this purpose, the stacking morphology of the splats in coatings made from different particle sizes was analyzed. The hardnesses and X-ray diffraction (XRD) patterns of the coatings were measured to evaluate the effect of particle size on mechanical properties and phase structure. Furthermore, the indentation technique was used to evaluate the influence of the lamellar interface morphology on the cracking resistance of the coatings. Based on these results, the failure modes of the coatings fabricated from different particle sizes under thermal cycling and erosion conditions were analyzed and discussed.

2. Materials and Methods

2.1. Sample Preparation

Samples were produced by air plasma spraying (APS) of YSZ powders (8 wt. % Y_2O_3) onto disk nickel-based super alloy (IN-738) substrates with thicknesses of 3 mm and diameters of 25.4 mm. Prior to coating deposition, the surface of the substrate was grit blasted to obtain a rough surface. Sand blasting was performed using corundum with a particle size of 60–80 mesh at an air pressure of 0.6 MPa. Before deposition of the YSZ coating, a commercially available NiCrAlY powder (45–106 μm, Beijing

SunSpraying New Material Co., Ltd., Beijing, China) was used to deposit the bond coat. To analyze the effect of lamellar interface morphology on the performance of YSZ coatings, YSZ powders with the size of 15–25 µm and 45–60 µm (Chengdu HuaYin Powder Technology Co., Ltd., Chengdu, China) were selected as the feedstocks for the deposited ceramic coat. Both the YSZ top coat and the bond coat were deposited by a commercial air plasma spray (APS) system (APS-2000, Beijing Aeronautical Manufacturing Technology Research Institute, Beijing, China) onto the substrate. Spraying parameters (such as plasma power, spray distance and powder feed rate) could influence the microstructure and mechanical properties of the coating. Therefore, in this study, except for the difference in particle size distribution of the YSZ powders, all process parameters were kept constant in order to introduce deliberate variations in the coating microstructures. During spraying, argon was used as the main gas and hydrogen as the auxiliary gas. The pressure was controlled at 0.4 MPa and 0.25 MPa, respectively. The main gas flow was controlled at 47 L/min. Argon was also used as the powder feed gas, with a flow rate of 9 L/min. The plasma power was maintained at approximately 36 kW (600 A/60 V) and 30 kW (500 A/60 V) to deposit the YSZ and the bond coat, respectively. The traverse speed of the spray gun was kept as 150 mm/s with a spray distance of 100 mm for the bond coat deposition and 70 mm for YSZ deposition. According to previous research experience, the YSZ powders with particle sizes of less than 60 µm were completely melted under these spray parameters [21,22]. Therefore, the spraying parameters in this study are suitable for the deposition of these two kinds of YSZ powders. The total thickness of the TBCs, including both the bond coat and ceramic coat, was about 450 µm, and the ceramic top coat was about 300 µm thick. For simplicity, in this study, the coatings fabricated from fine and coarse YSZ powders are labeled as F-TBC and C-TBC, respectively.

2.2. Sample Characterization

Microstructural analysis was conducted on the coating surfaces, polished cross-sections, and fracture surfaces using a scanning electron microscope (SEM, ZEISS EVO MA15, Carl Zeiss SMT Ltd., Cambridge, UK). Moreover, the 3D morphologies and roughness of coating surface were obtained using a non-contact profiler (Infinite Focus G4, Alicona, Graz, Austria). The phase analyses of the YSZ feedstocks and coatings before and after thermal cyclic testing were examined by X-ray diffraction (XRD, D/max2550VB/PC, RIGAKU, Tokyo, Japan) using filtered Cu Kα radiation at an accelerated voltage of 40 kV and a current of 100 mA. Diffraction angles were set in the range of 10°–80° with a step width of 0.02°. The Young's modulus and micro-hardness of YSZ coatings were determined using nano-indentation tests (Agilent Nano Indenter G200, Agilent Technologies Inc., Santa Clara, CA, USA). The indentation experiments were carried out using a Berkovich indenter in displacement-controlled mode with a constant strain rate of 0.05 s^{-1}. The maximum displacement was set to 1000 nm. Regions relatively devoid of cracks and surface defects were identified using a microscope and 15 indents made in those regions to obtain the Young's modulus and micro-hardness values of the coatings. It should be noted that measurements performed by nano-indentation give values approaching those of the local stiffness of the coatings, so the Vickers hardness of the coatings was also measured. The Vickers hardness of the coatings was measured on a polished cross-section at a load of 9.8 N with a loading duration of 15 s. 10 points were measured on each coating sample to obtain a reliable hardness value. The Vickers indentation fracture (VIF) test is a popular experimental technique for estimating of the fracture resistance of brittle ceramics. The cracking resistance of the YSZ coatings fabricated from the fine and coarse powders were investigated by analyzing crack behavior during the Vickers indentation test.

2.3. Thermal Shock and Erosion Test

Cyclic thermal shock and particle erosion testing were conducted to investigate the influence of cracking resistance on the failure modes of coatings fabricated from the fine and coarse YSZ powder. Thermal shock tests were conducted using a muffle furnace. When the temperature of the furnace reached 1050 °C, the samples were put into the furnace with a holding time of 10 min, then they were directly quenched in deionized water with a temperature of 20–30 °C until the samples had cooled to room temperature.

Erosion tests were carried out using a homemade erosive tester. The schematic of the erosion experiment is shown in Figure 1. The abrasive particles are fed into the high-speed steam and accelerated, then the particles are ejected from the steel tube nozzle to impact the coating. Irregular SiC powder with sizes in the range of 61–75 μm was used as abrasive. During the erosion process, the flow of the steam gas was fixed at 6.8 kg/h, and the feed rate of the abrasive was 5.12 g/min. A 6-cm nozzle-to-substrate distance was used, with an impingement angle of 90°. To produce an erosion scar with a well-defined geometry, the specimen surface was masked with a 20 mm diameter opening. Specimen mass was measured prior to erosion and after every 5.12 g of erodent feed.

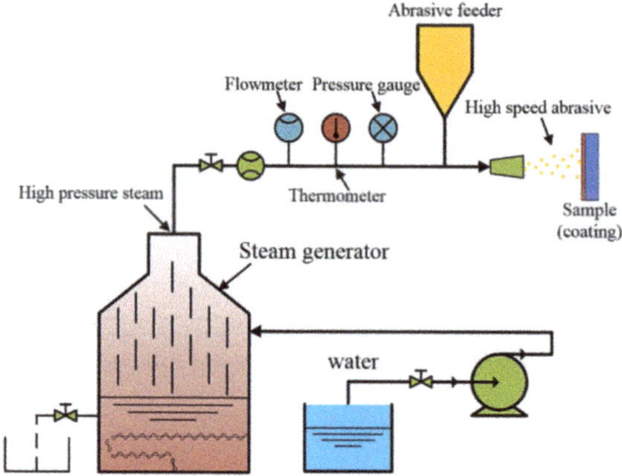

Figure 1. Schematic diagram of the erosion device.

3. Results and Discussion

3.1. Microstructural

Figure 2 shows the cross-section microstructure of as-sprayed coatings. It can be seen that more macro-pores are present in the coating prepared from the coarse YSZ powder (Figure 2d). The porosity of the coating prepared by the fine powder (Figure 2a) is lower than that by the coarse powder. Vertical cracks are the common in the structures of plasma-sprayed coatings and are caused by the release of residual stress during the deposition process [11,23]. In this study, vertical cracks were observed in the coatings prepared from both fine and coarse YSZ powders. However, some vertical cracks in F-TBC were accompanied by branching cracks, as shown in Figure 1b. Some branches were formed by cracking along the lamellar interface (Figure 2c). In contrast, no branching cracks were found in C-TBC. Instead, some intra-splat cracks were observed near the vertical cracks (Figure 2f).

Figure 3 shows the surface morphology and the splat stacking features of the coatings deposited using the two different YSZ powders. As shown in Figure 3a,c, the surface of the F-TBC is much smoother than that of the C-TBC. Babu et al. also reported that the surface roughness and porosity increased with an increase in powder particle size in the case of Ni-Cr and Al_2O_3 coatings [24]. As expected, the splats formed by the fine powder are much smaller than that those formed by the coarse powder, marked in Figure 3b,d. In addition, due to the tortuous lamellar interface of C-TBC, the splats formed by molten particles are stacked in waves in this coating. In contrast, the splats formed in F-TBC are much smoother.

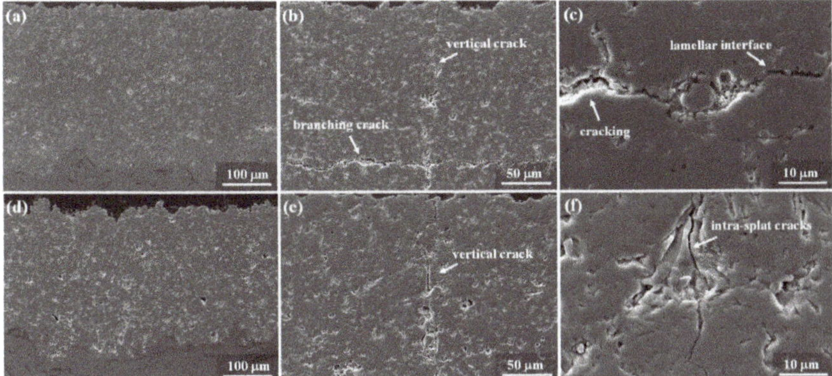

Figure 2. Cross-section microstructure of the as-sprayed YSZ coatings: (**a–c**) F-TBC; (**d–f**) C-TBC.

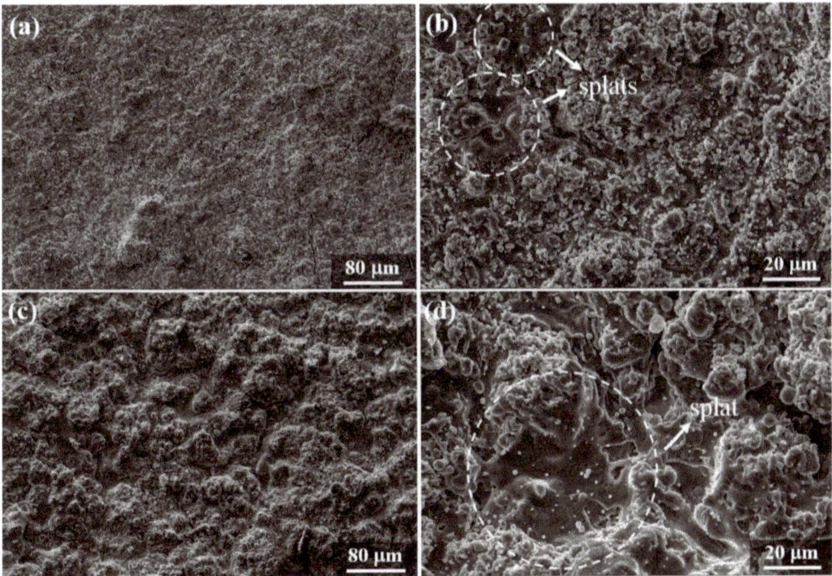

Figure 3. Surface morphology of the YSZ coatings: (**a,b**) F-TBC; (**c,d**) C-TBC.

The 3D images of the YSZ coating surface are shown in Figure 4a. The surface roughness presented in Ra was obtained by measuring the contour along the coating surface (shown in Figure 4b). The roughness of C-TBC is 2.71 μm, which is about three times that of F-TBC (0.99 μm). The maximum amplitude of the C-TBC surface profile exceeded 8 μm, whereas that of the F-TBC was less than 4 μm.

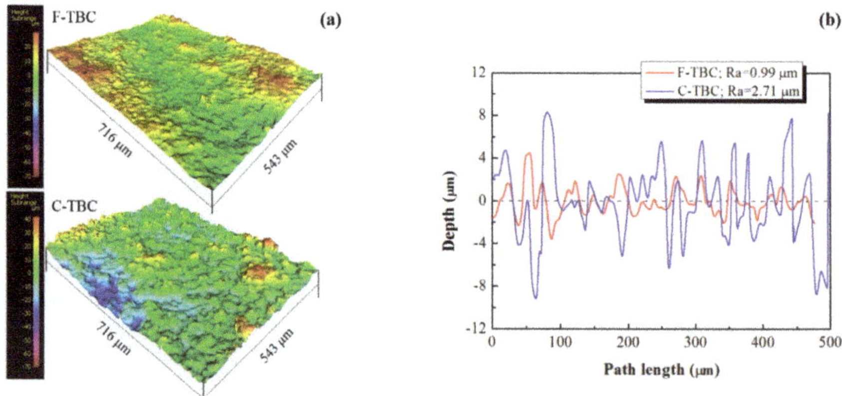

Figure 4. 3D morphological views (**a**) and roughness (**b**) of F-TBC and C-TBC.

3.2. Mechanical Properties and Phase Structure

The performance of plasma-sprayed coatings depends on the mechanical properties of the material itself and the defect structure formed during the deposition process. Table 1 shows the elastic modulus and hardness values of the coatings obtained by the nano- and micro-indentation method. The elastic modulus and hardness obtained through nano-indentation of the C-TBC is 213 and 15 GPa, respectively, which is larger than those of the F-TBC. However, the hardness of the C-TBC measured by Vickers indentation is slightly lower than that of F-TBC. There are some fine pores in the coating beneath the nano-indenter, which causes the Young's modulus of coatings measured by nano-indentation to be lower than that of the corresponding bulk material (~200–210 GPa [25]). The pore size in PS TBC usually reveals a bimodal distribution [11]. Due to the small size of the splats, the number of fine pores attributed to micro-cracks such as inter-splats gaps in F-TBC is relatively greater [21]. Therefore, the hardness and Young's modulus examined by nano-indentation are lower than those of C-TBC. There are more large voids in the C-TBC (show in Figure 2d), which leads to it having a lower Vickers hardness compared to F-TBC.

Table 1. Elastic modulus and hardness of the F-TBC and C-TBC.

Coatings	Elastic Modulus by Nano-Indentation Technique (GPa)	Hardness by Nano-Indentation Technique (GPa)	Vickers Hardness (Hv)
F-TBC	166 ± 16	10 ± 2	786 ± 28
C-TBC	213 ± 19	15 ± 2	765 ± 46

Phase composition is considered to be a crucial factor affecting the properties and durability of TBCs. The XRD patterns of the two kinds of YSZ feedstocks and the corresponding coatings before and after thermal cycling testing are shown in Figure 5. It can be seen that coatings made from coarse and fine YSZ powders have the same phase structure. Tetragonal zirconia is the main phase for the YSZ coatings before and after thermal cyclic test. From the detailed analysis in a diffraction angle range from 27° to 33° (Figure 5b), a very small fraction of monoclinic phase can be detected for both fine and coarse YSZ feedstocks. However, no monoclinic phase was present in any of the coatings after the plasma spraying process. No obvious changes in phase composition can be found for the coatings before and after the thermal cycling test. Consequently, the phase structure may not influence the cracking resistance and service performance of YSZ coatings prepared from particles with different sizes.

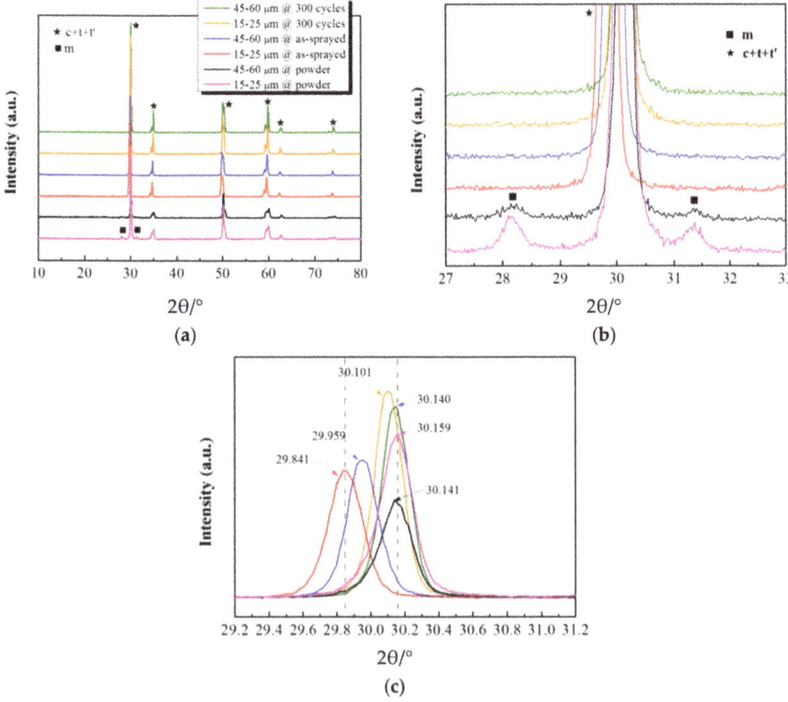

Figure 5. XRD patterns of YSZ feedstocks and coatings before and after the thermal cyclic test: (**a**) global analysis in a diffraction angle range from 10° to 80°; (**b**) detailed analysis in a diffraction angle range from 27° to 33° and (**c**) peak shift of XRD patterns of coatings during thermal cycling.

Besides the phase composition, residual stresses in coatings also have an important influence on the service behavior, performance, and lifetime of the TBCs. As shown in Figure 5c, the XRD peak positions of the as-sprayed F-TBC and C-TBC shifted to the left with respect to the YSZ powders, which indicates that both F-TBC and C-TBC contain tensile residual stress arising from the plasma spraying process during the coating deposition. In addition, the residual stress in F-TBC is higher than that of C-TBC, based on the peak position offset. After thermal cycling, the peak positions of both coatings shifted to the right, indicating that stress relaxation had occurred during the thermal cyclic test.

3.3. Cracking Resistance

It is known that YSZ spallation is often related to crack initiation and propagation in the vicinity of the YSZ/bond coat interface [26]. Therefore, the cracking behavior in the coating under the action of indentation was studied in this study. Figure 6 shows four typical indentation morphologies under the same load conditions: indentation with vertical cracks, indentation with horizontal cracks, crushed indentation, and indentation with inter-lamellar cracking. The difference between inter-lamellar cracking and horizontal crack is judged by the crack length. Horizontal cracks exceeding the diagonal length of the indentation are considered to be inter-lamellar cracking. It can be seen that there is a great difference in the morphology of cracks generated near the indentation, and some indentations do not even cause cracks. Due to the complexity of the crack morphology, the fracture toughness of the coatings in this study could not be measured by the indentation method. Other publications have also pointed out that indentation-based measurement techniques are not very reliable for coatings, as they do not provide legitimate toughness values [12,27].

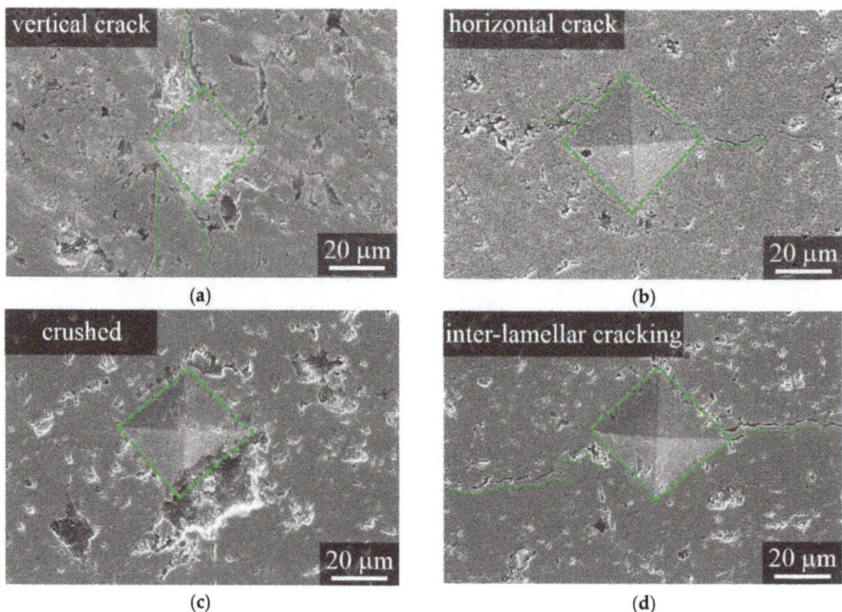

(a)　　　　　　　　　(b)

(c)　　　　　　　　　(d)

Figure 6. Four types of typical indentation morphology of F-TBC and C-TBC: (**a**) indentation with vertical cracks, C-TBC; (**b**) indentation with horizontal cracks, F-TBC; (**c**) crushed indentation, C-TBC and (**d**) indentation with inter-lamellar cracking, F-TBC.

Although the specific fracture toughness of the coatings cannot be obtained, the cracking resistance of F-TBC and C-TBC can be evaluated by analyzing the crack morphology formed by indentation. By sorting the indentations in the coating according to the features in Figure 6, the statistics of the indentation features of F-TBC and C-TBC are shown in Figure 7. The percentage of indentation is determined by the number of indentations containing the feature divided by the total number of indentations performed in the coating. The same indentation may belong to several features. For example, indentations with inter-lamellar cracking also belong to the indentation with horizontal cracks. It can be seen that more than 90% of the indentations in F-TBC are accompanied by horizontal cracks, and that more than 65% of the horizontal cracks belong to inter-lamellar cracking. However, in C-TBC, the indentations accompanied by horizontal cracks are fewer than 25%, and no inter-lamellar cracking can be observed. These data show that coatings prepared by fine powder tend to form large cracks parallel to the substrate direction, indicating that the cracking resistance of F-TBC is worse than that of C-TBC. In addition, the presence of delamination in the as-sprayed F-TBC (Figure 2b) also shows that coatings deposited by fine YSZ powders have poor cracking resistance.

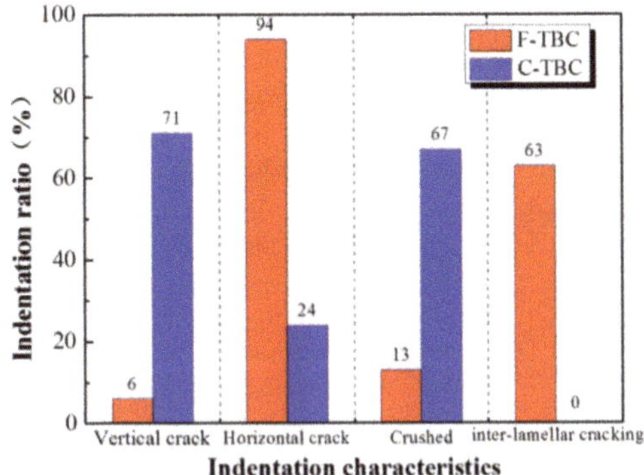

Figure 7. Statistics of the indentation features of F-TBC and C-TBC.

According to Dwivedi's report, coatings prepared by finer powders show higher fracture toughness due to the denser microstructure with lower defect density [12]. The proportion of the crushed indentation of C-TBC is about 5 times that of F-TBC, indicating that C-TBC contains more macro-pores, which is consistent with the microstructure observation. Therefore, prior to our study, the cracking resistance of F-TBC was considered to be superior to that of C-TBC. The results of this study show that the cracking resistance of plasma-sprayed coatings cannot be judged solely from the macroscopic pore structure. Besides the pore structure, the stacking morphology of the splats and the adhesion between the splats are also important factors for the cracking resistance of coatings. As illustrated by the surface morphologies shown in Figures 3 and 4, the particle size of YSZ powder affects the stacking morphology of the splats that form the coating. The lamellar interface roughness of C-TBC is about 3 times that of F-TBC. Furthermore, from the cracking propagation path in the F-TBC (Figure 2b,c and Figure 6d), it can be inferred that the cracking in the coating tends to propagate along the lamellar interface. Therefore, by considering the effect of pore structure, mechanical properties, phase composition, splat stacking morphology, and crack behavior of the coats all together, insufficient lamellar interface roughness is believed to be the major factor responsible for the poor cracking resistance of F-TBC in this study.

A bond coat with a roughness approximately in the range of Ra 5–10 μm provides good adhesion for the ceramic coating [17]. The roughness value Ra of C-TBC is about 3 μm, with a maximum amplitude of the surface profile of up to 8 μm. Therefore, such a rough interface facilitates the adhesion of the subsequent droplets with the previously formed coating surface during the plasma spraying process. Due to the good adhesion of the lamellar interface, it is difficult for the cracks to expand along the lamellar interface in C-TBC. In addition, the rough interface makes the crack propagation path twist and turn, which also increases the difficulty of crack expansion. As a result, intra-splat cracks perpendicular to the substrate (vertical cracks) are formed in C-TBC instead of cracking along the lamellar interface (horizontal cracks). From the statistics of the indentation characteristics in Figure 7, it can be seen that more than 70% of the indentations in C-TBC are accompanied by vertical cracks, and indentations accompanied by horizontal cracks are fewer than 25% with no inter-lamellar cracking. In contrast, the smooth lamellar interface in F-TBC could only provide limited mechanical bonding to accommodate splats. Thus, the cracking resistance of F-TBC is poor due to its susceptibility to cracking along the lamellar interface.

3.4. Thermal Shock Resistance

The macro images of F-TBC and C-TBC samples after 300 thermal cycles are shown in Figure 8a,d. It can be seen that more than 20% of the coating in F-TBC has peeled off, and the failure happens within the top coating (cohesive failure). Failure phenomena did not occur in C-TBC, even after

300 thermal cycles, and the coating remained intact. Figure 8b,e shows the fracture morphologies of F-TBC and C-TBC after 300 thermal cycles. From the fracture characteristics of F-TBC, it can be seen that the coating prepared by fine powder has formed substantial inter-lamellar cracks after the thermal shock test, due to the poor cracking resistance of the interface between the lamellas. It is interesting to note that the distance between these cracks is about 25 μm, exactly the thickness of the coating deposited for each pass. The crack path is consistent with the lamellar interface, indicating that the lamellar interface is prone to debonding. As vertical cracks inside the coating propagate to the lamellar interface, the coating will peel off along the interface, resulting in the partial spalling of the coating, as shown in Figure 8a. For C-TBC, no horizontal cracking along the interface occurs in the coating, so the coating will not fracture, even with vertical cracks.

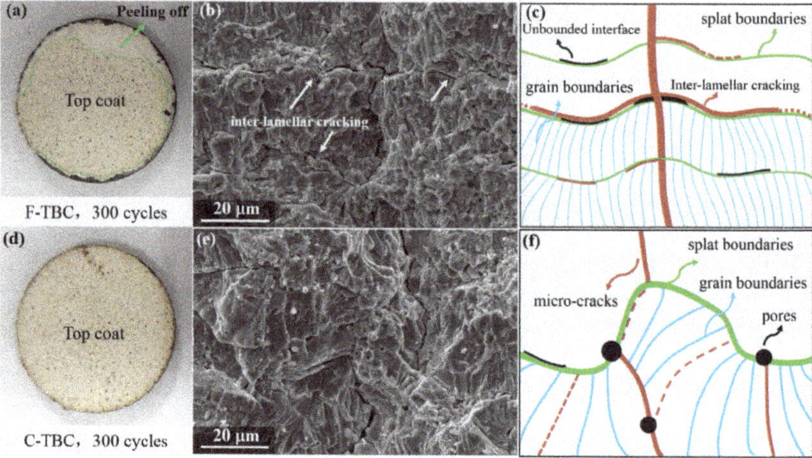

Figure 8. Illustration of thermal shock failure modes of two kinds of coatings. Macro image of F-TBC (**a**) and C-TBC (**d**) after 300 thermal cycles; fracture morphology of F-TBC (**b**) and C-TBC (**e**); schematic of crack behavior of F-TBC (**c**) and C-TBC (**f**).

The interfaces within the plasma-sprayed coating include the lamellar interface, splat interface, and grain boundary. In F-TBC, the lamellar interfaces are smooth, and most of the droplets are stacked on the surface in a flat manner. The cracking resistance of the lamellar interface is worse than that of the splat interface and grain boundary, so inter-lamellar cracking occurs in the coating. The schematic of the crack behavior in F-TBC is shown in Figure 8c. In C-TBC, due to the undulating lamellar interface, the splats in the coating are stacked and anchored to each other. Therefore, the resistance of the cracks propagating along the lamellar interface is too high for continuous transverse cracks to form. Only some scattered intra-splat cracks form in the coatings (cracking along grain boundary), as shown in the scheme in Figure 8f.

Previous study has found that the rate of crack propagation in coatings prepared with coarse YSZ powder is slower than that prepared with fine powder [22]. Based on the results of this study, it can be seen that the stacking morphology of splats has a great influence on the crack propagation behavior of coatings. The roughness of the lamellar interface may affect the cracking resistance of the coatings. A certain degree of roughness is conducive to increasing the resistance to transverse crack expansion along the lamellar interface. Therefore, in future studies, apart from the porosity, the effects of the splat interface on the performance of the coating should also be taken into consideration.

3.5. Particle Erosion Resistance

The erosion results are provided in Figure 9. From the ratio of coating mass loss to erodent exposure, it can be seen that the erosion rate of F-TBC is slightly higher than that of C-TBC. Since the erosion failure of the APS coating is mainly due to the delamination of the layered elements, the erosion performance of the coating mainly depends on the porosity, thickness of the lamellas and the bonding strength between the splats [28]. The pores and cracks in the coating may provide the starting sources for the peeling of the coating, and thus the erosion resistance of the coating decreases with increasing porosity [28,29]. In addition, according to Schmitt's research, smaller splat size provides a lower surface roughness, which has been shown to reduce erosion rate by yielding smaller failure regions [30,31]. In this study, the porosity of F-TBC is lower than that of C-TBC, and the splat size of F-TBC is smaller than that of C-TBC due to the fine feedstock. However, the erosion resistance of F-TBC is worse than that of C-TBC, indicating that in addition to the size of splat and porosity, there are other more important factors that affect the erosion rate of the coating.

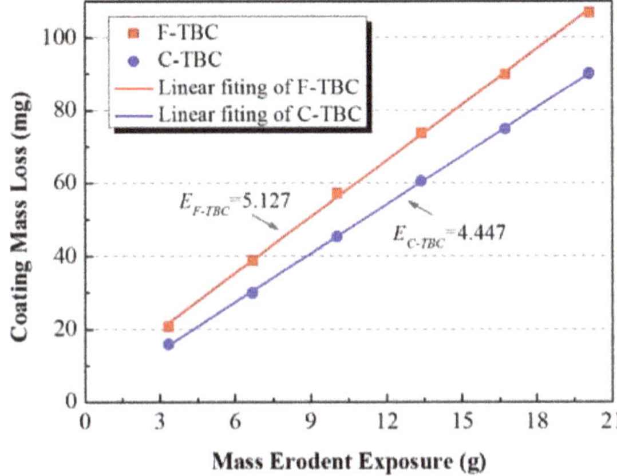

Figure 9. Erosion results in terms of coating mass loss as a function of erodent exposure.

Figure 10 shows the typical surface morphology of the coatings after the erosion test. It was found that the number of splat interfaces (marked in Figure 10b) in the fracture of F-TBC was much higher than that of C-TBC. These interfaces mean that the cracks in F-TBC tend to propagate along the splat boundary under the impact of abrasives. For C-TBC, a large number of splat sections are present in the fracture surface, indicating that the expansion of cracks has to cross the splat. Figure 11 shows the cross-sectional microstructure of the coatings after erosion. Lateral cracking occurred in both F-TBC and C-TBC, and the cracks in F-TBC are larger than those in the C-TBC in terms of both crack length and width. By observing the surface of the cracks at a high magnification, one can see that the cracking in F-TBC is torn along the splat interface, while it is elusive in C-TBC.

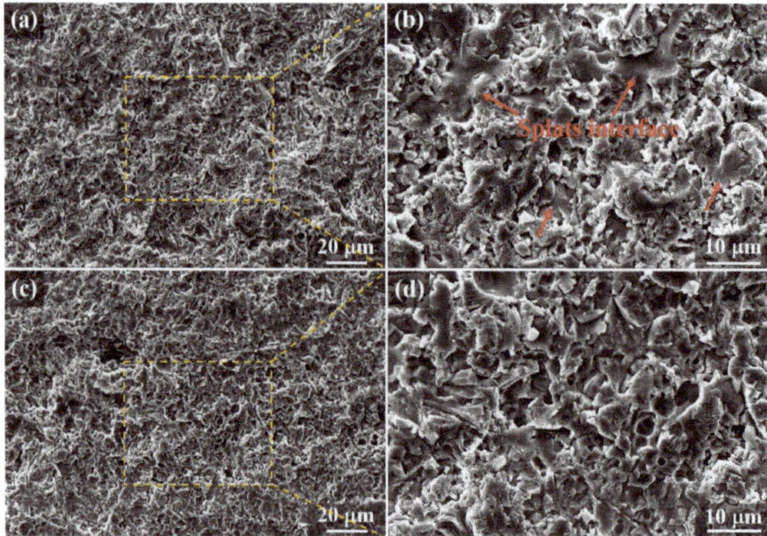

Figure 10. Surface morphology of the coatings after the erosion test. (**a**,**b**) F-TBC; (**c**,**d**) C-TBC.

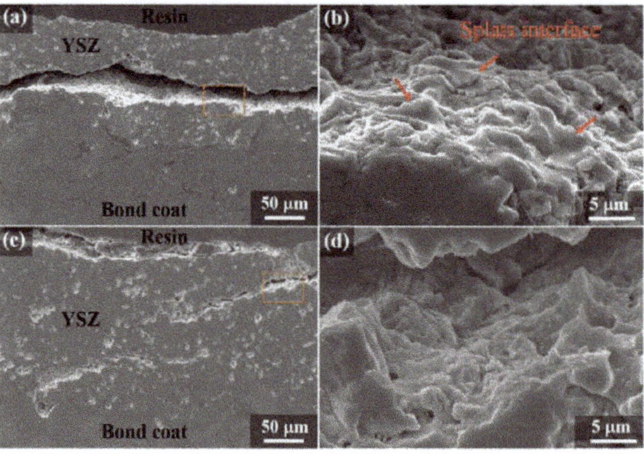

Figure 11. Cross-sectional microstructure of the coatings after erosion. (**a**,**b**) F-TBC; (**c**,**d**) C-TBC.

Li et al. have reported that the erosion of plasma-sprayed coatings occurred through spalling of the lamella exposed to the coating surface resulting from cracking along the lamellar interface [28]. In this study, the erosion of the F-TBC occurred from the lamellar or splat interfaces, whereas cracking in C-TBC tends to occur inside the splat. The difference in crack propagation mode may be an important factor that affects the erosion resistance of F-TBC and C-TBC. Figure 12 illustrates the behavior of cracks in the coating under the impact of abrasives. As described in Section 3, the cracking resistance of the lamellar interface in F-TBC is poor, so it is easy for cracks to expand through the whole interface of the lamella and consequently lead to the spalling of the coating. For C-TBC, the stack of splats is undulating, and the adhesion between splats is strong. Therefore, the expansion of the cracks needs to cross the splats, instead of cracking along the lamellar interface. It is well known that cracking within the splat is more difficult than that along the splat or lamellar interface, so from the perspective of the anti-cracking performance of the coating, the erosion resistance of C-TBC should be better than that of F-TBC. Considering the effect of porosity, splat size and cracking resistance on the erosion performance of the coatings together, although the splat size and porosity of F-TBC are smaller than those of C-TBC, the poor crack resistance of the F-TBC has a more significant impact on decreasing the erosion resistance of the coating.

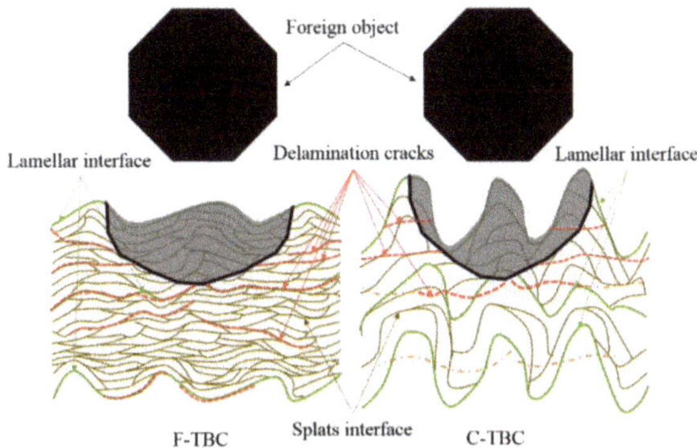

Figure 12. Illustration of particle erosion failure modes of two kinds of coatings.

4. Conclusions

In this study, YSZ coatings with different lamellar interface morphologies were prepared by using feedstocks with different particle sizes. The influence of lamellar interface roughness on cracking resistance of the coatings was investigated. Furthermore, the effect of cracking resistance on thermal shock and erosion resistance of the coatings was evaluated. The major useful conclusions can be summarized as follows:

- The cracking resistance of plasma-sprayed coatings cannot be judged solely from the macroscopic pore structure. Besides the pore structure, the stacking morphology of the splats and adhesion between the splats are also important factors for the cracking resistance of coatings.
- The particle size of the feedstock powders affects the stacking morphology of the splats that form the coating. The splat stack in C-TBC is undulating, while that in F-TBC is much smoother. Coatings fabricated from the coarse YSZ powders show a relatively rough inter-lamellar surface.
- The cracking resistance of F-TBC is worse than that of C-TBC, and the insufficient lamellar interface roughness is the major factor responsible for the poor cracking resistance of F-TBC. The smooth lamellar interface in F-TBC can only provide limited mechanical bonding to accommodate splats. Thus, the cracking resistance of F-TBC is poor due to its tendency to crack along the lamellar interface. In C-TBC, due to the undulating lamellar interface, the splats in the coating are stacked and anchored to each other. Therefore, the resistance of cracks propagating along the lamellar interface is large.
- The stacking morphology of splats have a great influence on the crack propagation behavior of coatings. During the thermal shock and erosion test, the cracking of F-TBC tends to occur from the lamellar or splat interfaces, whereas that in C-TBC tends to occur inside the splat. The difference in crack propagation mode is an important factor that affects the cracking resistance of the coatings. Coatings prepared using the coarse powder show better thermal shock and erosion resistance than those using the fine one due to their higher cracking resistance.

Author Contributions: J.H., W.W. and C.L. conceived and designed the experiments; J.H., X.L. and S.L. performed the experiments; and J.H. and W.W. analyzed the data and wrote the paper.

Funding: This research was funded by National Natural Science Foundation of China (No. 51775189), Science and Technology Commission of Shanghai Municipality Project (16DZ2260604), Aviation funding (2015ZES7001) and Shanghai Pujiang Program (15PJD009).

Conflicts of Interest: The authors declare no conflict of interest.

References

1. Darolia, R. Thermal barrier coatings technology: Critical review, progress update, remaining challenges and prospects. *Int. Mater. Rev.* **2013**, *58*, 315–348. [CrossRef]
2. Padture, N.P.; Gell, M.; Jordan, E.H. Thermal barrier coatings for gas-turbine engine applications. *Science* **2002**, *296*, 280–284. [CrossRef] [PubMed]
3. Bakan, E.; Vaßen, R. Ceramic top coats of plasma-sprayed thermal barrier coatings: Materials, processes, and properties. *J. Therm. Spray Technol.* **2017**, *26*, 992–1010. [CrossRef]
4. Fauchais, P.; Vardelle, M.; Goutier, S. Latest researches advances of plasma spraying: From splat to coating formation. *J. Therm. Spray Technol.* **2016**, *25*, 1534–1553. [CrossRef]
5. Mutter, M.; Mauer, G.; Mücke, R.; Guillon, O.; Vaßen, R. Correlation of splat morphologies with porosity and residual stress in plasma-sprayed ysz coatings. *Surf. Coat. Technol.* **2017**, *318*, 157–169. [CrossRef]
6. Li, D.; Zhao, H.; Zhong, X.; Liu, C.; Wang, L.; Yang, K.; Tao, S. Effect of the bond coating surface morphology on ceramic splat construction. *J. Therm. Spray Technol.* **2015**, *24*, 1450–1458. [CrossRef]
7. Song, X.; Liu, Z.; Suhonen, T.; Varis, T.; Huang, L.; Zheng, X.; Zeng, Y. Effect of melting state on the thermal shock resistance and thermal conductivity of aps ZrO_2–7.5 wt. % Y_2O_3 coatings. *Surf. Coat. Technol.* **2015**, *270*, 132–138. [CrossRef]
8. Syed, A.; Denoirjean, A.; Hannoyer, B.; Fauchais, P.; Denoirjean, P.; Khan, A.; Labbe, J. Influence of substrate surface conditions on the plasma sprayed ceramic and metallic particles flattening. *Surf. Coat. Technol.* **2005**, *200*, 2317–2331. [CrossRef]
9. Ercan, B.; Bowman, K.J.; Trice, R.W.; Wang, H.; Porter, W. Effect of initial powder morphology on thermal and mechanical properties of stand-alone plasma-sprayed 7 wt. % Y_2O_3–ZrO_2 coatings. *Mater. Sci. Eng. A* **2006**, *435*, 212–220. [CrossRef]
10. Kulkarni, A.; Vaidya, A.; Goland, A.; Sampath, S.; Herman, H. Processing effects on porosity-property correlations in plasma sprayed yttria-stabilized zirconia coatings. *Mater. Sci. Eng. A* **2003**, *359*, 100–111. [CrossRef]
11. Guo, H.; Murakami, H.; Kuroda, S. Effect of hollow spherical powder size distribution on porosity and segmentation cracks in thermal barrier coatings. *J. Am. Ceram. Soc.* **2006**, *89*, 3797–3804. [CrossRef]
12. Dwivedi, G.; Viswanathan, V.; Sampath, S.; Shyam, A.; Lara-Curzio, E. Fracture toughness of plasma-sprayed thermal barrier ceramics: Influence of processing, microstructure, and thermal aging. *J. Am. Ceram. Soc.* **2014**, *97*, 2736–2744. [CrossRef]
13. Li, C.-J.; Ohmori, A. Relationships between the microstructure and properties of thermally sprayed deposits. *J. Therm. Spray Technol.* **2002**, *11*, 365–374. [CrossRef]
14. Zheng, Z.; Luo, J.; Li, Q. Mechanism of competitive grain growth in 8YSZ splats deposited by plasma spraying. *J. Therm. Spray Technol.* **2015**, *24*, 885–891. [CrossRef]
15. Li, C.J.; Li, Y.; Yang, G.J.; Li, C.X. Evolution of lamellar interface cracks during isothermal cyclic test of plasma-sprayed 8YSZ coating with a columnar-structured YSZ interlayer. *J. Therm. Spray Technol.* **2013**, *22*, 1374–1382. [CrossRef]
16. Wang, Z.; Kulkarni, A.; Deshpande, S.; Nakamura, T.; Herman, H. Effects of pores and interfaces on effective properties of plasma sprayed zirconia coatings. *Acta Mater.* **2003**, *51*, 5319–5334. [CrossRef]
17. Vaßen, R.; Kerkhoff, G.; Stöver, D. Development of a micromechanical life prediction model for plasma sprayed thermal barrier coatings. *Mater. Sci. Eng. A* **2001**, *303*, 100–109. [CrossRef]
18. Eriksson, R.; Sjöström, S.; Brodin, H.; Johansson, S.; Östergren, L.; Li, X.H. TBC bond coat-top coat interface roughness: Influence on fatigue life and modelling aspects. *Surf. Coat. Technol.* **2013**, *236*, 230–238. [CrossRef]
19. Gupta, M.; Skogsberg, K.; Nylén, P. Influence of topcoat-bondcoat interface roughness on stresses and lifetime in thermal barrier coatings. *J. Therm. Spray Technol.* **2014**, *23*, 170–181. [CrossRef]
20. Cheng, B.; Zhang, Y.M.; Yang, N.; Zhang, M.; Chen, L.; Yang, G.J.; Li, C.X.; Li, C.J. Sintering-induced delamination of thermal barrier coatings by gradient thermal cyclic test. *J. Am. Ceram. Soc.* **2017**, *100*, 1820–1830. [CrossRef]

21. Huang, J.; Wang, W.; Lu, X.; Hu, D.; Feng, Z.; Guo, T. Effect of particle size on the thermal shock resistance of plasma-sprayed ysz coatings. *Coatings* **2017**, *7*, 150. [CrossRef]

22. Huang, J.; Wang, W.; Yu, J.; Wu, L.; Feng, Z. Effect of particle size on the micro-cracking of plasma-sprayed ysz coatings during thermal cycle testing. *J. Therm. Spray Technol.* **2017**, *26*, 755–763. [CrossRef]

23. Guo, H.; Kuroda, S.; Murakami, H. Microstructures and properties of plasma-sprayed segmented thermal barrier coatings. *J. Am. Ceram. Soc.* **2006**, *89*, 1432–1439. [CrossRef]

24. Babu, P.S.; Rao, D.S.; Rao, G.V.N.; Sundararajan, G. Effect of feedstock size and its distribution on the properties of detonation sprayed coatings. *J. Therm. Spray Technol.* **2007**, *16*, 281–290. [CrossRef]

25. Paul, S. Stiffness of plasma sprayed thermal barrier coatings. *Coatings* **2017**, *7*, 68. [CrossRef]

26. Ahmadian, S.; Jordan, E.H. Explanation of the effect of rapid cycling on oxidation, rumpling, microcracking and lifetime of air plasma sprayed thermal barrier coatings. *Surf. Coat. Technol.* **2014**, *244*, 109–116. [CrossRef]

27. Quinn, G.D.; Bradt, R.C. On the vickers indentation fracture toughness test. *J. Am. Ceram. Soc.* **2010**, *90*, 673–680. [CrossRef]

28. Li, C.J.; Yang, G.J.; Ohmori, A. Relationship between particle erosion and lamellar microstructure for plasma-sprayed alumina coatings. *Wear* **2006**, *260*, 1166–1172. [CrossRef]

29. Janos, B.Z.; Lugscheider, E.; Remer, P. Effect of thermal aging on the erosion resistance of air plasma sprayed zirconia thermal barrier coating. *Surf. Coat. Technol.* **1999**, *113*, 278–285. [CrossRef]

30. Schmitt, M.P.; Schreiber, J.M.; Rai, A.K.; Eden, T.J.; Wolfe, D.E. Development and optimization of tailored composite tbc design architectures for improved erosion durability. *J. Therm. Spray Technol.* **2017**, *26*, 1062–1075. [CrossRef]

31. Schmitt, M.P.; Harder, B.J.; Wolfe, D.E. Process-structure-property relations for the erosion durability of plasma spray-physical vapor deposition (PS-PVD) thermal barrier coatings. *Surf. Coat. Technol.* **2016**, *297*, 11–18. [CrossRef]

Article

Hot Corrosion Resistance of Laser-Sealed Thermal-Sprayed Cermet Coatings

Lidia Baiamonte [1], Cecilia Bartuli [1,*], Francesco Marra [1], Annamaria Gisario [2] and Giovanni Pulci [1]

[1] LIMS, INSTM Reference Laboratory for Engineering of Surface Treatments, Department of Chemical Engineering Materials Environment, Sapienza University of Rome, Via Eudossiana 18, 00184 Rome, Italy; lidia.baiamonte@uniroma1.it (L.B.); francesco.marra@uniroma1.it (F.M.); giovanni.pulci@uniroma1.it (G.P.)

[2] Department of Mechanical and Aerospace Engineering, Sapienza University of Rome, Via Eudossiana 18, 00184 Rome, Italy; annamaria.gisario@uniroma1.it

* Correspondence: cecilia.bartuli@uniroma1.it; Tel.: +39-6-4458-5633

Received: 12 March 2019; Accepted: 25 May 2019; Published: 28 May 2019

Abstract: Hot corrosion affects the components of diesel engines and gas turbines working at high temperatures, in the presence of low-melting salts and oxides, such as sodium sulfate and vanadium oxide. Thermal-sprayed coatings of nickel–chromium-based alloys reinforced with ceramic phases, can improve the hot corrosion and erosion resistance of exposed metals, and a sealing thermal, post-treatment can prove effective in reducing the permeability of aggressive species. In this study, the effect of purposely-optimized high-power diode laser reprocessing on the microstructure and type II hot corrosion resistance of cermet coatings of various compositions was investigated. Three different coatings were produced by high velocity oxy-fuel and was tested in the presence of a mixture of Na_2SO_4 and V_2O_5 at 700 °C, for up to 200 h: (i) Cr_3C_2–25% NiCr, (ii) Cr_3C_2–25% CoNiCrAlY, and (iii) mullite nano–silica–60% NiCr. Results evidenced that laser sealing was not effective in modifying the mechanism, on the basis of the hot corrosion degradation but could provide a substantial increase of the surface hardness and a significant decrease of the overall coating material consumption rate (coating recession), induced by the high temperature corrosive attack.

Keywords: type II hot corrosion; diode laser; sealing; thermal spray; HVOF; cermet

---

1. Introduction

Hot corrosion is a severe degradation mechanism that occurs on steels and superalloys exposed to aggressive oxidizing environments, at high temperatures, in the presence of low-melting salts or oxides, which induce accelerated damage to the protective surface oxide layers [1].

The most commonly affected components are the nozzle guide vanes and rotor blades in the hot sections of gas turbines, which can get covered with deposits of sodium sulfate (formed by the reaction of sulfur from the fuel and sodium chloride from sea water) and vanadium oxide (formed from the oxidation of the vanadium present as an impurity in the combustion fuel) [2].

Two types of hot corrosion (known as type I and type II) occur in different and specific temperature ranges—type I attack is observed at 800–900 °C, in the presence of totally molten salt deposits that act to disrupt the surface oxide with a sudden and important increase in the corrosion rate; type II mechanism is active at lower temperatures (670–750 °C), and is characterized by a more localized attack, with little or no material loss underneath the pit [2].

Type II hot corrosion, rarely observed in aeroengines, is more frequently experienced by marine and industrial gas turbines that generally operate at lower temperatures.

Type II corrosion mechanism has often been proposed to be a liquid-phase attack by the mixed sulfates in the molten state [3,4], initiating corrosion by fluxing the protective oxide layer and rapidly propagating degradation by transporting the reactive species through the liquid, in the corrosion pits. However, more recent studies [5], do not support the molten sulfate attack theory, and rather indicate a solid reaction process involving the cooperative formation of nanosized sulfides and oxides, during corrosion front propagation.

The selection of materials for the construction of components operating in such extreme environments requires specific attention [1–8]. Nickel-base alloys are more resistant to type II hot corrosion than cobalt-base alloys; and NiCrAlY and NiCoCrAlY coatings are generally more resistant than CoCrAlY coatings. Additionally, increasing the chromium content improves the resistance of the material to both type I and type II hot corrosion attack [2].

Erosion resistance is often an additional important requirement for the coatings operating in the hot sections of gas turbines. Thermal-sprayed [9], chromium-based cermet coatings represent a viable option when it is necessary to provide high hardness and corrosion resistance at temperatures up to 900 °C [10–12]. Corrosion resistance and toughness are provided by the metallic matrix, while the wear resistance is mainly offered by the dispersed (usually carbide-based), ceramic phase.

Coating microstructure plays a fundamental role in the combined corrosion-wear resistance [13]. Porosity is certainly the main feature to be controlled, especially in the case of a hot corrosion degradation phenomenon where molten salts characterized by extremely low viscosities can easily penetrate through the permeable coatings. Partial decarburation of the ceramic phase can also represent a relevant issue, being potentially responsible for a decrease in the coating hardness [14–17], with detrimental consequences on the erosion/corrosion resistance, especially at higher temperatures.

In thermal-sprayed cermets, the specific type of spraying process and the selected deposition parameters can induce substantial variations in the composition and the microstructure of the coatings [18]. Coatings deposited by high velocity oxy-fuel (HVOF) generally exhibit dense microstructures, due to a high powder particle velocity and impact energy, even if relatively low temperatures are achieved in the plume [19,20]. Ni–Cr-based HVOF cermet overlays have, therefore, been proposed as the best-performing coatings, to overcome the serious consequences of hot corrosion in the presence of molten salts, e.g., for applications in power plant boilers [21–23].

However, in the most demanding operating conditions, residual porosities of a few percent units can still present a significant limitation, and the coatings applied for corrosion protection need to be made impervious by means of appropriate surface sealing treatments [24,25].

Laser technology has become the object of increasing interest as a viable, affordable, and reliable sealing and densification post-treatment process, in thermal spray coatings, for improving the performance of the coatings [26–35].

Gisario et al. [30] found consistent improvements in the mechanical properties of Diamalloy coatings deposited by high velocity oxy-fuel (HVOF), by reprocessing them by diode laser at a higher power and a reduced scan speed. Similarly, appropriate combinations of laser parameters have been shown to improve the structure density, hardness, and wear endurance of the WC–Co/NiCr coatings, using diode lasers, post-treatment [31].

Morimoto et al. [32] proposed a study on HVOF Cr_3C_2–25% NiCr cermet coatings, reprocessed by direct diode laser system. They showed an improvement in the micro-hardness and a material loss of the laser-treated Cr_3C_2–25% NiCr coatings, as evaluated by the sand erosion testing, and found it to be as low as 50% of that measured on the as-sprayed coatings.

While research on the effect of the CoCrAlY coatings densification processes on type I hot corrosion resistance have been reported in the recent literature [36], no study is currently available that has specifically investigated the effect of laser sealing on hot corrosion resistance of cermet coatings. Efficient densification of cermets is, in fact, expected to be hindered by the thermal coefficients mismatch of the metallic and ceramic phases coexisting in the coating, making the optimization of laser treatment and the formation of a protective densified layer more difficult. In [37], laser densification of HVOF WC–Co NiCr coatings has been reported for oxidation resistance, but surface treatment optimization procedures have not been discussed in detail, nor has the microstructure of densified layers been specifically investigated.

The present work investigated high power diode laser reprocessing of metal–ceramic coatings of various compositions, deposited by HVOF: (a) Cr_3C_2–NiCr coatings, widely applied on the industrial components working at a high temperature, to improve erosion, corrosion, and oxidation resistance; (b) Cr_3C_2–MCrAlY (with M = Ni, Co or a combination, thereof) overlay coatings, commonly used in gas turbines to protect the surface of blades against oxidation and hot corrosion; and (c) an original composite coating consisting of a Ni–Cr matrix, containing a combined dispersion of large mullite $(3Al_2O_3 \cdot 2SiO_2)$ particles and silica nanoparticles. The original composite coating was developed in [38], on the basis of studies that have reported that the presence of silicate minerals in Ni-based metallic matrix could be beneficial for the resistance to hot corrosion phenomena, in heavy oil-fueled diesel engines [39,40].

The optimal laser treatment conditions were identified for each type of coating, and the effect of coating densification on composition, microstructure and type II hot corrosion resistance was investigated. The main purpose of the proposed experiment was to produce experimental evidence of the potential beneficial effect of laser sealing procedures on hot corrosion resistance of coatings, rather than offering novel insight for the type II hot corrosion mechanism. Very dissimilar compositions were in fact tested, and no common mechanism could be envisaged for the different investigated coatings.

2. Materials and Methods

2.1. Starting Powders and Coatings Deposition

Two commercially available Cr_3C_2-based cermet powders were selected for the coatings deposition (composition, size distribution, and suppliers are reported in Table 1)—a standard Cr_3C_2–NiCr formulation (designated as CRCZ), and a Cr_3C_2–CoNiCrAlY (CRCY) expressly designed for an improved high-temperature-corrosion-resistance. The fraction of the metal matrix was 25 wt % for both CRCZ and CRCY.

Table 1. Designation and composition of powders used in the experimental activity.

Name	Composition	Particle Size	Trade Name
CRCZ	75 wt % Cr_3C_2 25wt % Ni20Cr	$-45 + 15$ μm	Sulzer-WOKA 7302
CRCY	75 wt % Cr_3C_2 25wt % CoNiCrAlY (Co 9.50 Ni 7.50 Al 1.75 Y 0.20 C10 Cr bal.)	$-45 + 15$ μm	FST-K880.23
MSN	10wt % nano-SiO_2 30wt % mullite $(3Al_2O_3\text{-}2SIO_2)$ 60wt % Ni20Cr	$-53 + 20$ μm (sieved)	Purposely produced

A non-commercial mullite nano–silica–NiCr powder (designated as MSN), originally proposed and developed by the authors in [36], was also included in the experimental campaign, in order to investigate the influence of post-deposition laser treatment on cermet coatings characterized by unconventional ceramic reinforcements, and by a remarkably higher fraction (60 wt %) of the metallic matrix. The powder was obtained by mixing, metallic (NiCr–HC Starck-Amperit 251.051, H.C. Starck GmbH, Goslar, Germany) and ceramic (mullite–Nabaltec Symulox M 72 K0C, Nabaltec AG Schwandorf, Germany) powders, in aqueous suspension, with addition of the nano–silica particles (Grace LUDOX TM50, W.R. Grace, Columbia, MD, USA), with an average diameter of 22 nm. The suspensions were spray-dried to obtain a granulated powder that was sieved to $-53 + 20$ μm size distribution.

All particle size specifications reported in Table 1 were provided by powder suppliers and were confirmed by sieving and laser scattering granulometry (Malvern Mastersize 2000, Malvern Panalytical Ltd, Malvern, UK).

All materials were deposited onto 3 mm thick, 20 mm in diameter, Nimonic disks, previously sand-blasted with alumina grit, to improve adhesion by mechanical interlocking.

Coatings were deposited by HVOF, using a Praxair Tafa JP-5000 kerosene–oxygen gun (Praxair, Danbyry, CT, USA). The process parameters were previously optimized by the Design of Experiment (DoE) procedures [38], with the aim of producing dense and hard coatings, offering the maximum protection against both penetration of corrosive media and erosion by solid particles; summarized in Table 2.

Table 2. Deposition parameters of high velocity oxy-fuel (HVOF) thermal-sprayed coatings.

Composition	CRCZ	CRCY	MSN
Kerosene feed rate (gph)	6.5	6.5	6
O_2 feed rate (scfh)	2000	2000	1850
Spray distance (mm)	350	370	350

2.2. Laser Post-Treatment

A diode laser (Rofin Sina DL015), with a wavelength of 940 nm and an elliptical spot size of 3.8 mm × 1.2 mm, with a maximum power of 1500 W, was used for the post-deposition laser treatment of all coatings. Argon was used as an assist gas.

Two laser scanning trajectories were initially tested—a circular spiral-like trajectory (circumference-to-center), characterized by high scan rates, V, of 13–15 mm/s (depending on the position), and a raster trajectory with lower scan rates (3–8.5 mm/s) but which allowed for a better precision and a limited overlapping between subsequent scans. Final laser treatments were performed in raster geometry, mainly for reasons correlated with the requirement of optimal control of the overlapping passes in the central area of the samples, to be used for the hot corrosion resistance experiments.

Based on preliminary "trial and error"-driven tests—aimed at identifying the conditions for a surface sealing without total remelting of the coatings or massive interdiffusion with the substrate—the following variation ranges for the laser treatment parameters (power, W, and scan rate, mm/s) for the three coating compositions were identified—200–800 W and 2–6.5 mm/s for CRCZ, 700–800 W and 3–4 mm/s for CRCY, 200–400 W and 2–4 mm/s for MSN.

A finer tuning of the optimal treatment conditions was then performed investigating the power/scan rate combinations reported in Tables 3–5, for CRCZ, CRCY, and MSN, respectively, where the corresponding value of fluence, or radiant exposure, H, was also reported. Fluence represents the radiant energy absorbed by the surface per unit area and was defined as

$$H = P \cdot t_{\text{interaction}}/A_{\text{spot}} \tag{1}$$

where H = fluence (J/mm^2); P = power (W); $t_{\text{interaction}}$ = interaction time (s) between laser and surface.

Table 3. Laser treatment parameters for the Cr_3C_2–NiCr (CRCZ) coatings.

Test No.	Power P (W)	Scan Rate V (mm/s)	Fluence H (J/mm^2)
1	800	4	53.08
2	800	3	70.77
3	700	3	61.92
4	200	2	26.54
5	300	2	39.81
6	400	2	53.08
7	500	2.5	52.08
8	600	2.5	63.69
9	300	3	26.54
10	400	4	26.54
11	500	5	26.54
12	300	4	20.51
13	400	5	20.51
14	500	6.5	20.51

Table 4. Laser treatment parameters for the Cr_3C_2–CoNiCrAlY (CRCY) coatings.

Test No.	Power P (W)	Scan Rate V (mm/s)	Fluence H (J/mm^2)
1	800	4	53.08
2	800	3	70.77
3	700	4	46.44
4	700	3	61.92

Table 5. Laser treatment parameters for the mullite nano–silica–NiCr (MSN) coatings.

Test No.	Power P (W)	Scan Rate V (mm/s)	Fluence H (J/mm^2)
1	200	2	26.54
2	200	4	13.27
3	300	2	39.81
4	300	4	20.51
5	400	2	53.08
6	400	4	26.54

It was observed that the same value of H could be obtained with different combinations of P and V (as in tests 9, 10, and 11, for example). These different combinations were expressly included in the test matrix to investigate the correlation between laser parameters and the kinetics of physico-chemical processes occurring during material/laser interaction.

The selected values of fluence for tests on different materials were strongly dependent on the composition and thermal properties of the metallic matrix and ceramic reinforcement. Fluence for the treatment of MSN coatings was generally much lower, taking into account the limited fraction of ceramic phase, in this last composition.

2.3. Hot Corrosion Tests

Hot corrosion tests simulating type II hot corrosion were carried out in air, at 700 °C, on both the as-sprayed and the laser-treated samples. A paste made of a mixture of 40 wt % Na_2SO_4–60 wt % V_2O_5 dispersed in distilled water, was applied on the central portion of the coated discs (0.8 cm^2), obtaining a total salt concentration of 3 mg/cm^2. To avoid salt dispersion during high temperature exposure, a refractory cement was used to mask the outer circular area of the disks. The slurry was then dried, and the sample was placed in the furnace for a high temperature cycle. At the end of each cycle, a sample for every coating, in both, the as-sprayed and laser-treated condition, was analyzed, as described below, to investigate the corrosion scale evolution. At least 20 measures of corrosion layer thickness were performed for each observed sample. On the remaining samples, the salt mixture was refurbished to perform a new cycle. Evaluation analyses were carried out after 50, 100, and 200 h of exposure.

2.4. Coatings Characterization

Characterization of the laser-treated coatings was carried out by visual inspection of the top view, by optical microscopy (Nikon Eclipse L150, Nikon Instruments Europe B.V., Amsterdam, the Netherlands), assisted by image analysis software Leica Qwin V. 2.2 and Lucia™ 4.80), and by scanning electron microscopy (Philips SEM XL40, FEI, Hillsboro, OR, USA), assisted by EDS microanalysis EDAX Octane SDD) of the cross sections.

Optimal laser treatment conditions were identified based on the impact on surface quality, coating thickness and microstructure, and micro and macro-porosity measured by image analysis, hardness, presence of vertical cracks, depth of remelting, and interdiffusion with the substrate.

Hot corrosion resistance of the as-sprayed and laser-sealed coatings was evaluated, using the above-mentioned instrumental techniques, assessing the microstructure and the composition of the corrosion scale, formed after 0, 50, 100, and 200 h corrosion treatments, and estimating the total recession of the coating (difference between the thickness of original coating and the residual thickness of the hot corroded coatings, under the corrosion scale).

Vickers microhardness measurements (HV0.05) were performed, both, on the cross section and on the top surface of the samples, following the ASTM E-384-89 standard [41] using a Leica VMHT microhardness tester (min least 30 measures per sample).

The identification of the crystalline phases in the corrosion scales was carried out by X-ray diffraction analysis (XRD), using a PANalytical X'Pert3 diffractometer (Malvern Panalytical Ltd, Malvern, UK). XRD measurements were carried out at 40 kV and 40 mA, with Cu Kα radiation ($\lambda_{K\alpha 1}$ = 1.540598 Å, $\lambda_{K\alpha 2}$ = 1.544426 Å), with a scan range of 15°–70° (2θ), step size 0.02°, and an acquisition time of 20 s/step.

3. Results and Discussion

3.1. Optimization of the Post-Deposition Laser Treatment

Different laser treatment conditions were tested (three tests for each power/scan rate combination) on the CRCZ, CRCY, and MSN samples. Examples of the laser patterns are shown in Figure 1a–c, respectively.

Different undesired effects (Figure 2) was induced on the coating microstructure and morphology, following the post-deposition laser treatment:

- Vertical cracks originating within the coating, as a consequence of tensile stresses caused by excessive densification (too high initial porosity or too high fluence); Figure 2a, CRCZ coating, track 6, 400 W, 2 mm/s and Figure 2d, CRCY coating, track 4, 700 W, 3 mm/s;
- Iincorporation of large amounts of the shielding gas inside the molten material, with formation of residual macro-porosities in the coating (mainly caused by too high a laser power, in the presence of high scan rates, where the viscosity of the molten material was drastically reduced and the interaction time was not sufficient to allow for proper degassing); Figure 2b, CRCZ coating, track 1, 800 W, 4 mm/s;
- Selective remelting or evaporation and recondensation in the shape of surface droplets of the lower melting fraction of the material (too long interaction times); Figure 2c, MSN coating, track 3, 300 W, 2 mm/s.

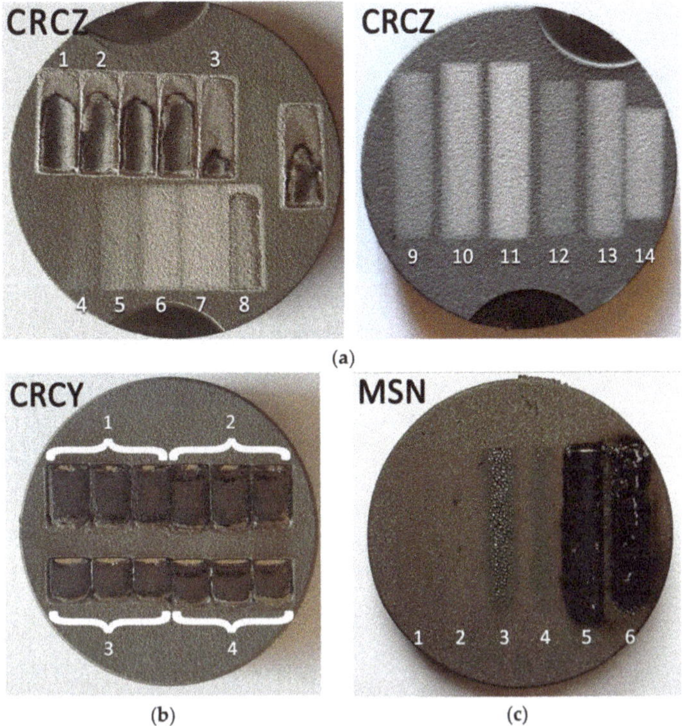

Figure 1. Laser treatment optimization tests: (**a**) CRCZ, (**b**) CRCY, and (**c**) MSN coating composition. Test numbers conditions are reported in Tables 3–5, respectively.

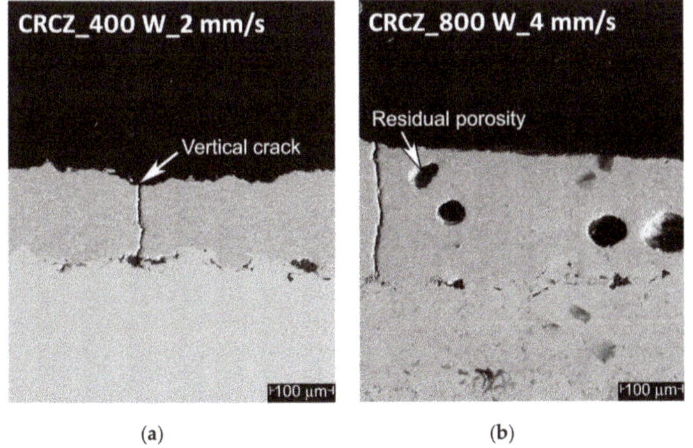

Figure 2. *Cont.*

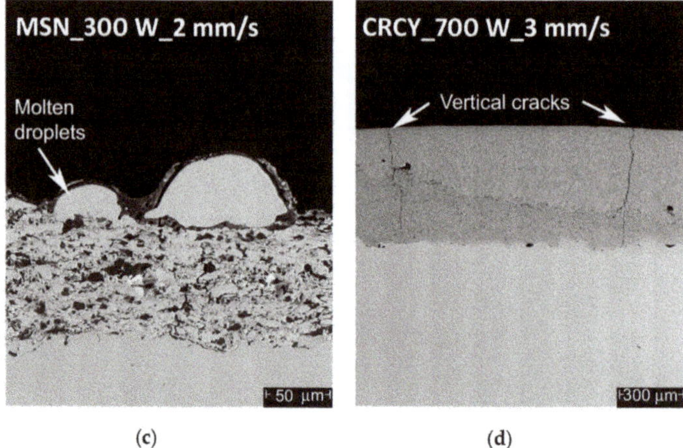

(c)   (d)

Figure 2. Potential undesired effects of the laser interaction with the coating: (**a**) Vertical cracks; (**b**) incorporation of the assist gas inside the molten material and formation of the residual macro-porosity; (**c**) selective remelting or evaporation and recondensation of the low melting material in surface droplets; and (**d**) partial remelting of the coating with the formation of vertical cracks.

Optimal microstructures were identified for the laser treatments, indicated in Tables 3–5 with bold character: 200 W-2 mm/s for CRCZ, 700 W-4 mm/s for CRCY, and 300 W-4 mm/s for MSN.

The cross-sections of the as-sprayed coatings are shown in Figure 3a, Figure 4a, and Figure 5a, while the samples treated by the optimal laser process are shown in Figure 3d, Figure 4d, and Figure 5d. It should be noted that the dark areas appearing in the MSN coatings cross-sections corresponded to large mullite grains and are not to be mistaken for porosity.

For Cr_3C_2–CoNiCrAlY, the desired effect of surface sealing was completely achieved (Figure 4d), and a depth of about half the original coating thickness was effectively densified.

Cr_3C_2–NiCr coatings exhibited a higher sensitivity to vertical cracking. For this reason, a higher number of laser treatment conditions were evaluated. The optimal treatment parameters were, therefore, selected to minimize the coating fragmentation (Figure 3d). Analogous laser treatment conditions for the Cr_3C_2–NiCr coatings have also been reported in the literature [32].

Finally, mullite nano–silica–NiCr coatings were found to be highly susceptible to nickel–chrome evaporation and redeposition in spherical droplets of about 200 μm diameter. On the other hand, in coatings treated with intermediate values of power and high scan rates, densification induced by laser interaction with the coating was limited to a thin (about 15 μm) surface layer, as shown in Figure 5d.

For the selected laser-treated coatings, microhardness and porosity were compared with the same features in the as-sprayed conditions. Results are shown in Figures 6 and 7, respectively.

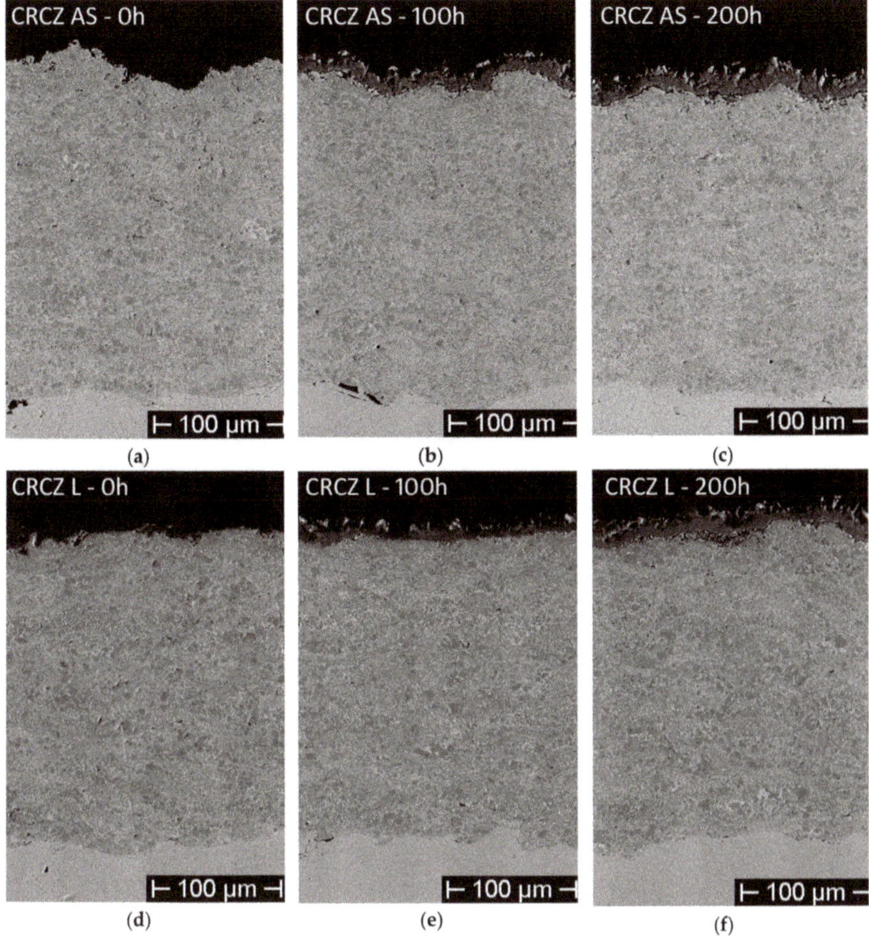

Figure 3. Cross-sections of the CRCZ samples: (**a**) As-sprayed; (**b**) as-sprayed after 100 h of hot corrosion test; (**c**) as-sprayed after 200 h of hot corrosion test; (**d**) laser-treated; (**e**) laser-treated after 100 h of hot corrosion test; and (**f**) laser-treated after 200 h of hot corrosion test.

In terms of densification, the best results were achieved for the CRCY coatings, whose porosity was reduced by an order of magnitude (from 3.2% to 0.15%), with a corresponding 50% improvement in hardness. CRCZ and MSN coatings showed negligible differences in terms of porosity (Figure 7) and hardness was measured in the sample cross-section (Figure 6a), highlighting a lack of in-depth microstructure modification, induced by the laser treatments. However, the analysis of the microhardness measured on the top surface of the treated samples (Figure 6b), revealed an increase in the coating hardness, as compared to, both, the top surface of the as-sprayed coatings and cross-section of the laser-treated samples. Specifically, CRCZ–L and MSN–L coatings exhibited an improvement in hardness, as measured on the top surface, by about 20% and 50%, respectively, compared to the untreated samples. These results suggest the presence of thin, harder, surface layers modified by laser treatments, which were barely distinguishable in the cross-section micrographs, and were certainly characterized by a locally-sealed and denser microstructure.

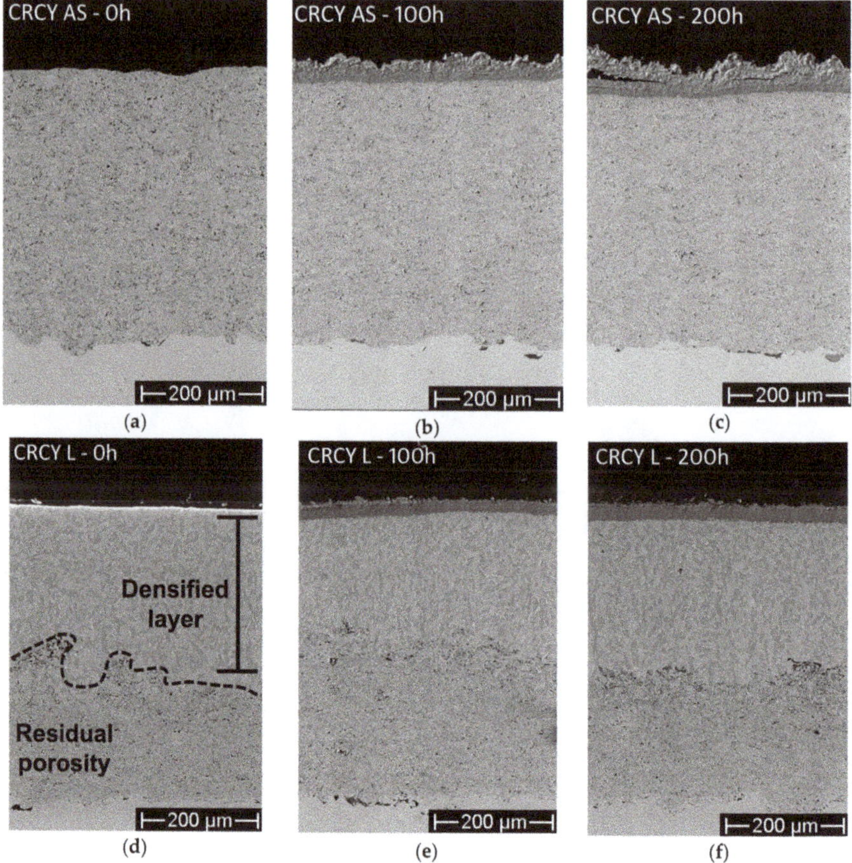

Figure 4. Cross-sections of the CRCY samples: (**a**) As-sprayed; (**b**) as-sprayed after 100 h of hot corrosion test; (**c**) as-sprayed after 200 h of hot corrosion test; (**d**) laser-treated; (**e**) laser-treated after 100 h of hot corrosion test; and (**f**) laser-treated after 200 h of hot corrosion test.

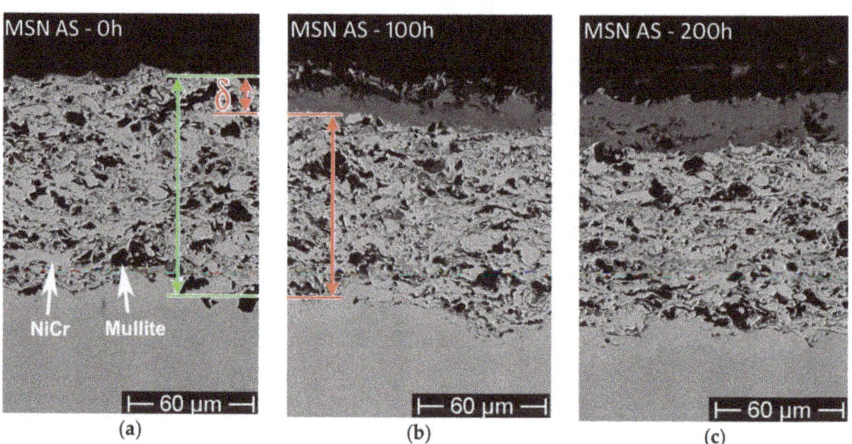

Figure 5. *Cont.*

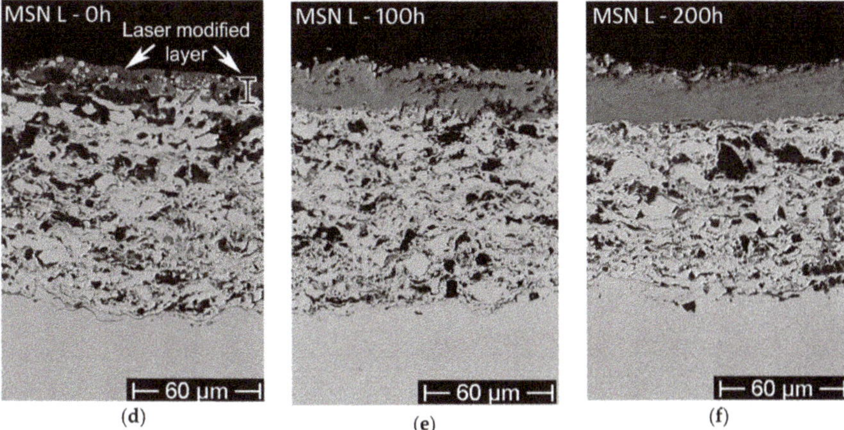

Figure 5. Cross-sections of the MSN samples: (**a**) As-sprayed; (**b**) as-sprayed after 100 h of hot corrosion test; (**c**) as-sprayed after 200 h of hot corrosion test; (**d**) laser-treated; (**e**) laser-treated after 100 h of hot corrosion test; and (**f**) laser-treated after 200 h of hot corrosion test.

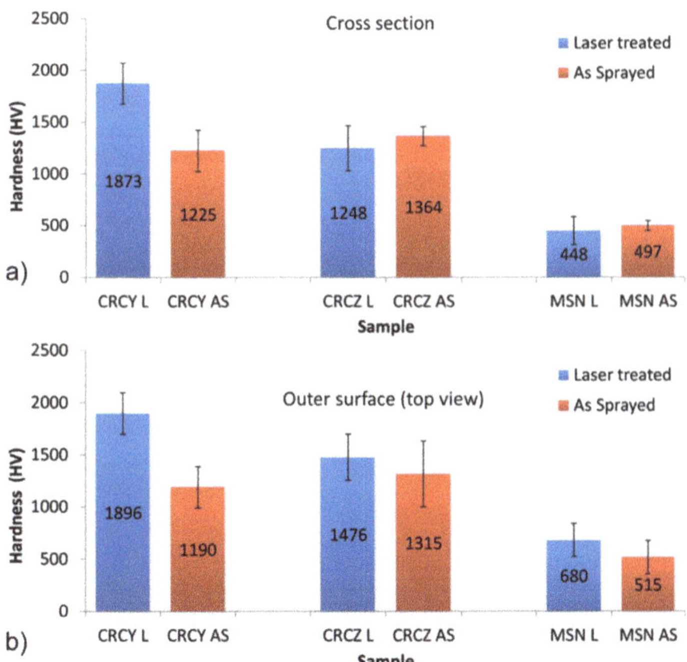

Figure 6. Microhardness of the as-sprayed (AS) and laser-treated (L) samples measured on the cross-section (**a**) and the top surface (**b**) (error bars indicate +/− standard deviation).

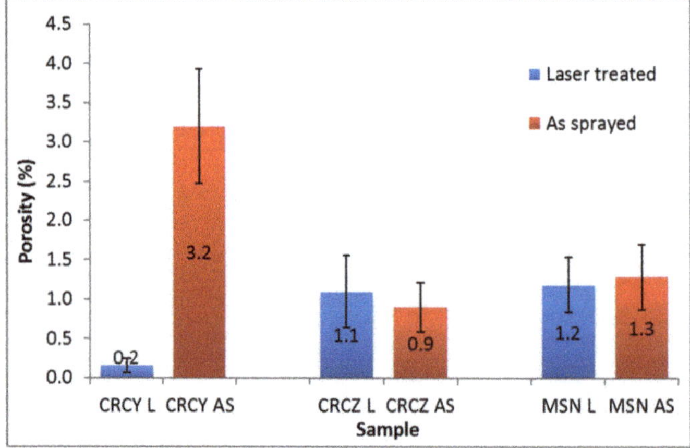

Figure 7. Porosity of the as-sprayed (AS) and laser-treated (L) samples. (error bars indicate +/− standard deviation).

Results of the XRD analyses of the top surfaces are presented in Figures 8–10 for the as-sprayed and laser-treated coatings (see left-hand patterns). No significant variations in phase composition were induced by the laser treatment for the CRCZ and MSN compositions, while a certain degree of oxidation of Al and Y was evidenced for the CRCY laser-treated coatings, most probably as a consequence of the higher energy input of the reprocessing conditions.

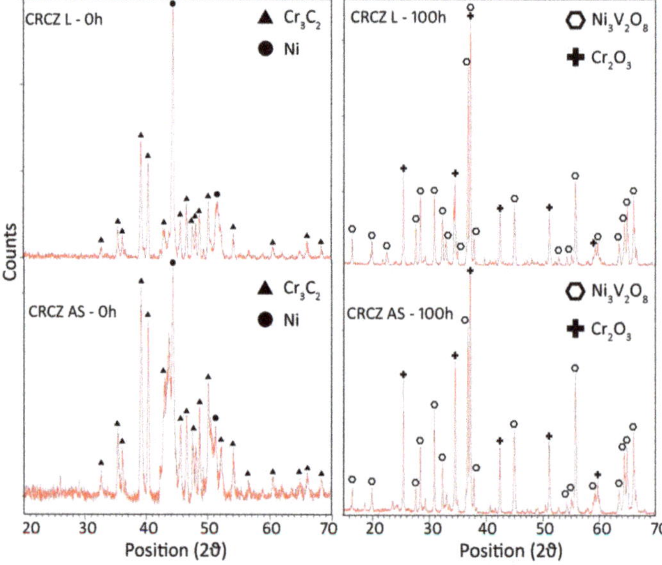

Figure 8. XRD patterns for the CRCZ samples: AS—as sprayed (lower left); L—laser treated (upper left); AS-100 h—as sprayed, exposed for 100 h (lower right); and L-100h—laser treated, exposed for 100 h (upper right).

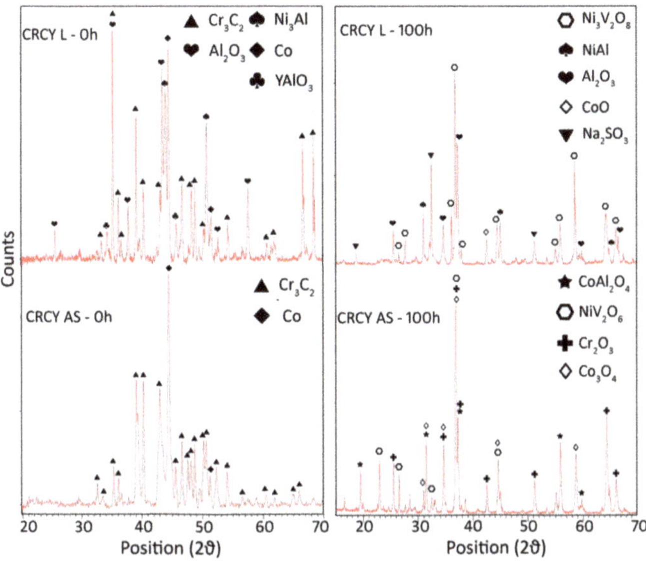

Figure 9. XRD patterns for the CRCY samples: AS—as-sprayed (lower left); L—laser-treated (upper left); AS-100 h—as-sprayed, exposed for 100 h (lower right); L-100h—laser-treated, exposed for 100 h (upper right).

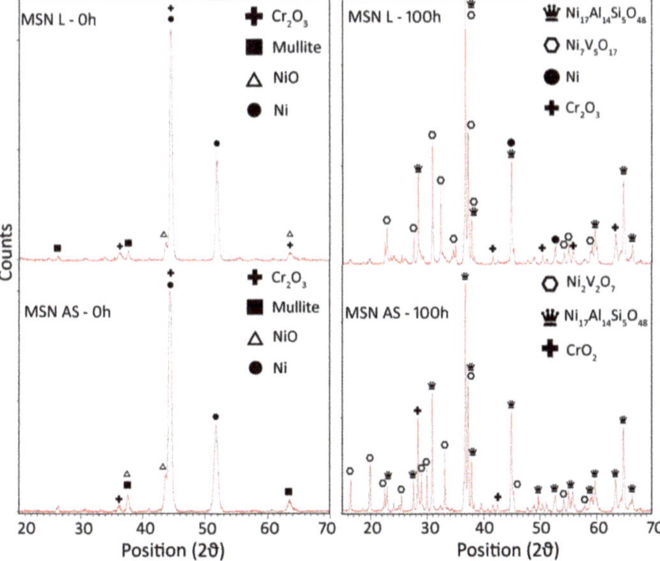

Figure 10. XRD patterns for the MSN samples: AS—as-sprayed (lower left); L—laser-treated (upper left); AS-100 h—as-sprayed, exposed for 100 h (lower right); L-100h—laser-treated, exposed for 100 h (upper right).

3.2. Hot Corrosion Results

Following exposure of the as-sprayed and laser-densified samples to the action of aggressive salts at 700 °C for 50, 100, and 200 h, the cross-sections were examined by SEM and the microstructure, thickness, and composition of the scale were evaluated. Cross-sections of the samples exposed for 100 and 200 h are reported in Figure 3b,c, Figure 4b,c and Figure 5b,c, for Cr_3C_2–NiCr, Cr_3C_2–CoNiCrAlY and mullite-nanosilica-NiCr, respectively.

3.2.1. Microstructure

Hot corrosion scales appeared generally continuous and free from relevant fragmentation, horizontal cracking, detachment, or spallation, for all coating materials, with the exception of the as-sprayed CRCY exposed for 200 h.

The following evidence was gathered for the individual compositions:

- For the CRCZ as-sprayed coatings (Figure 3a–c), a scale of about 25 µm was formed after 100 h, made of an inner layer of about 10 µm, compact and continuous, and an outer, less dense, coat, characterized by the presence of both acicular and sharp-cornered non-cohesive phases. The same corrosion scale morphology was highlighted in the CRCZ laser-treated samples, after 100 h of exposure test (Figure 11a).

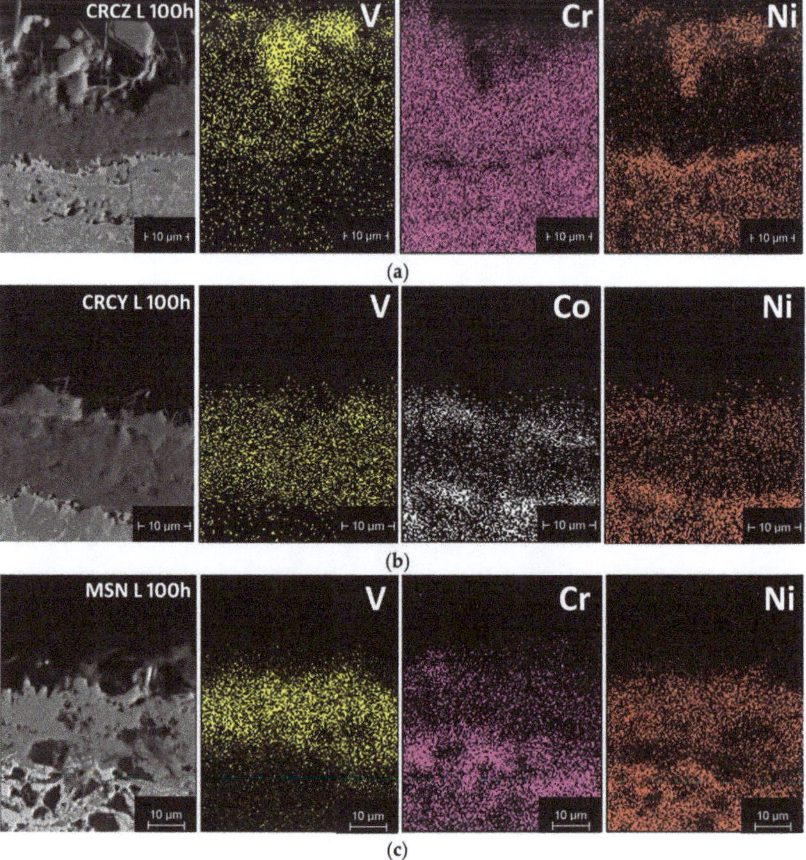

Figure 11. EDS analysis of the corrosion scale for laser-treated samples, after 100 h of the corrosion test: (a) CRCZ; (b) CRCY; and (c) MSN.

After 200 h, the scale thickness was slightly increased, but the inner compact layer had reached a total thickness of almost 20 µm (Figure 3c).

The laser-treated coatings (Figure 3d–f, Figure 11a) exhibited the same scale microstructure but for each exposure time the thickness of the corrosion layer was lower than the as-sprayed coatings.

- For CRCY, the thickness of the scale increased with the number of hot corrosion cycles for, both, the as-sprayed (Figure 4a–c) and laser-treated coatings (Figure 4d–f), but in this case, a more evident beneficial effect of the sealing treatment could be observed—after 200 h the as-sprayed coatings exhibited a spalled, detached, and thick (≈50 μm) corrosion scale, whereas, the laser-treated samples showed thinner and more compact surface layers formed by hot corrosion products.
- The as-sprayed (Figure 5a,c) and laser-treated (Figure 5d–f) MSN samples did not show significant differences in terms of hot corrosion behavior—the scale thickness was comparable for each exposure time, while the corrosion layers of the laser-treated samples seemed to be slightly more compact and cohesive than the untreated.

3.2.2. Composition of the Scale

Elemental maps of the cross-sections of the laser-treated coatings, after 100 h of hot corrosion exposure, were acquired, to identify the distribution of the original coating elements, throughout the scale thickness, evidencing any diffusion or enrichment phenomenon that had occurred during high temperature exposure, and the presence of elements from the salt mixture, in the corrosion products.

In the laser-treated CRCZ-L coatings (Figure 11a), a clear evidence of the conforming distributions of nickel and vanadium confirmed the formation of a nickel vanadate in the reaction products, concentrated in the superficial layer (about 10 μm thick), where the presence of acicular and cuboidal structures was previously reported. Chromium remained uniformly distributed in the rest of the scale, whereas sodium and sulfur (not shown in Figure 11) were only present in small traces.

EDS microanalysis of the scale formed on the CRCY-L coatings (Figure 11b), evidenced an analogous behavior, but the microstructure of the surface vanadate layer appeared denser and more compact. Moreover, the analysis revealed the presence of cobalt in the outer part of the scale and traces of aluminum, sulfur, and sodium (not reported in Figure 11b), thus, confirming the mechanism involving the migration of cobalt ions proposed in [3].

The MSN-L coatings showed the presence of Ni, Cr, and V, in the corrosion scale, as reported in Figure 11c.

The corrosion products of the hot corroded samples were also analyzed by means of X-ray diffraction—the results are reported in Figures 8–10 for the as-sprayed and laser-treated coatings (see right-hand patterns). The composition of the scale was unchanged for the CRCZ compositions in the sealed and unsealed conditions. Sodium sulfate residues were only identified in the scale of laser-treated CRCY coatings, together with cobalt oxides of various stoichiometry and mixed nickel vanadium oxides. The formation of mixed nickel aluminum silicon oxides and nickel vanadium oxides was confirmed for the MSN coatings, with small differences in the stoichiometry of the oxidized phases. These results suggested, in general, that the hot corrosion mechanism was not modified by the coating sealing treatment.

3.2.3. Recession

Results in terms of the coating recession (marked as δ in Figure 5a and measured as the difference between the original thickness of the coating and the thickness of the remaining unaffected coating after the hot corrosion treatment) are shown in Figure 12. Standard deviations were not reported to facilitate the graph readability, and they were all contained between ±10% of the mean value.

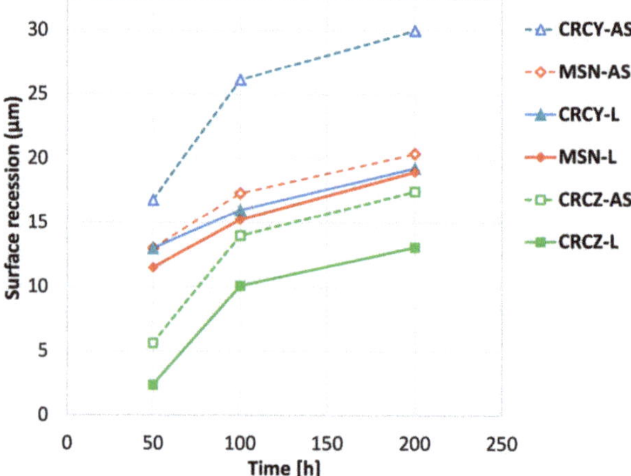

Figure 12. Coating recession caused by hot corrosion exposure of increasing duration for the as-sprayed (AS, dashed line) and laser-treated (L, solid line) coatings.

Among the as-sprayed coatings (dashed lines), the lowest recession ($\approx$17 µm at 200 h) was exhibited by the CRCZ deposits, while the highest coating consumption affected the CRCY samples ($\approx$30 µm after 200 h). However, the different effectiveness of the laser surface treatment on the coatings of different compositions modified the protection ability of the deposits—Figure 12 shows a significant improvement induced by the laser on the CRCZ coatings (with a final recession of only 13 µm at 200 h) and an almost negligible effect on the MSN coatings. On the other hand, it must be noted that the highest improvement of hot corrosion behavior was shown by the CRCY coatings, for whom the recession decreased from 30 µm before treatment to 19 µm after laser sealing.

4. Conclusions

Ni–Cr based cermet coatings, with different ceramic content and variable metal matrix composition, were treated after the HVOF deposition by a high-power diode laser to induce a surface sealing effect aimed to improve their hot corrosion resistance.

Cermet coatings proved extremely sensitive to the surface treatment parameters (in terms of specific laser power–scan rate combinations), and the identification of optimal laser reprocessing conditions required extensive experimental activities, in order to induce appropriate surface sealing, while avoiding concurrent coating fragmentation, gas incorporation, or substrate remelting.

The results of the corrosion resistance characterization of the as-sprayed and laser-treated coatings (carried out in air, at 700 °C, for up to 200 h in the presence of sodium sulfate and vanadium pentoxide mixture) evidenced that:

- The mechanism of the degradation occurring in typical type II hot corrosion conditions was not substantially altered by the formation of a surface-laser-densified layer, as confirmed by the analysis of the corrosion products of the treated and as-sprayed samples exhibiting an analogous microstructure and composition;
- The formation of surface compact layers (with overall thickness and microstructure very much dependent on coating composition) was responsible for a considerable increase of surface hardness of all coatings, and for a consistent improvement in the hot corrosion resistance, promoting the formation of thinner and more compact corrosion scales, and considerably reducing the surface recession rate (up to 60%, after 200 h exposure for Cr_3C_2–25% CoNiCrAlY coatings).

Author Contributions: Conceptualization: L.B., G.P., F.M. and A.G.; Investigation: L.B., F.M., A.G. and G.P.; Data Curation and Formal Analysis: C.B., F.M. and G.P.; Writing—Original Draft Preparation, Review and Editing: C.B.

Funding: This research received no external funding.

Conflicts of Interest: The authors declare no conflict of interest.

References

1. Eliaz, N.; Shemesh, G.; Latanision, R.M. Hot corrosion in gas turbine components. *Eng. Fail. Anal.* **2002**, *9*, 31–43. [CrossRef]
2. Lai, G.Y. High-Temperature Corrosion of Engineering Alloys. ASM International: Metals Park, OH, USA, 1990.
3. Pettit, F. Hot corrosion of metals and alloys. *Oxid. Met.* **2011**, *76*, 1–21. [CrossRef]
4. Bornstein, N.S. Reviewing sulfidation corrosion—Yesterday and today. *J. Met.* **1996**, *11*, 37–39.
5. Zhang, W.-J.; Sharghi-Moshtaghin, R. Revisit the type II corrosion mechanism. *Metall. Mater. Trans. A* **2018**, *49*, 4362–4372. [CrossRef]
6. Birks, N.; Meier, G.H.; Pettit, F.S. *Introduction to the High Temperature Oxidation of Metals*, 2nd ed.; Cambridge University Press: Cambridge, UK, 2006.
7. Young, D.J. *High Temperature Oxidation and Corrosion of Metals*, 1st ed.; Elsevier: Amsterdam, The Netherlands, 2008.
8. Rajendran, R. Gas turbine coatings–An overview. *Eng. Fail. Anal.* **2008**, *26*, 355–369. [CrossRef]
9. Pawlowski, L. *The Science and Engineering of Thermal Spray Coatings*, 2nd ed.; John Wiley & Sons: Hoboken, NJ, USA, 2008.
10. Saini, H.; Kumar, D.; Shukla, V.N. Hot corrosion behaviour of nanostructured cermet based coatings deposited by different thermal spray techniques: A review. *Mater. Today Proc.* **2017**, *4*, 541–545. [CrossRef]
11. Valente, T.; Bartuli, C.; Tului, M. Thermal Sprayed Hard Cr_3C_2-Ni Cr Coatings for Wear Protection. In *Thermal Spray 2001: New Surfaces for a New Millennium*; ASM International: Materials Park, OH, USA, 2001; pp. 1075–1084.
12. Kamal, S.; Jayaganthan, R.; Prakash, S. Evaluation of cyclic hot corrosion behaviour of detonation gun sprayed Cr_3C_2-25%NiCr coatings on nickel- and iron-based superalloys. *Surf. Coat. Technol.* **2009**, *203*, 1004–1013.
13. Bluni, S.T.; Marder, A.R. Effects of thermal spray coating composition and microstructure on coating response and substrate protection at high temperatures. *Corrosion* **1996**, *52*, 213–218. [CrossRef]
14. Di Girolamo, G.; Pilloni, L.; Pulci, G.; Marra, F. Tribological characterization of WC-Co plasma sprayed coatings. *J. Am. Ceram. Soc.* **2009**, *92*, 1118–1124. [CrossRef]
15. Bartuli, C.; Valente, T.; Cipri, F.; Bemporad, E.; Tului, M. Parametric study of an HVOF process for the deposition of nanostructured WC-Co coatings. *J. Therm. Spray Technol.* **2005**, *14*, 187–195. [CrossRef]
16. Bartuli, C.; Cipri, F.; Valente, T.; Verdone, N. CFD Simulation of an HVOF Process for the Optimization of WC-Co Protective Coatings. *WIT Trans. Eng. Sci.* **2003**, *39*, 71–83.
17. Bartuli, C.; Valente, T.; Cipri, F.; Bemporad, E. Nanostructured Wear Resistant Coatings Deposited by HVOF. In *Surface Engineering: Coatings and Heat Treatments*; Popoola, O., Dahotre, N.B., Midea, S., Kopech, H., Eds.; ASM International: Materials Park, OH, USA, 2003; pp. 480–488.
18. Rickerby, D.S.; Eckold, G.; Scott, K.T.; Buckley-Golder, I.M. The interrelationship between internal stress, processing parameters and microstructure of physically vapour deposited and thermally sprayed coatings. *Thin Solid Films* **1987**, *154*, 125–141. [CrossRef]
19. Lih, W.-C.; Yang, S.H.; Su, C.Y.; Huang, S.C.; Hsu, I.C.; Leu, M.S. Effects of process parameters on molten particle speed and surface temperature and the properties of HVOF CrC/NiCr coatings. *Surf. Coat. Technol.* **2000**, *133*, 54–60. [CrossRef]
20. Sidhu, T.S.; Prakash, S.; Agrawal, R.D. State of the art of HVOF coating investigations-A review. *Mar. Technol. Soc. J.* **2005**, *39*, 53–64. [CrossRef]
21. Sreenivasulu, V.; Manikandan, M. Hot corrosion studies of HVOF sprayed carbide and metallic powder coatings on alloy 80A at 900 °C. *Mater. Res. Express* **2019**, *6*, 036519. [CrossRef]
22. Zhou, W.; Zhou, K.; Deng, C.; Zeng, K.; Li, Y. Hot corrosion behaviour of HVOF-sprayed Cr3C2-NiCrMoNbAl coating. *Surf. Coat. Technol.* **2017**, *309*, 849–859. [CrossRef]

23. Sidhu, T.S.; Prakash, S.; Agrawal, R.D. Hot corrosion studies of HVOF sprayed Cr3C2-NiCr and Ni-20Cr coatings on nickel—based superalloy at 900 °C. *Surf. Coat. Technol.* **2006**, *201*, 792–800. [CrossRef]

24. Chatha, S.S.; Sidhu, H.S.; Sidhu, B.S. The effects of post-treatment on the hot corrosion behavior of the HVOF-sprayed Cr_3C_2-NiCr coating. *Surf. Coat. Technol.* **2012**, *206*, 4212–4224. [CrossRef]

25. Houdková, S.; Smazalová, E.; Vostřák, M.; Schubert, J. Properties of NiCrBSi coating, as sprayed and remelted by different technologies. *Surf. Coat. Technol.* **2014**, *253*, 14–26. [CrossRef]

26. Smurov, I.; Uglov, A.; Krivongov, Y.; Sturlese, S.; Bartuli, C. Pulsed laser treatment of plasma sprayed thermal barrier coatings: Effect of pulse duration and energy input. *J. Mater. Sci.* **1992**, *27*, 4523–4530.

27. Garcia-Alonso, D.; Serres, N.; Demian, C.; Costil, S.; Langlade, C.; Coddet, C. Pre-/during-/post-laser processes to enhance the adhesion and mechanical properties of thermal-sprayed coatings with a reduced environmental impact. *J. Therm. Spray Technol.* **2011**, *20*, 719–735. [CrossRef]

28. Koivuluoto, H.; Milanti, A.; Bolelli, G.; Latokartano, J.; Marra, F.; Pulci, G.; Vihinen, J.; Lusvarghi, L.; Vuoristo, P. Structures and properties of laser-assisted cold-sprayed aluminum coatings. *Mater. Sci. Forum* **2017**, *879*, 984–989. [CrossRef]

29. Liu, Z.; Liu, H.; Viejo, F.; Aburas, Z.; Rakhes, M. Laser-induced microstructural modification for corrosion protection. *Proc. Inst. Mech. Eng. Part C J. Mech. Eng. Sci.* **2010**, *224*, 1073–1085. [CrossRef]

30. Gisario, A.; Barletta, M.; Veniali, F. Laser surface modification (LSM) of thermally-sprayed Diamalloy 2002 coating. *Opt. Laser Technol.* **2012**, *44*, 1942–1958. [CrossRef]

31. Gisario, A.; Puopolo, M.; Venettacci, S.; Veniali, F. Improvement of thermally sprayed WC-Co/NiCr coatings by surface laser processing. *Int. J. Refract. Met. Hard Mater.* **2015**, *52*, 123–130. [CrossRef]

32. Morimoto, J.; Sasaki, Y.; Fukuhara, S.; Abe, N.; Tukamoto, M. Surface modification of Cr_3C_2-NiCr cermet coatings by direct diode laser. *Vacuum* **2006**, *80*, 1400–1405. [CrossRef]

33. Ghasemi, R.; Shoja-Razavi, R.; Mozafarinia, R.; Jamali, H.; Hajizadeh-Oghaz, M.; Ahmadi-Pidani, R. The influence of laser treatment on hot corrosion behavior of plasma-sprayed nanostructured yttria stabilized zirconia thermal barrier coatings. *J. Eur. Ceram. Soc.* **2014**, *34*, 2013–2021. [CrossRef]

34. Ahmadi, M.S.; Shoja-Razavi, R.; Valefi, Z.; Jamali, H. Evaluation of hot corrosion behavior of plasma sprayed and laser glazed YSZ–Al_2O_3 thermal barrier composite. *Opt. Laser Technol.* **2019**, *111*, 687–695. [CrossRef]

35. Yi, P.; Mostaghimi, J.; Pershin, L.; Xu, P.; Zhan, X.; Jia, D.; Yi, H.; Liu, Y. Effects of laser surface remelting on the molten salt corrosion resistance of yttria-stabilized zirconia coatings. *Ceram. Int.* **2018**, *44*, 22645–22655. [CrossRef]

36. Cai, J.; Gao, C.; Lv, P.; Zhang, C.; Guan, Q.; Lu, J.; Xu, X. Hot corrosion behaviour of thermally sprayed CoCrAlY coating irradiated by high-current pulsed electron beam. *J. Alloy. Compd.* **2019**, *784*, 1221–1233. [CrossRef]

37. Dharamendara, M.; Jegadeeswaran, N. Development of laser heat treatment HVOF coating to combat oxidation on gas turbines. *Mater. Today Proc.* **2018**, *5*, 24937–24943. [CrossRef]

38. Baiamonte, L.; Marra, F.; Gazzola, S.; Giovanetto, P.; Bartuli, C.; Valente, T.; Pulci, G. Thermal sprayed coatings for hot corrosion protection of exhaust valves in naval diesel engines. *Surf. Coat. Technol.* **2016**, *295*, 78–87. [CrossRef]

39. Verlotski, V.; Stanglmaier, R.H.; Moormann, G.; Geraets, R. Mineral-metal, multiphase coatings to protect combustion chamber components against hot-gas corrosion and thermal loading. *J. Eng. Gas Turbines Power* **2011**, *133*, 1–5. [CrossRef]

40. Verlotski, V. Thermally sprayed gastight protective layer for metal substrates. US patent No. US8784979B2, 30 May 2018.

41. *ASTM E384-89 Standard Test Method for Microhardness of Materials*; ASTM International: West Conshohocken, PA, USA, 1989.

Article

Oxidation Behavior of MoSi$_2$-Coated TZM Alloys during Isothermal Exposure at High Temperatures

Kwangsu Choi [1], Young Joo Kim [1], Min Kyu Kim [1], Sangyeob Lee [1], Seong Lee [2] and Joon Sik Park [1],*

[1] Department of Materials Science and Engineering, Hanbat National University, Daejeon 34158, Korea; rkskejfj@naver.com (K.C.); youngjoo.kim@ekosa.or.kr (Y.J.K.); hykim1207@hanbat.ac.kr (M.K.K.); seongl@add.re.kr (S.L.)

[2] Agency for Defense Development, Yuseong P.O. Box 35-44, Daejeon 34186, Korea; seongl@add.re.kr

*   Correspondence: jsphb@hanbat.ac.kr; Tel.: +82-42-821-1276; Fax: +82-42-821-1592

Received: 15 May 2018; Accepted: 8 June 2018; Published: 11 June 2018

Abstract: Coating properties and oxidation behaviors of Si pack cementation-coated TZM (Mo-0.5Ti-0.1Zr-0.02C) alloys were investigated in order to understand the stability of the coating layer at high temperatures up to 1350 °C in an ambient atmosphere. After the pack cementation coatings, MoSi$_2$ and Mo$_5$Si$_3$ layers were formed. When MoSi$_2$-coated TZM alloys were oxidized in air at high temperatures, the Si in the outer MoSi$_2$ layer diffused and formed SiO$_2$. Also, due to the diffusion of Si, the MoSi$_2$ layer was transformed into a columnar shaped Mo$_5$Si$_3$ phase. During isothermal oxidation, the Mo$_5$Si$_3$ phase was formed both within the coated MoSi$_2$ layer and between the MoSi$_2$ and the substrate. The coating properties and the oxidation behavior of the Si pack-coated TZM alloys were discussed along with the identification of growth kinetics.

Keywords: TZM alloys; pack cementation; oxidation

1. Introduction

TZM alloys are traditional Mo-based alloys that are typically used for high temperature applications due to their superior material properties at high temperatures. The phase constituting TZM alloys are composed of mainly Mo matrix with precipitates of TiC and/or ZrC. Detailed material properties have been reported elsewhere [1–5]. In particular, since the alloys are stable with liquid state metals, they have been applied in high temperature components such as atomic plant components. However, since the alloys are mainly composed of Mo, the low stability of Mo during high temperature exposure under an ambient atmosphere is a serious limitation of these alloys. Specifically, when the alloys are exposed in air at high temperatures above 400 °C, a volatile and non-protective Mo oxide, MoO$_3$, is formed. This ruins the usefulness of the alloys due to the nature of the non-protective oxidation behaviors of Mo [6,7].

Mo-based alloys have been studied in the past to improve oxidation resistance through surface coatings. Surface protection coatings such as thermal plasma spray, sputtering coatings, or pack cementation coatings have been studied in order to improve oxidation resistance at the exposure to high temperatures under an ambient atmosphere [8–11]. Among the coating routes, the pack cementation coating process has the capability of producing a uniform coating layer on complex shaped substrates, and is effective in preventing oxygen penetration [2,4,7,12–17]. According to the literature, the formation of protective coating layers can be achieved by Si or Al pack cementation coatings [2,4,11]. Upon the application of Si pack cementation coatings on the alloy, MoSi$_2$ phase, a highly oxidation resistant material, is formed as a result of the diffusion of Si into Mo in the matrix. When the MoSi$_2$ phase is exposed in air at high temperatures, SiO$_2$ is formed on the surface of the MoSi$_2$ layer, implying that the coated TZM alloys exhibit excellent oxidation resistant behaviors [1].

Furthermore, the effectiveness of Al-Si silicide coatings has been reported. The coating layers composed of $Mo(Si, Al)_2$ can provide protection to the TZM alloys through the formation of an Al_2O_3 layer on the surface [7]. Besides these, systematic reports have also been published which focus on the formation of pack cementation coating layers with various compositions and coating kinetics using different halide activators [7].

In order to apply the coated TZM alloys on structural parts that require reproducible credibility, such as defense systems, the coating layers should have properties such as (i) strong bonding with the substrate; (ii) a defect-free coating layer in complex-shaped structures; and (iii) a defined lifetime of the coating layer with respect to exposure at high temperatures. While the nature of the pack cementation coating process suitably matches with the aforementioned two requirements, the lifetime kinetics of $MoSi_2$-coated TZM alloys have not been systematically investigated at high temperatures during oxidation exposure. In reality, the lifetime analyses of the coating layers are a critical issue for the practical application of the coated TZM alloys, such as structural parts for defense systems, in order to provide reproducible service time and temperatures.

In this study, the growth kinetics of the $MoSi_2$ coating layer were investigated to identify the lifetime of the coated TZM alloys during isothermal exposure under aerobic conditions. The disintegrated kinetics of the $MoSi_2$ phase was also investigated during the isothermal oxidation exposure for various exposure times up to 1350 °C in air. The formation of columnar Mo_5Si_3 inside the $MoSi_2$ layer was observed for the first time. The coating layer kinetics and disintegrated kinetics were discussed via microstructural observations and kinetic estimations.

2. Experimental Details

Bar-shaped TZM alloy with a diameter of 10 mm (purchased from Hansh®, Seoul, Korea), was cut into small pieces of 10 mm thickness. Each piece was polished using a fine Al_2O_3 powder and cleaned ultrasonically. For Si pack cementation coatings, the powder mixture was composed of 25 wt % coating source (Si), 5 wt % activator (NaF), and 70 wt % anti-sintering material (Al_2O_3). The powder mixture was blended in a jar of a milling machine for 24 h. The powder mixture with the TZM alloys in an alumina pack were put inside a furnace under an Ar atmosphere, and the coatings were carried out at temperatures of 900–1100 °C for various time ranges (up to 24 h). The process involved the deposition of Si vapor carried by volatile halide species on the substrate embedded in a mixed powder pack at elevated temperatures. In order to identify the thickness of the coating layer, the coated TZM alloys were cut perpendicular to the surface. The TZM alloys coated at 1100 °C for 6 h were selected for oxidation tests, since the coating layer thickness of ~30 μm was a desirable thickness for practical application. For the oxidation tests, a coated sample disc placed in an alumina boat was inserted into a tube furnace which was initially set at a high temperature (up to 1350 °C) in air. After the oxidation time reached the selected exposure time, the specimens were pulled out of the furnace promptly and cooled in air. The coated alloys and oxidized specimens were cut into cross-sections for microstructure observations. The sectioned specimens were polished with fine Al_2O_3 powder and cleaned ultrasonically. The cross-sections of the coating layers were examined by scanning electron microscope (SEM, JEOL-6300, JEOL, Ltd., Tokyo, Japan) equipped with energy dispersive spectrum (EDS, TEAM XP, Tokyo, Japan) and electron backscatter diffraction pattern systems (EBSD, TSL-Hikari Super, Tokyo, Japan).

3. Results and Discussion

In the case of the Si diffusion coatings, chemical reactions such as $Si + Al_2O_3 + NaF \rightarrow SiF_2 + NaF + Al_2O_3$ may progress during the coating heat treatments. Al_2O_3 is a ceramic material that is an anti-sintering agent, and the ceramic material does not participate in chemical reactions. When a high-temperature SiF_2 activating gas atmosphere is formed through the heat treatment, Si is coated on TZM alloys through gas and solid state diffusion. The additional detailed coating procedure was described elsewhere [7,12–15]. The outlook images of (a) as-received TZM; (b) the

MoSi$_2$-coated TZM; (c) the oxidized TZM without coatings; and (d) the oxidized TZM with Si pack cementation coatings exposed in air at 1350 °C for 50 h are shown in Figure 1. When the coatings were not applied, the TZM alloy turned yellowish in color and lost its initial shape due to the non-protective MoO$_3$ phase formation after oxidation tests. An additional analysis of the oxidation behaviors of the uncoated TZM was not investigated, since the loss of Mo was so obvious [6,7]. However, when the coatings were applied on the TZM alloy, the surface turned dark grey and the initial shape was well maintained after oxidative exposure, implying that the coatings are needed for practical applications. Figure 2 shows a typical cross-section of the coating layer with Si pack cementation coatings at 1100 °C for 6 h. The formation of the coating layer of MoSi$_2$ has been mentioned in previous studies [7,14]. After the Si pack cementation coatings, MoSi$_2$ formed on the surface and formation of a Mo$_5$Si$_3$ layer was observed between the MoSi$_2$ and the TZM alloy (Figure 2b). However, no Mo$_3$Si phase was observed in the current study. It can be argued that the Mo$_3$Si phase might have formed between Mo$_5$Si$_3$ and TZM, but this was not observed possibly due to the nature of low growth kinetics and/or the limitation of the SEM, which corresponds to previous documents by other researchers [18,19]. In order to identify the growth behaviors of the coating layers, the cross-sections of the MoSi$_2$-coated TZM alloys annealed at 1100 °C for 24 h were observed with EBSD, as shown in Figure 3. In order to show a thick coated layer, the specimen coated for the longest time was selected. The MoSi$_2$ grains underwent the preferred directional grain growth with a ~1 μm width, and Mo$_5$Si$_3$ grains developed beneath the MoSi$_2$ grains. The direction of the MoSi$_2$ grain grew towards the surface, and the length of the MoSi$_2$ grain was not consistent, ranging from 10 to 40 μm. Figure 4 shows the growth kinetics of the MoSi$_2$ and the Mo$_5$Si$_3$ phases underneath. The thicknesses of the synthesized MoSi$_2$ and Mo$_5$Si$_3$ phases increased as the coating time was increased. Moreover, the thickness of the coating layer linearly increased with respect to the annealing time. When the coating layer thickness was plotted as $x^2 = kt$, where x denotes thickness (m), t denotes time (s), and the k represents the kinetic parameters. The k values of the MoSi$_2$ phases were larger than that of Mo$_5$Si$_3$. The k values for MoSi$_2$ were estimated to be 2.18×10^{-11}, 1.17×10^{-10}, and 5.60×10^{-10}, and those for Mo$_5$Si$_3$ were evaluated to be 2.25×10^{-14}, 3.03×10^{-13}, and 2.40×10^{-12} at annealing temperatures of 900, 1000, and 1100 °C, respectively. It was noted that the orders of the current values were similar to the previous results obtained from solid state diffusion annealing [18]. To note, the kinetic parameter (k) also followed the equation, $k = k_o \exp(-Q/RT)$, where Q is the activation energy (kJ/mol), R is the gas constant (J/K mol), and T (K) is the absolute temperature. The evaluated Q values of MoSi$_2$ and Mo$_5$Si$_3$ phases were 217 and 313 kJ/mol, respectively, which are similar to the results of the previous documents [19].

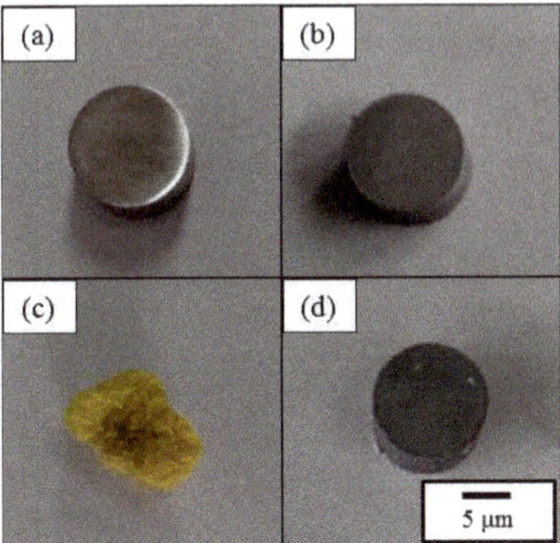

Figure 1. Surface image of (**a**) the as-received TZM alloy; (**b**) after coating the TZM alloy at 1100 °C for 6 h; (**c**) oxidized TZM alloy at 1300 °C for 50 h; and (**d**) MoSi$_2$-coated TZM alloy oxidized at 1350 °C for 50 h.

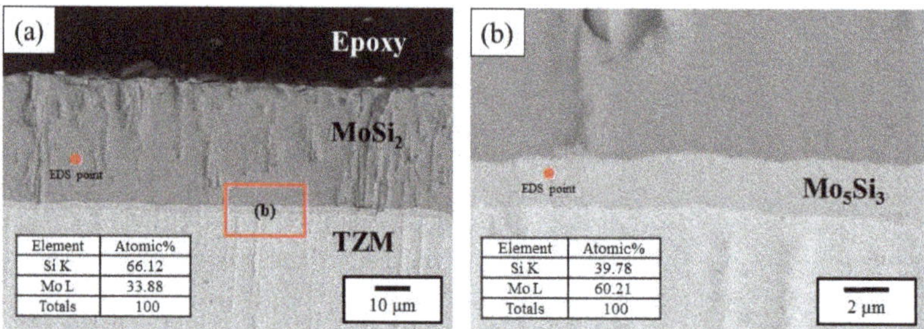

Figure 2. (**a**) Cross-section of MoSi$_2$-coated TZM alloys at 1100 °C for 6 h; (**b**) shows the enlargement of the inset in (**a**).

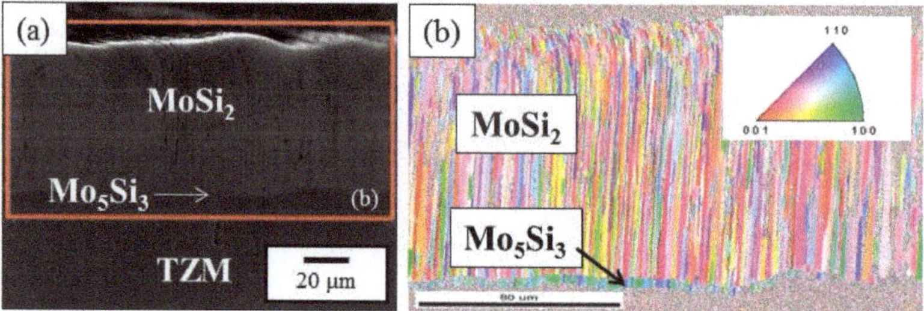

Figure 3. (**a**) Cross-section of MoSi$_2$-coated TZM alloys at 1100 °C for 24 h; (**b**) shows the electron backscatter diffraction (EBSD) micrographs of the marked area in (**a**).

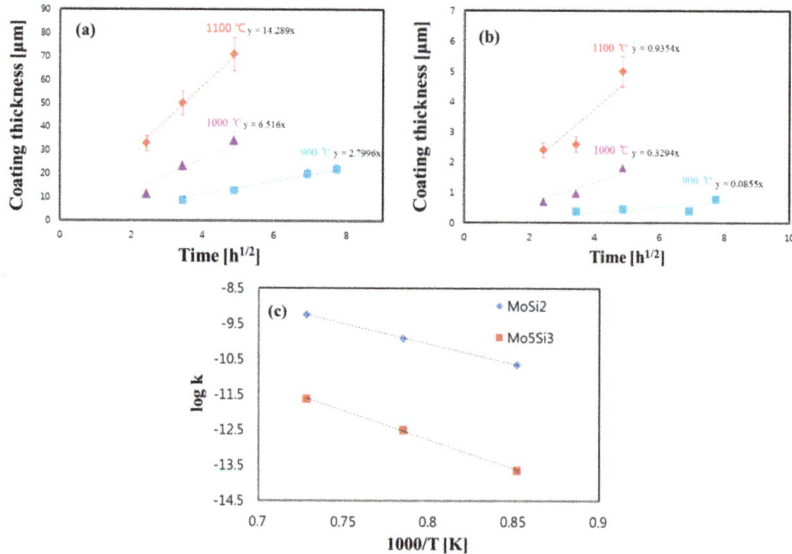

Figure 4. Growth kinetics of (**a**) MoSi$_2$; (**b**) Mo$_5$Si$_3$ layer; and (**c**) kinetic parameters (k) with respect to exposure temperature.

Figure 5 shows SEM micrographs of the surface of the TZM alloy after oxidation at 1350 °C for 5, 20, and 50 h. When the $MoSi_2$ phase-coated TZM was exposed in air, an SiO_2 phase formed on the surface. However, with increase in time (i.e., when the thickness of the SiO_2 phase was increased), cracks were observed on the surface of the SiO_2 layer, possibly due to the coefficient of thermal expansion (CTE) difference between SiO_2 and the coated $MoSi_2$ layer [7,19]. It was observed that porous mushroom-shaped oxides were formed at the location of the cracks. The insets in Figure 5a–c are shown in Figure 5d–f, respectively.

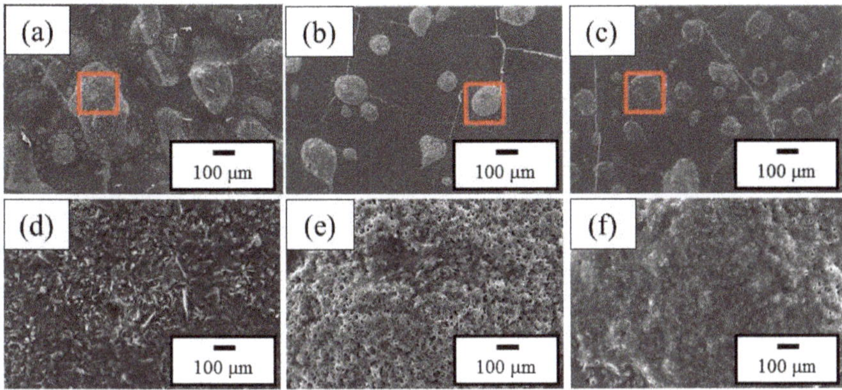

Figure 5. Surface morphology of $MoSi_2$-coated TZM alloys oxidized at 1350 °C for (**a**) 5, (**b**) 20, and (**c**) 50 h; (**d–f**) show the enlargements of the marked areas in (**a–c**), respectively.

In order to identify the effect of the coating layer on the oxidation exposure, the cross-section of the oxidized specimen was prepared and EBSD was carried out for the coated specimen oxidized at 1350 °C for 20 h. The SEM is shown in Figure 6a, and the EBSD micrograph of the marked large box in Figure 6a is shown in Figure 6b, in which the thickness of Mo_5Si_3 was about 50 µm, located between the surface $MoSi_2$ and the TZM substrate. Also, the marked small box in Figure 6a is shown in Figure 6c,d. The micrographs show that columnar grains of Mo_5Si_3 were located inside the coated $MoSi_2$ layer. It was clear that some grains of vertical Mo_5Si_3 were connected to the Mo_5Si_3 layer located between the surface $MoSi_2$ and the TZM substrate, as shown in Figure 6d.

In order to identify the disintegrated kinetics of $MoSi_2$ and the growth kinetics of the Mo_5Si_3 layer during oxidation exposure, the thicknesses of each phase were measured with respect to the oxidation time, as shown in Figure 7. AS can be seen in the figure, the thickness of the $MoSi_2$ phase decreased, and the reduction rate increased when the oxidation temperature reached the temperature of 1350 °C. At the same time, it was noted that the thickness of the Mo_5Si_3 phase located between the surface $MoSi_2$ and the TZM substrate increased as the temperature and time of the oxidation exposure was increased. Again, during the oxidation, the k values for the $MoSi_2$ phases (thickness reduction rate) were estimated to be 2.36×10^{-11}, 2.77×10^{-11}, and 3.86×10^{-10} (Figure 7a), and the values for Mo_5Si_3 were evaluated to be 6.17×10^{-10}, 3.93×10^{-10}, and 2.20×10^{-10} (Figure 7b), at oxidation temperatures of 1250, 1300, and 1350 °C, respectively. The Q value for the decomposition of $MoSi_2$ was estimated as 567 kJ/mol, and that for the Mo_5Si_3 phase was evaluated as 210 kJ/mol. Figure 8 shows the mass change of the $MoSi_2$-coated TZM alloy after oxidation. Although an increase in mass was observed, the amount of increment was marginal, i.e., the increased values were less than 1 wt %. This is possibly due to the formation of SiO_2 at the surface after oxidation tests.

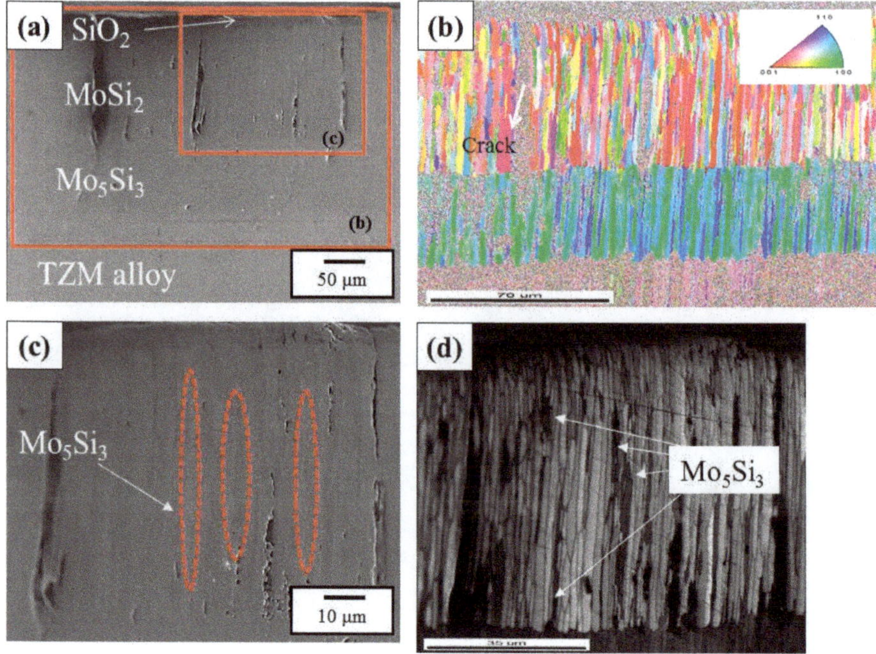

Figure 6. Cross-sectional SEM of the MoSi$_2$-coated TZM alloy oxidized at 1350 °C for 20 h. (**a**) SEM; (**b**) EBSD micrograph of the large area marked in (**a**); (**c**) the enlargement of the small inset in (**a**); and (**d**) image with black and white contrast of (**c**).

Regarding the columnar growth of Mo$_5$Si$_3$ inside the surface of MoSi$_2$, a possible mechanism is shown in Figure 9. According to a previous report, the growth of the MoSi$_2$ phase is usually observed as a columnar manner, since the MoSi$_2$ phase is synthesized via diffusion reactions [2]. At the same time, when the coated MoSi$_2$ layer was exposed to high temperatures in the presence of air, the surface MoSi$_2$ phase underwent a disintegration reaction, viz. (5/7) MoSi$_2$ (s) + O$_2$ (g) → (1/7) Mo$_5$Si$_3$ + SiO$_2$ (s) [15]. This reaction shows that when the surface SiO$_2$ layer was formed, an Mo$_5$Si$_3$ phase should be produced due to the loss of Si from MoSi$_2$. In this regard, it is possible that the Mo$_5$Si$_3$ phase inside the MoSi$_2$ phase could grow, and Si should move through the grain boundaries of the MoSi$_2$ layer, as shown in Figure 6.

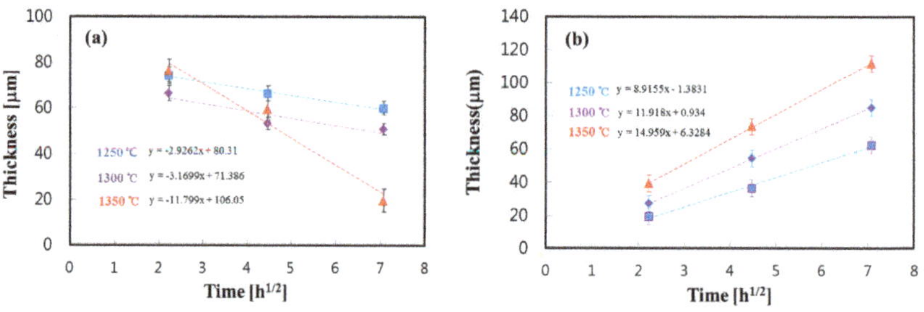

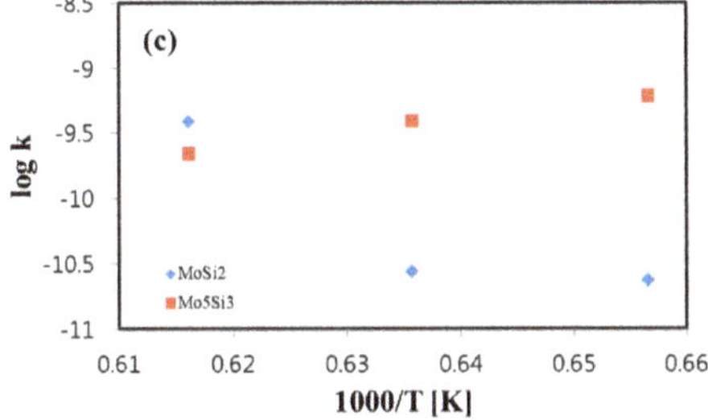

Figure 7. Layer kinetics of (**a**) $MoSi_2$ layer, (**b**) Mo_5Si_3 layer during oxidation exposure with respect to the exposure time; (**c**) shows the log k vs. inverse temperature plot.

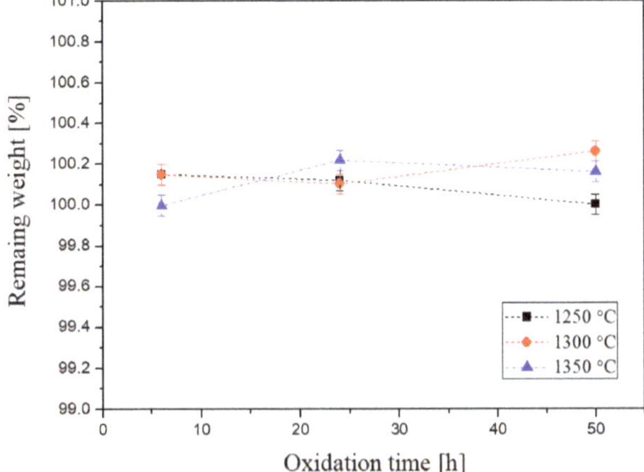

Figure 8. Remaining weight percent of TZM alloys after oxidation exposure at the designated temperature and time.

Since phase development during oxidation is related to the diffusion process, it was useful to examine the diffusion pathways with respect to the sequence of the product phases. In regard to the diffusion pathway, some approaches have been proposed under simplified conditions [20–22]. The analysis methods included two requirements of the diffusion path. First, the pathway should cross the line connecting the end member compositions in order to satisfy mass balance. Second, a stable path should follow the isothermal phase diagram and coincide with two-phase field tie lines to satisfy local equilibrium. Furthermore, the reaction pathway analyses with finite terminal members and infinite terminal members are totally different situations [23]. For the current cases, probably the diffusion pathway was somehow transient due to the presence of the unstable $MoSi_2$. When the coated $MoSi_2$ phase was exposed to air, two separated reactions of (i) oxygen/$MoSi_2$ ((1/7) $MoSi_2$ + O_2 → (1/7) Mo_5Si_3 + SiO_2) and (ii) $MoSi_2$/TZM ($2MoSi_2$ + $6Mo$ → Mo_5Si_3 + (Mo_3Si)) may occur during oxidation exposure. Then, the $MoSi_2$ layer would eventually disappear as the oxidation time increased due to the loss of Si, i.e., the current case is a diffusion reaction between finite terminal members, and the diffusion pathway might change when the $MoSi_2$ phase disappears.

The observed diffusion pathway of the $MoSi_2$-coated TZM after the oxidation tests was O (MoO_3)/SiO_2/$MoSi_2$/Mo_5Si_3/TZM, as shown in Figure 10. When the $MoSi_2$ coating layer was oxidized, Mo initially evaporated to the surface. Then, a SiO_2 layer formed on the surface of $MoSi_2$ due to the reaction of oxygen that diffused into $MoSi_2$ and reacted with the Si [7]. According to

Figure 10, the diffusion pathway was O $(MoO_3)/SiO_2/MoSi_2/Mo_5Si_3/TZM$ during oxidation (Path I). When further oxidation occurred, the $MoSi_2$ layer may undergo loss of Si due to the diffusion of Si towards the surface and inside the alloy. In that case, the diffusion pathway may change to O $(MoO_3)/SiO_2/Mo_5Si_3/TZM$ (Path II). Furthermore, when the oxidation exposure time is increased, the $MoSi_2$ phase eventually disappears.

While further investigation into factors such as the grain orientations of Mo_5Si_3 inside the $MoSi_2$ layer is needed, the current observations clearly showed that the disintegration of the outer $MoSi_2$ layer occurs together with both the formation of Mo_5Si_3 inside the $MoSi_2$ layer and the formation of Mo_5Si_3 located between the outer $MoSi_2$ layer and the TZM alloy. Also, the lifetime (decomposition of the $MoSi_2$ layer) of the $MoSi_2$ coating layer stability was identified during the exposure of high temperatures under an ambient atmosphere.

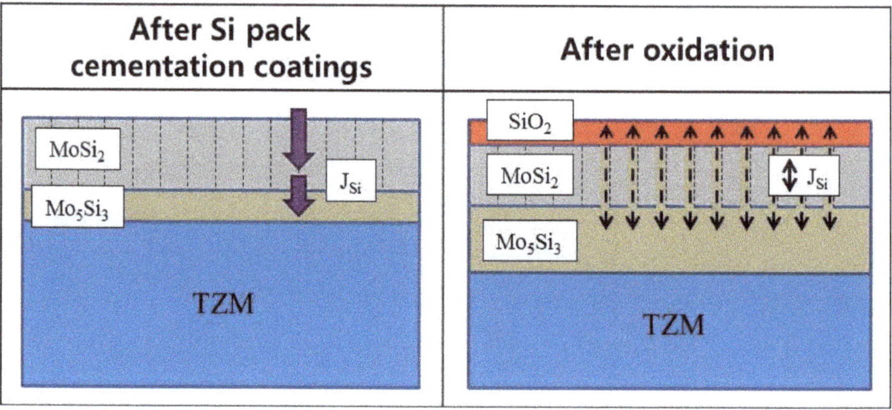

Figure 9. Schematic illustration of $MoSi_2$ coatings and the formation of Mo_5Si_3 phase inside the $MoSi_2$ layer after oxidation.

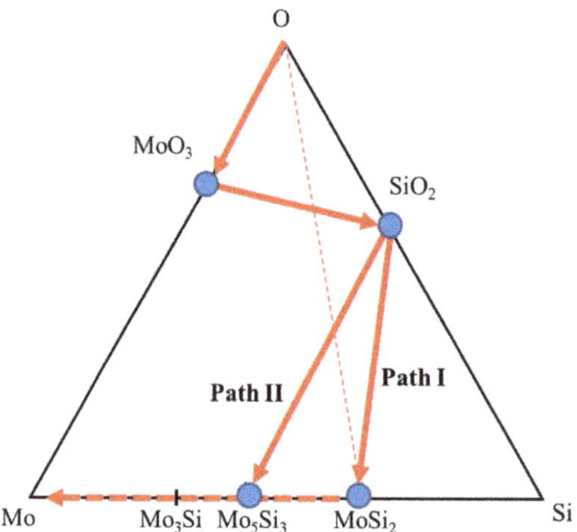

Figure 10. Diffusion pathways of the oxidation of $MoSi_2$ (diffusion path I: O $(MoO_3)/SiO_2/MoSi_2/Mo_5Si_3/TZM$; diffusion path II: O $(MoO_3)/SiO_2/Mo_5Si_3/TZM$).

4. Conclusions

The growth kinetics and the lifetime kinetics of the coating layers during oxidation were investigated for the $MoSi_2$-coated TZM alloys. When $MoSi_2$ coatings were laid on TZM alloys, the columnar $MoSi_2$ and the Mo_5Si_3 phases located between the surface of $MoSi_2$ and the substrate TZM alloy were formed. During the coating process, the activation energies of the kinetic parameters of $MoSi_2$ and Mo_5Si_3 were estimated to be 217 and 313 kJ/mol, respectively. When the $MoSi_2$-coated TZM was oxidized up to 1350 °C, the thickness of the $MoSi_2$ layer decreased, whereas that of the Mo_5Si_3 phase constantly increased during oxidation exposure. The activation energy of the disintegration parameter of $MoSi_2$ and the growth parameter of Mo_5Si_3 was estimated to be 567 and 210 kJ/mol, respectively. Also, during the oxidation, the Mo_5Si_3 phase was also observed inside the surface $MoSi_2$ layer as a columnar shape due to the loss of Si, which formed a surface SiO_2 layer. The formation of the SiO_2 layer can accelerate the disintegration of the $MoSi_2$ coating layer. The present observations show that the disintegration kinetics of the coated outer $MoSi_2$ layer is meaningful for the estimation of the lifetime of TZM alloys.

Author Contributions: J.S.P. and S.L. conceived and designed the experiments; K.C., Y.J.K., M.K.K. and S.L. performed the experiments; and K.C. and J.S.P. wrote the paper.

Funding: This research was funded by the Agency for Defense Development (contract No. UE105099GD) and Basic Science Research Program through the National Research Foundation of Republic of Korea (NRF) funded by the Ministry of Education, Science, and Technology (contract No. 2016R1D1A1A0991905).

Conflicts of Interest: The authors declare no conflict of interest.

References

1. Chakraborty, S.P. Studies on the development of TZM alloy by aluminothermic coreduction process and formation of protective coating over the alloy by plasma spray technique. *Int. J. Refract. Met. Hard Mater.* **2011**, *29*, 623–630. [CrossRef]
2. Majumdar, S.; Sharma, I.G.; Raveendra, S.; Samajdar, I.; Bhargava, P. In situ chemical vapour co-deposition of Al and Si to form diffusion coatings on TZM. *Mater. Sci. Eng. A* **2008**, *492*, 211–217. [CrossRef]
3. Chakraborty, S.P.; Banerjee, S.; Sharma, I.G.; Suri, A.K. Development of silicide coating over molybdenum based refractory alloy and its characterization. *J. Nucl. Mater.* **2010**, *403*, 152–159. [CrossRef]
4. Chakraborty, S.P.; Banerjee, S.; Singh, K.; Sharma, I.G.; Grover, A.K.; Suri, A.K. Studies on the development of protective coating on TZM alloy and its subsequent characterization. *J. Mater. Proc. Technol.* **2008**, *207*, 240–247. [CrossRef]
5. Cockeram, B.V. The mechanical properties and fracture mechanisms of wrought low carbon arc cast (LCAC), molybdenum–0.5pct titanium–0.1pct zirconium (TZM), and oxide dispersion strengthened (ODS) molybdenum flat products. *Mater. Sci. Eng. A* **2006**, *418*, 120–136. [CrossRef]
6. Smolik, G.R.; Petti, D.A.; Schuetz, S.T.J. Oxidation and volatilization of TZM alloy in air. *J. Nucl. Mater.* **2000**, *283*, 1458–1462. [CrossRef]
7. Majumdar, S.; Sharma, I.G. Oxidation behavior of $MoSi_2$ and Mo(Si, Al)$_2$ coated Mo–0.5Ti–0.1Zr–0.02C alloy. *Intermetallics* **2011**, *19*, 541–545. [CrossRef]
8. Park, J.S.; Sakidja, R.; Perepezko, J.H. Coating designs for oxidation control of Mo-Si-B alloys. *Scr. Mater.* **2002**, *46*, 765–770. [CrossRef]
9. Nomura, N.; Suzuki, T.; Yoshimi, K.; Hanada, S. Microstructure and oxidation resistance of a plasma sprayed Mo-Si-B multiphase alloy coating. *Intermetallics* **2003**, *11*, 735–742. [CrossRef]
10. Lange, A.; Braun, R. Magnetron-sputtered oxidation protection coatings for Mo-Si-B alloys. *Corros. Sci.* **2014**, *84*, 74–84. [CrossRef]
11. Riedl, H.; Vieweg, A.; Limbeck, A.; Kalaš, J.; Arndt, M.; Polcik, P.; Euchner, H.; Bartosik, M.; Mayrhofer, P.H. Thermal stability and mechanical properties of boron enhanced Mo–Si coatings. *Surf. Coat. Technol.* **2015**, *280*, 282–290. [CrossRef]
12. Park, J.S.; Kim, J.M.; Son, Y.I. Oxidation response of Al coated TZM alloys under dynamic plasma flame. *Int. J. Refract. Met. Hard Mater.* **2013**, *41*, 110–114. [CrossRef]

13. Sakidja, R.; Rioult, F.; Werner, J.; Perepezko, J.H. Aluminum pack cementation of Mo-Si-B alloys. *Scr. Mater.* **2006**, *55*, 903–906. [CrossRef]

14. Park, J.S.; Kim, J.M.; Kim, H.Y.; Lee, J.S.; Oh, I.H.; Kang, C.S. Surface protection effect of diffusion pack cementation process by Al-Si powders with chloride activator on magnesium and its alloys. *Mater. Trans.* **2008**, *49*, 1048–1051. [CrossRef]

15. Park, J.S.; Kim, J.M.; Cho, S.H.; Kim, Y.I.; Kim, D.S. Oxidation of $MoSi_2$-coated and uncoated TZM (Mo0.5Ti0.1Zr0.02C) alloys under high temperature plasma flame. *Mater. Trans.* **2013**, *54*, 1517–1523. [CrossRef]

16. Schilephake, D.; Gombola, C.; Kauffmann, A.; Heilmaier, M.; Perepezko, J.H. Enhanced oxidation resistance of Mo-Si-B-Ti alloys by pack cementation. *Oxid. Met.* **2017**, *88*, 267–277. [CrossRef]

17. Perepezko, J.H. High temperature environmental resistant Mo-Si-B based coatings. *Int. J. Refract. Met. Hard Mater.* **2018**, *71*, 246–254. [CrossRef]

18. Tortorici, P.C.; Dayananda, M.A. Diffusion structures in Mo vs Si solid-solid diffusion couples. *Scr. Mater.* **1998**, *38*, 1863–1969. [CrossRef]

19. Yoon, J.K.; Byun, J.Y.; Kim, G.H.; Kim, J.S.; Choi, C.S. Growth kinetics of three Mo-Silicide layers formed by chemical vapor deposition of Si on Mo substrate. *Surf. Coat. Technol.* **2002**, *155*, 85–95. [CrossRef]

20. Yanagihara, K.; Przybylski, K.; Maruyama, T. The role of microstructure on pesting during oxidation of $MoSi_2$ and $Mo(Si,Al)_2$ at 773 K. *Oxid. Met.* **1997**, *47*, 277–293. [CrossRef]

21. Van Loo, F.J. Multiphase diffusion in binary and ternary solid-state systems. *Prog. Solid State Chem.* **1990**, *20*, 47–99. [CrossRef]

22. Kirkaldy, J.S.; Young, D.J. *Diffusion in the Condensed State*; The Institute of Metals: London, UK, 1988; pp. 361–400.

23. Perepezko, J.H.; da Silva Bassani, M.H.; Park, J.S.; Edelstein, A.S.; Everett, R.K. Diffusional reactions in composite synthesis. *Mater. Sci. Eng. A* **1995**, *195*, 1–11. [CrossRef]

Article

Enhancement of the Corrosion Resistance of 304 Stainless Steel by Cr–N and Cr(N,O) Coatings

Mihaela Dinu [1], Emile S. Massima Mouele [2], Anca C. Parau [1], Alina Vladescu [1,3], Leslie F. Petrik [2] and Mariana Braic [1,*]

[1] National Institute for Optoelectronics, 409 Atomistilor St., 077125 Magurele, Romania; mihaela.dinu@inoe.ro (M.D.); anca.parau@inoe.ro (A.C.P.); alinava@inoe.ro (A.V.)
[2] Department of Chemistry, Environmental and Nano Sciences, University of the Western Cape, Robert Sobukwe Road, Bellville 7535, South Africa; 2916096@myuwc.ac.za (E.S.M.M.); lpetrik@uwc.ac.za (L.F.P.)
[3] National Research Tomsk Polytechnic University, Lenin Avenue 43, Tomsk 634050, Russia
* Correspondence: mariana.braic@inoe.ro; Tel.: +4-21-457-57-59

Received: 27 February 2018; Accepted: 2 April 2018; Published: 5 April 2018

Abstract: Chromium nitride and oxynitride coatings were deposited as monolayers ((Cr–N), Cr(N,O)) and bilayers (Cr–N/Cr(N,O), Cr(N,O)/Cr–N) on 304 steel substrates by reactive cathodic arc method. The coatings were characterised by X-ray diffraction (XRD), scanning electron microscopy (SEM), energy dispersive X-ray spectrometry (EDS), surface profilometry, and scratch tester. The anticorrosive properties of the coatings were assessed by electrochemical tests in 0.10 M NaCl + 1.96 M H_2O_2, carried out at 24 °C. Cr_2N, CrN, and Cr(N,O) phases were identified in the coatings by grazing incidence X-ray diffraction (GI-XRD) measurements. The measured adhesion values ranged from 19 N to 35 N, the highest value being obtained for the bilayer with Cr(N,O) on top. Electrochemical tests showed that Cr(N,O) presence in both mono- and bilayered coatings determined the lowest damage in corrosive solution, as compared to the Cr–N coatings. This improvement was ascribed to the more compact structure, lower coatings porosity, and smoother surface.

Keywords: chromium nitride; chromium oxynitride; multilayer; cathodic arc deposition; corrosion resistance; coating adhesion

1. Introduction

For the past decades, material science has emerged as one of the major scientific fields aiming at the improvement of the physical and chemical properties of materials, taking their projected purpose into consideration. The stainless steel (SS) family represents one of the most utilized alloys, being used in a wide range of applications, such as cutting tools, convenient support in biomedicine, rotor blades of gas turbines, aircraft parts, automobiles, pipelines, and naval vessels [1–20]. The selection of the specific SS to be used in a certain application is done considering the economic aspects as well as its physical and chemical properties. An excessive amount of transition metals (TMs) such as Cr and Fe in SS alloys may influence the chemical stability. The abundance of Cr in SS alloy may largely contribute to the formation of a passive Cr_2O_3 layer, acting as a corrosion-protective layer. However, an excessive amount of Fe in SS alloy may induce its oxidation to FeO, which accelerates SS rusting [21]. More often, the chemical stability of alloys is either temperature- or pH-dependent [20,22,23]. The exposure of SS to such harsh environments may result in corrosion that further limits its performance and durability [6,20,24]. Even though various approaches to corrosion prevention have been proposed in the literature [16], they might be costly.

The use of coatings containing carbon, oxygen, or nitrogen (e.g., carbides, nitrides, carbonitrides, or oxynitrides) has become a practical method used to improve the performance of SS [15,17–20,24]. Coatings designed to withstand corrosion are already in use in various fields, such as orthopaedics, dentistry, tribology, and photocatalysis. In order to enhance the performance of SS, numerous anticorrosion solutions have been developed based on coating application by the pulse laser deposition (PLD) or cathodic arc deposition (CAE) [14]. While large area coatings are still difficult to produce economically by PLD, CAE is a high-productivity method, producing an intense plasma flux at the cathode spots generated by the electric arc. The arc plasma comprises an important fraction of single or multiple ionized atoms. This peculiarity is of great importance, since the bombardment of the growing film by energetic ions and neutrals creates a highly adhesive coating, which can ensure the integrity of the coated structure in the long term [25,26]. Another particularity of the CAE method consists of the significant roughness of the coated surfaces, resulting from droplets ejected by the cathode due to its local melting. Considering this issue, different techniques have been developed, including venetian blinds, magnetic shielding, increased deposition pressure, as well as the development of more sophisticated methods consisting mainly of short-pulsed voltage applied on the substrate [27–31].

Therefore, understanding the corrosion behaviour of coated materials is an important step in achieving the maximum protection of coated materials and hence boosting their life service in chemically aggressive environments [12–14,16,32]. Based on the reported results, various corrosion-resistant coatings have been developed to improve the corrosion resistance of SS. Different TM nitride coatings have been synthesised and investigated due to their high hardness, good oxidation resistance, and low wear rate in dry atmosphere, as well as their phase stability in corrosive environments [10,15,33–37]. Oxynitride coatings have also been studied due to their plasmonic properties, relatively low electrical resistivity, good biocompatibility, and high oxidation resistance [25,38–46].

The aim of the present study was to develop rough, large surface area, corrosion- and erosion-resistant coatings for 304 stainless steel mesh, used as support for powder photocatalysts, utilised in water remediation technologies [47]. The first step was to ascertain the corrosion behaviour of such coatings deposited on solid pieces of 304 stainless steel in relation to other properties, such as crystalline structure, mechanical properties, and surface roughness.

In the current study, we report on the electrochemical behaviour of Cr nitride and oxynitride coatings deposited on 304 SS substrates by CAE technique, as there is abundant scientific literature about the use of Cr-based coatings in severe environments, due to their superior corrosion resistance [45,48–54]. The corrosion tests of the plain and coated SS were carried out in saline solution by potentiodynamic polarization technique. The corrosion performance of the coatings was examined in relation to the coatings features, such as elemental and phase composition, surface morphology, hardness, reduced elastic modulus, and adhesion to SS substrate.

2. Materials and Methods

The Cr-based coatings (nitride and oxynitride) were prepared on 304 SS discs ($\Phi = 20$ mm) and Si pieces (20 × 20) mm^2 by the reactive CAE, using a Cr cathode (99.5% purity, Cathay Advanced Materials Ltd., Guangdong, China).

The chemical composition of the 304 SS, as given by the manufacturer in wt.% and at.%, is presented in Table 1.

Each 304 SS substrate was sanded using a SiC abrasive paper (grit 800), polished ($R_a = 60$ nm), ultrasonically washed in isopropyl alcohol and water for 10 min, then dried at 120 °C for 1 h.

For controlling the coating uniformity, the samples were placed on a rotating sample holder. Moreover, before deposition, the substrates were sputter etched with Ar$^+$ for 5 min to remove any contaminant layer. The residual pressure in the system was 2×10^{-3} Pa. The total gas pressure during the deposition was 8×10^{-2} Pa. For the nitride coatings, nitrogen was introduced in the chamber at a mass flow rate of 60 sccm, while for the oxynitrides, the same value of the nitrogen mass flow rate was used, and 17 sccm of the oxygen was added. The arc current applied on the Cr cathode was 90 A, and the substrates were biased at -200 V. This value was selected based on our previous studies [55], aiming for the enhancement of coatings' corrosion resistance. As previously reported, by selecting the appropriate value for the bias voltage, the coatings' properties can be tuned, mainly related to their crystallinity and surface roughness [56]. For the bilayer coatings, the same deposition parameters were used. However, the deposition time of each individual layer was modified in order to obtain almost the same thickness for all the coatings, around 1 μm.

Table 1. Chemical composition of 304 L stainless steel (wt.% and at.%).

Composition	Element										
	Fe	C	Mn	Si	Cr	Ni	P	S	Co	Mo	Cu
wt.%	70.976	0.004	1.220	0.208	17.746	8.524	0.020	0.014	0.160	0.589	0.539
at.%	70.380	0.020	1.230	0.410	18.900	8.040	0.036	0.024	0.150	0.350	0.470

The thickness and the surface roughness of the coatings were investigated using a Dektak 150 surface profilometer (Bruker, Billerica, MA, USA) equipped with a 2.5 μm radius stylus. The thickness of the coatings was measured by surface profilometry according to the standard DD ENV1071-1:1994 [57], as follows. Part of a Si substrate was masked, such that an edge was formed during the deposition. Ten lines transverse to this edge were scanned at a scan rate of 20 μm/s, resulting in different heights; the averaged value was considered as the mean thickness of the film. The roughness of each SS sample was determined before and after the corrosion testing using the same 2.5 μm radius stylus, moving over a length of 10 mm at a scan rate of 50 μm/s. The roughness of all the investigated specimens (substrate and coatings) was evaluated based on two roughness parameters: the arithmetic average deviation from the mean line (R_a) and the root mean square average of the profile heights over the evaluation length (R_q). Moreover, the asymmetry of the profile about the mean line was also considered by calculating the skewness parameter (S_k). The presented roughness values represent the average data obtained from 5 measurements (10 mm length), performed on different areas of each specimen. After the corrosion tests, the profilometry measurements were also carried out on the substrate and all the coatings for visualisation of the significant profiles of the resulted pits.

A scanning electron microscope (SEM) (Hitachi TM3030 Plus, Tokyo, Japan) coupled with an energy dispersive X-ray spectrometer (EDS) (Bruker, Billerica, MA, USA), operated at an accelerating voltage of 15 kV, was used for the surface morphology and elemental composition investigation of the bare and coated substrate. For EDS, measurements were performed on 10 different areas of (298 × 217) μm², before and after the corrosion tests. The arithmetic mean and the standard deviation were then calculated. Images of the surface morphology for each specimen were recorded at both 30× and 100× magnification. Moreover, for the identification of corrosion products, images of element mapping were also acquired in different areas of each specimen's surface.

The phase composition of the samples was investigated by X-ray diffraction (XRD) using a SmartLab diffractometer (Rigaku, Tokyo, Japan), with a Cu Kα radiation (λ = 0.15405 nm). Grazing incidence scans in the range 30°–70° at 2° incidence, with a step size of 0.02° were obtained for the coatings deposited on Si and SS substrates.

A Hysitron TI Premier nanoindenter, equipped with a Berkovich diamond tip (100 nm radius), was used for the nanoindentation measurements, for obtaining the hardness (H) and the reduced modulus (E_r) values. The normal force used was 5 mN to obtain individual indents for penetration depths in the 50–80 nm range. In order to take the possible imperfections of the indenter into account, the system was calibrated before indentation measurements using a standard fused quartz piece. Five indentations were done at the same force for each measurement, which were positioned at least 12 μm apart in order to prevent the effects of any possible interference between the indentation points. The Oliver–Pharr method was used to extract hardness (H) and reduced elastic modulus (E_r) values from the load–displacement curves [58].

The corrosion resistance was evaluated by the potentiodynamic techniques in 0.10 M NaCl + H_2O_2 (pH = 4), at room temperature (24 °C), using a VersaStat 3 Potentiostat/Galvanostat (Princeton Applied Research, Oak Ridge, TN, USA). A typical three-electrode cell was used, with a Pt counter-electrode and an Ag/AgCl reference electrode. The coated and uncoated (used as control) stainless steels were used as working electrode, being placed in a Teflon holder with a working area of 1 cm^2. Firstly, the open circuit potential (E_{OC}) was monitored for 15 h after immersion. To identify the polarisation resistance (R_p), a linear polarisation technique was used by applying a perturbation potential of -0.01 to 0.01 V vs. E_{OC} at a scanning rate of 1 mV/s. R_p parameter was determined as the slope of the linear region of the $\Delta E - \Delta i$ curve at corrosion potential (E_{corr}). Further, Tafel plots were recorded from -0.25 V to 0.25 V vs. E_{OC} at a scanning rate of 1 mV/s. The E_{corr} parameter, anodic (β_a) and cathodic (β_c) slopes, and corrosion current density (i_{corr}) were extracted from Tafel plots. The porosity (P) was estimated from Elsener's empirical equation (Equation (1)) based on the polarisation resistance of the uncoated (R_{ps}) and coated (R_{pc}) SS specimen, the difference between the corrosion potentials of the coated and uncoated SS specimen (ΔE_{corr}) and corresponding anodic slope (β_a).

$$P = \left(\frac{R_{ps}}{R_{pc}} \right) \times 10^{\frac{-|\Delta E_{corr}|}{\beta_a}} \tag{1}$$

The protective efficiency (P_e) was also calculated (Equation (2)) based on the corrosion current densities of the coatings (i_{corr_c}) and the SS specimen (i_{corr_s}):

$$P_e = \left(1 - \frac{i_{corr\_c}}{i_{corr\_s}} \right) \times 100 \tag{2}$$

Scratch tests under standard conditions (indenter—0.2 mm radius diamond tip, load—continuous increase from 0 to 100 N, scratching speed—10 mm/min, scratching distance—10 mm) were undertaken to determine the coating adhesion, using a laboratory system. The critical loads values at which the film flaking starts (L_1) and at which the delamination is completed (L_2) were determined by optical microscopy.

The coatings were labelled considering the left written coating, being near the substrate, such as in Cr–N/Cr(N,O) bilayer, the top layer being Cr(N,O).

3. Results and Discussion

3.1. Surface Morphology and Elemental Composition before Corrosion Measurements

Representative SEM micrographs of Cr–N and Cr(N,O) coatings are presented in Figure 1. As expected in CAE deposition, microdroplets can be noticed over the surface of all the investigated specimens—more evident in the Cr–N coating, while the Cr(N,O) coating presents a smoother surface, as seen in the magnified images. The same results were reported by Li Ming-sheng on the reactive CAE deposition of chromium in nitrogen and oxygen atmosphere [59]. It was documented by Munz that a cathode material with low melting temperature (T_m) generates an increased number of droplets with larger size [60]. The observed differences in roughness may be explained by the difference between the T_m of Cr and its compounds: $T_m^{Cr_2N}$ (1923 K) $< T_m^{CrN}$ (2043 K) $< T_m^{Cr}$ (2143 K) $< T_m^{Cr_2O_3}$ (2708 K), as documented by the binary phase diagrams of Cr–N and Cr–O [61]. The process is related to the temporary formation of small islands of the reactive compounds on the cathode surface due to the dense plasma arc condition. The islands are melted by the steering arc, such that the resulting microdroplets are propelled to the substrate. In a nitrogen atmosphere, the surface of the metallic Cr cathode is covered with Cr$_2$N and CrN, and mainly with CrN or even Cr$_2$O$_3$ if the deposition atmosphere consists of a mixture of oxygen and nitrogen, as a direct result of the higher reactivity of oxygen compared to the nitrogen molecule. It is reasonable to believe that the observed difference in surface roughness of Cr–N and Cr(N,O) coatings is related to the different T_m of the reactive compounds formed on the cathode as a result of metal target poisoning in reactive CAE.

Despite this peculiarity, the surface of both coatings was uniform, without major morphological defects such as pores, pinholes, and voids, proven as deleterious for corrosion protection of the steel substrate.

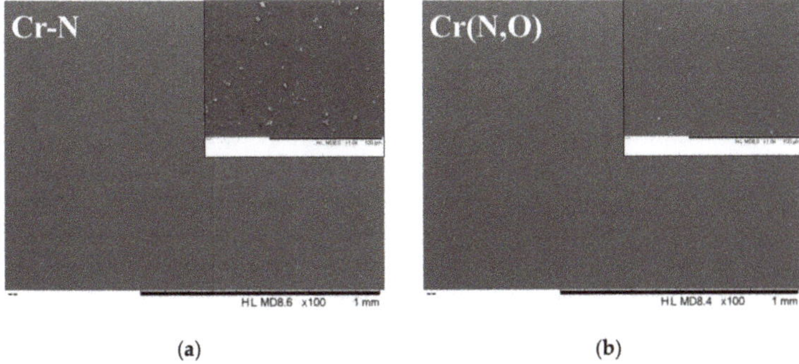

(a)   (b)

Figure 1. SEM micrographs of the investigated monolayers before the corrosion tests: (**a**) Cr–N; (**b**) Cr(N,O).

For each type of coating, the averaged values of the atomic concentrations, together with O/N, O/Cr, and (O + N)/Cr concentration ratios, are presented in Table 2. As standard samples were not available, the accepted precision in EDS measurement was defined as the relative standard deviation: RDS (%) = $(\delta_{at}/C_{am}) \times 100$, where δ_{at} represents the calculated SD (%) of 10 measurements, and C_{am} represents the arithmetic mean concentration of the constituent element. In order to improve the accuracy of the results, long counting (acquisition) times were used [46,47]. To improve the accuracy of the light elements concentration, the low-energy domain of each spectrum was deconvoluted, considering the constituent elements of the substrate and of the coatings. The calculated RDS values are presented in Table 2. One can notice the presence of oxygen and nitrogen in the coatings, as well as a small amount of carbon, probably due to the external handling of samples. A high oxygen content was measured on Cr(N,O) coatings. The low oxygen concentration found in the Cr–N coatings should be treated as contamination, as already reported [62].

Chromium oxynitride coating has a complex structure, its composition and properties being controlled by the deposition parameters, which should ensure the reproducibility of each deposition run. The deposition of chromium oxynitride coating, performed in a mixture of two reactive gases such as O_2 and N_2, is a complex process due to the different reactivity of the Cr with nitrogen and the more reactive oxygen atoms and ions. The observed increase of O/N ratio—about five-fold—demonstrates the successful deposition of chromium oxynitride coating. The observed significant increase can be ascribed to the higher affinity of Cr for oxygen compared to nitrogen [63].

Table 2. The elemental composition of the monolayered coatings deposited on Si substrates.

Coating	Elemental Composition (at.%)				O/N	O/Cr	(O + N)/Cr
	N	O	Cr	C			
Cr–N	30.5 ± 1.2	2.6 ± 0.2	63.4 ± 2.5	3.5 ± 0.2	0.085	0.041	0.522
Cr(N,O)	35.6 ± 1.7	14.9 ± 0.8	46.8 ± 2.2	2.6 ± 0.1	0.419	0.318	1.079

3.2. Phase Composition

Due to the significant overlapping of 304 SS peaks and those of the deposited coatings (Figure 2a), Figure 2b shows the XRD patterns of the coatings deposited on silicon wafers, discussed as follows. The diffractogram of the Si/Cr–N monolayer presents a shallow peak at 37.7° corresponding to (111) cubic CrN phase (JCPDS 11-0065, at 37.57°) or to the hexagonal (110) Cr_2N phase (JCPDS 35-0803, at 37.35°). Additionally, a second small peak situated at 43.38° was attributed more probably to the (200) plane of hexagonal Cr_2N phase (JCPDS 35-0803, at 43.40°) than to (200) plane in B1 phase (JCPDS 11-0065, at 43.77°). The high-intensity peak observed at 67.6° was ascribed to the hexagonal Cr_2N

phase (JCPDS 35-0803, at 67.3°), suggesting that the deposition conditions favour the formation of a highly-stressed Cr_2N coating. Our results are in good agreement with those reported by Rebholz et al., who showed that in chromium nitride with nitrogen content of about 30 at.%, only Cr_2N phase is present [64], similar to our result (30.5 at.% N). They also showed that both Cr_2N and CrN phases coexist in coatings if the nitrogen content is raised to about 40 at.% [64]. This result is also consistent with the phase diagram of the Cr–N system [65]. Considering the XRD pattern of the coating deposited on SS, one can observe a first broad peak, most probably representing the superposition of the (111) cubic CrN phase and (110) hexagonal Cr_2N phase. The second peak is also broad, higher than the one observed on Si substrate. As such, the presence of (200) planes of the hexagonal Cr_2N and cubic CrN phases is evident from the peak asymmetry. A distinctive feature of the coatings grown on SS is the presence of a third broad peak at about 63°, indicating the overlap of the (211) plane of Cr_2N (JCPDS 35-0803, at 62.44°) and the (220) plane of CrN (JCPDS 11-0065, at 63.60°). Note also that the (300) Cr_2N maximum observed on the Si deposited coating is no longer visible.

In the Cr(N,O) coating deposited on Si, three peaks were observed at 37.9°, 44.2°, and 63.6°, shifted to higher 2θ angles compared to the cubic CrN phase (JCPDS 11-0065, at 37.60°, 43.77°, and 63.60°). The observed shift is due to oxygen incorporation, as also reported by Suzuki et al. for PLD grown Cr(N,O) with different oxygen contents [66]. The observed decrease in the lattice constant and peak shift to higher angles suggests the formation of a solid solution which stabilises the single-phase cubic solid solution of Cr(N,O) [67]. This conclusion is supported by the almost unity value of (N + O)/Cr ratio obtained by EDS. The diffractogram of this coating deposited SS substrate shows the clear signatures of the (200) and (220) planes of Cr(N,O), while the peak ascribed to the (111) plane is not visible. As previously reported, the overall energy contains surface energy and strain energy, and the main coating orientation is correlated with the lower overall energy direction of the films. The presence of only (200) and (220) maxima might be an indication of the surface energy minimisation in Cr(N,O) coating deposited on SS when compared to the one deposited on Si, in which the plane (111) was also present [68,69].

Considering the bilayer coatings, one may observe that the peaks' location was shifted in between the positions observed in Cr–N and Cr(N,O) monolayers, indicating the overlapping of the peaks specific for each monolayer. On the Si/Cr–N/Cr(N,O) diffractogram, one can observe the (111) preferred orientation, as well as the presence of the (300) Cr_2N peak. As the Cr–N layer is located closest to the substrate the peak has a lower intensity, especially in grazing incidence measurement set-up. The diffractogram of SS/Cr–N/Cr(N,O) presents the same peaks as the Si/Cr–N/Cr(N,O), except for the Cr_2N peak (300), which was also missing from that of the SS/Cr–N monolayer.

As expected, in Si/Cr(N,O)/Cr–N bilayer, the Cr_2N peak is more intense, and a (111) significant orientation is present. Moreover, in this bilayer, two diffraction lines were found at 43.3° and 44.3°, ascribed to Cr_2N and Cr(N,O), respectively. As expected, the diffractogram of the SS/Cr(N,O)/Cr–N presents all peaks observed on both SS/Cr–N and SS/Cr(N,O) diffractograms. Summarising the crystalline structures observed in Cr–N/Cr(N,O) coatings deposited on both Si and SS substrates, they are accurately depicting the two composing monolayers. However, the crystalline structures of Si/Cr(N,O)/Cr–N and SS/Cr(N,O)/Cr–N are more complex, as the crystalline Cr(N,O) layer underneath seems to promote the growth of CrN and Cr_2N crystallites in the Cr–N top layer, as it results from the better separated maxima of the two types on chromium nitride.

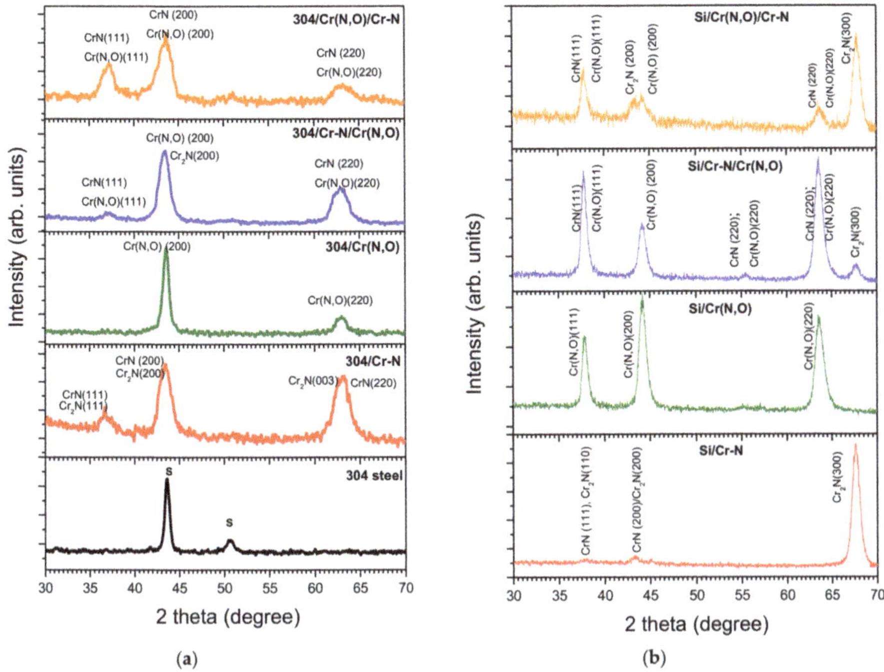

Intensity (arb. units)

2 theta (degree)

(a)

(b)

Figure 2. Grazing incidence X-ray diffraction (GI-XRD) patterns of the Cr–N and Cr(N,O) coatings: (**a**) deposited on stainless steel (SS); (**b**) deposited on Si.

3.3. Coating Thickness and Roughness

Figure 3 presents, as an example, a profile line of the edge measured on Cr(N,O) coating deposited on the masked Si piece. Additionally, the thickness values of all the investigated coatings are shown.

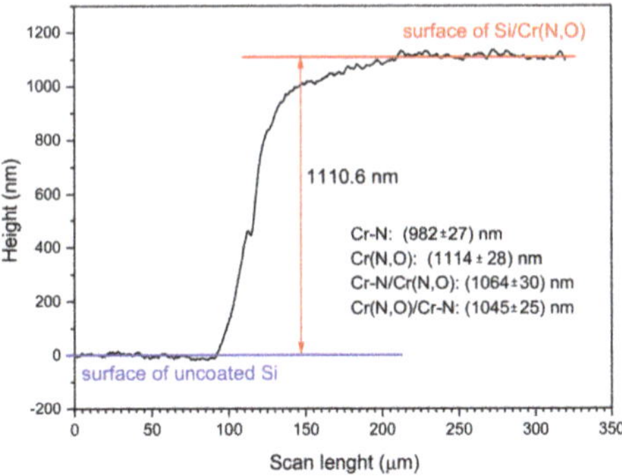

Figure 3. Thickness of the Cr(N,O) coating deposited on Si, measured by surface profilometry, according to the standard DD ENV1071-1:1994, along with all coatings' thickness values.

The values of the roughness parameters, R_a and R_q, determined for the SS substrate and all deposited coatings, are presented in Figure 4a,b, respectively. One may notice an increase of coating roughness (by a factor of up to approximately 8.5) compared to the bare substrate (R_a = 48 nm, R_q = 60 nm). In accordance with SEM images, the highest values were obtained for the Cr–N coating:

$R_a = 281.2$ nm, $R_q = 470.6$ nm. This result is due to the observed microdroplets generated from the target material during the reactive cathodic arc deposition, determining the presence of the most numerous peaks. The oxygen addition in the deposition atmosphere slightly decreased the surface roughness, as observed in Cr(N,O) monolayers. In the case of bilayers, considering the errors, the surface roughness is in the same range as that of the oxynitride coating. The observed decrease of the roughness for both bilayers, as compared to the nitride coating, may be ascribed to the lower roughness of Cr(N,O) layer in the bilayer, and to the thinner dimensions of the composing individual layers in the bilayers compared to the monolayer. The trend evidenced for the R_a parameter (Figure 4a) was also preserved for the R_q parameter (Figure 4b).

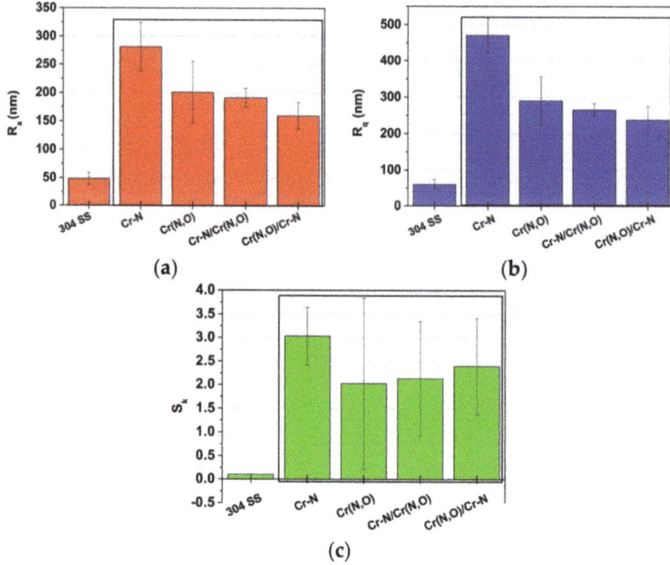

Figure 4. The roughness of Cr-based coatings before the corrosion test: (**a**) R_a parameter; (**b**) R_q parameter; (**c**) S_k parameter.

Since R_a and R_q parameters are not sufficient to evaluate the surface quality, the skewness parameter (S_k) values are presented in Figure 4c. Noe the value close to 0 for the SS substrate, indicating a relatively uniform distribution of the peaks and valleys on the surface. The S_k values corresponding to the coatings are in the 2–3 range, indicating the presence of high peaks, since a positive number relates to a higher percentage of profiles situated above the mean line. The presence of these peaks on the surface of coatings is the result of the droplets, as observed by SEM. As mentioned above, this is a characteristic topography of the coatings obtained by the CAE deposition method. It was also reported that other hard coatings (CrN [70], TiN/CrN [71]) deposited by CAE present structural defects on the surface in the form of overgrown droplets, leading to an increased surface roughness, which can affect the corrosion behaviour.

3.4. Coating Mechanical Properties

A coating's functionality is dependent on its superior mechanical properties, which might be the warrant of a prolonged lifetime.

The SEM micrographs of the scratch traces are illustrated in Figure 5. The first sign of coating delamination (L_1) was measured around 10 N loads on the coatings. The highest value, 11.2 N, was obtained for the Cr–N/Cr(N,O) coating.

The Cr(N,O) monolayer was completely delaminated for a load L_2 higher than 19.1 N, while Cr–N coating withstood the gradual increasing loading force up to 23.7 N, denoting an excellent adhesion to the metallic substrate.

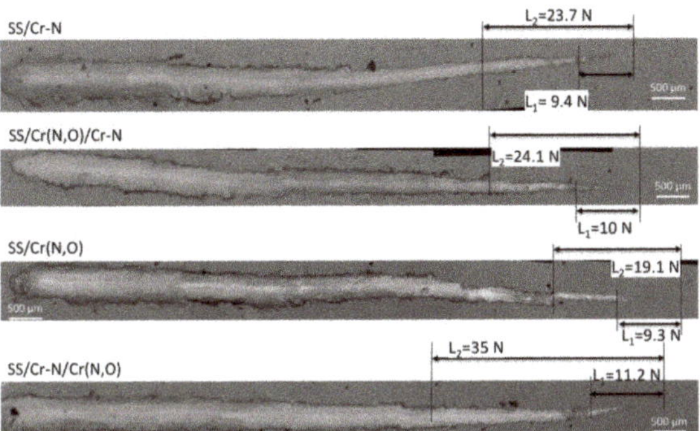

Figure 5. Scratch traces of Cr-based coatings deposited on 304 SS substrate.

Considering the two bilayered coatings, the one containing Cr(N,O) layer at the metal interface presented lower values for both loads (L_1 and L_2), but notably higher than the ones corresponding to Cr(N,O) monolayer. As expected, the presence of Cr–N layer at the metal interface determined the highest values for both L_1 and L_2 (35 N). The best adhesion was measured for the coating Cr–N/Cr(N,O), as a result of the good adhesion of the Cr–N monolayer to the substrate (L_2 = 23.7), in comparison with the one exhibited by the Cr(N,O) monolayer (L_2 = 19.1 N).

To sum up, both types of bilayer exceeded the delamination load values obtained for monolayers: $L_{1,2}{}^{Cr(N,O)} < L_{1,2}{}^{Cr-N} < L_{1,2}{}^{Cr(N,O)/Cr-N} < L_{1,2}{}^{Cr-N/Cr(N,O)}$. The higher adhesion of the two bilayers may be ascribed to the lower internal stress in the bi- and multi-layers, as a reduced intrinsic stress is commonly associated with an enhanced adhesion. Additionally, the presence of a supplementary interface between the two monolayers hinders the crack propagation, determining a better adhesion of the coating to the substrate.

The hardness and reduced elastic modulus of the coatings as determined by nanoindentation are presented in Table 3, and a typical force–displacement curve, with an indentation depth of 67.9 nm—less than 10% of the coating thickness—is presented in Figure 6. As expected, the highest hardness values were obtained for the Cr–N monolayer and for the SS/Cr(N,O)/Cr–N double layer, due to the superior mechanical characteristics of nitrides compared with oxynitrides. However, the aim of the study was to develop corrosion-resistant coatings working under soft erosive conditions. As such, we also looked for superior mechanical properties of the coating, which are related to the various ratios of hardness to reduced elastic modulus [72–74]. It has been reported that H/E_r is related to wear resistance, H^2/E_r to the coating's resilience, and $H^3/E_r{}^2$ to plastic deformation resistance, thus it can also predict erosion resistance. As can be observed in Table 3, all these ratios present the same trend as critical load L_2. The obtained values for H, E_r, and $H^3/E_r{}^2$ for Cr–N coating are in agreement with the data previously reported in the literature [75–77].

Table 3. Hardness (H), reduced elastic modulus (E_r), H/E_r, $H^3/E_r{}^2$, H^2/E_r ratios, and the critical loads L_2, as determined for the Cr-based coatings deposited on 304 SS substrate.

Coating	H (GPa)	ΔH (GPa)	E_r (GPa)	ΔE_r (GPa)	H/E_r	$H^3/E_r{}^2$	H^2/E_r	L_2 (N)
SS/Cr–N	24.53	±1.17	227.12	±6.71	0.1080	0.286	2.649	23.7
SS/Cr(N,O)	21.43	±1.70	203.34	±7.95	0.1054	0.238	2.259	19.1
SS/Cr(N,O)/Cr–N	25.28	±1.56	233.57	±8.26	0.1082	0.296	2.736	24.1
SS/Cr–N/Cr(N,O)	22.75	±1.03	175.45	±5.31	0.1296	0.382	2.949	35

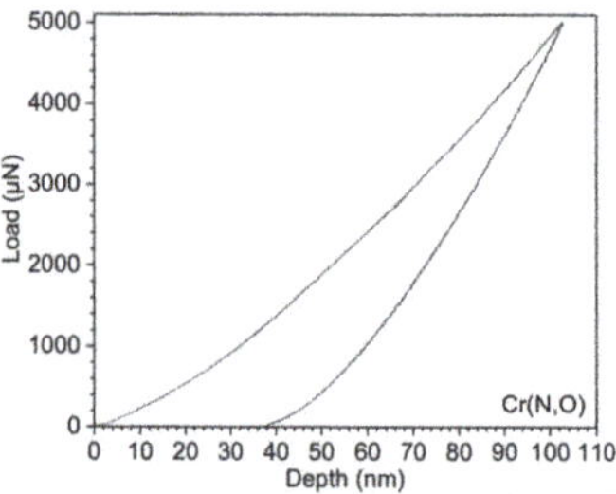

Figure 6. Typical force–displacement curves of Cr(N,O) coating.

3.5. Corrosion Measurements

The open circuit potential evolution during 15 h immersion in 0.10 M NaCl + 1.96 M H_2O_2 is presented in Figure 7. In the first hours of immersion, a slight decrease of potential was observed for all analysed specimens, more pronounced for 304 SS. However, all the coated specimens reached a steady state, the corresponding E_{OC} being in the range 0.02–0.3 V. The stable E_{OC} of the coated specimens is a sign of the probable formation of a stable passivation layer. The slow decrease of the potential corresponding to 304 SS indicates the instability of the passive layer. For the Cr(N,O)/Cr–N bilayer, the initial rapid decrease of the E_{OC} value was followed by a steady increase, such that after 15 h of immersion the value exceeded the one obtained for 304 SS, displaying an increasing tendency up to 117 mV, proving the protective nature of this coating. The most passive layers were formed on the surfaces of the Cr–N/Cr(N,O) and Cr(N,O) coatings, both presenting almost the same variation tendency for E_{OC}, leading to the conclusion that the presence of Cr(N,O) at the point of contact with the corrosive environment has a beneficial effect on the corrosion resistance. In contrast with the above result, the Cr–N monolayer exhibited a better passivation of its surface compared to the Cr(N,O)/Cr–N bilayer, showing an E_{OC} value as low as 0.1 V. This result may be ascribed to the poorer adhesion measured for the Cr(N,O)/Cr–N bilayer. Han and coworkers also studied the corrosion resistance of chromium nitride deposited on low alloy steel (AISI 4140) in an aerated 3% NaCl solution, and reported small negative E_{OC} values (~−0.7 V), which might be ascribed mainly to the different corrosive environment [78].

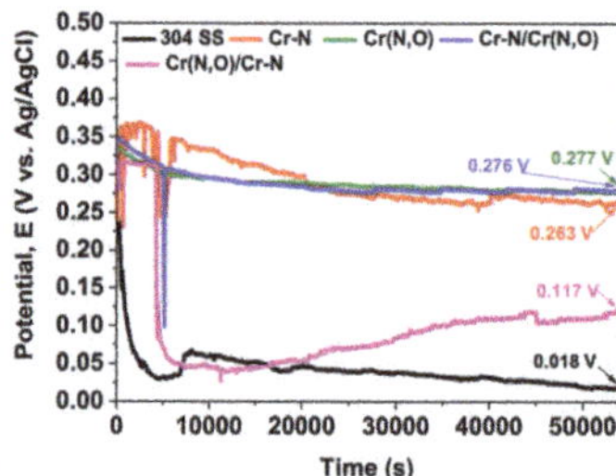

Figure 7. Open circuit potential evolution of Cr-based coatings deposited on 304 SS.

The R_p parameter presented in Table 4 indicates the resistance of the investigated coatings when only a small perturbation was applied (±10 mV). When compared with the uncoated substrate, all the coatings had a higher R_p value (with a factor of 40 to 106). The bilayer Cr–N/Cr(N,O) exhibited higher polarisation resistance value, followed by the Cr(N,O) monolayer. This result can be explained first by the double layer structure which acts as an enhanced barrier to the ingress of the electrolyte through the surface defects, and secondly by the lower porosity of the Cr(N,O) layer as indicated in Table 4, which further blocked the electrolyte ingress. According to Inoue et al., the oxygen atoms in the Cr(N,O) crystallites diffuse outwards, forming Cr_2O_3 layers which surround the crystallites and slow down the oxidation speed, such that the oxidation resistivity of Cr(N,O) is increased [67]. No significant differences were found between the Cr–N monolayer and Cr(N,O)/Cr–N bilayer.

Furthermore, E_{corr} and corrosion current density (i_{corr}) were extracted from Tafel plots, as frequently reported (e.g., [79]). Figure 8 shows the Tafel plots for the plain SS substrate and all the coatings deposited on SS. The inset presents the fitting lines for the Cr(N,O) coating. It can be stated that the behaviour of a material is nobler in a corrosive solution, when the corrosion potential (E_{corr}) value is more electropositive. According to Table 4, all coatings were nobler compared to the uncoated substrate. When the Cr(N,O) is used as top layer, the sample has a more electropositive corrosion potential, indicating that the corrosive solution had less influence on these surfaces. Comparing the coating with Cr–N in top, the most noble corrosion potential was measured for Cr–N monolayer, suggesting a good corrosion resistance.

R_p was determined at the open-circuit potential by linear polarisation tests performed by applying a small perturbation (±10 mV vs. E_{OC}). The R_p values were determined as the slope of the linear region of the $\Delta E - \Delta i$ curve near E_{corr}. The i_{corr} was extracted based on Tafel plots which were recorded from ±0.25 V vs. E_{OC}, using the corrosion test software (VersaStudio) and performing a numerical fit to the Butler–Volmer equation, as we considered that both a cathodic and an anodic reaction occur on the same electrode. As also indicated by the current density parameter, the coated surfaces are less inclined to allow current to flow, as the i_{corr} parameter was two orders of magnitude lower in value than the SS substrate. Considering this parameter, the Cr(N,O) monolayer showed the lowest i_{corr}, followed by Cr–N/Cr(N,O), Cr(O,N)/Cr–N, and Cr–N coatings. The corrosion rate is proportionally related to the i_{corr}. Thus, we can conclude that the Cr(N,O) monolayer has the lowest corrosion rate. It is interesting to note that the presented Cr–N-based coatings exhibited i_{corr} values lower than the multilayered NbN/CrN coatings with 2 to 10 bilayers, immersed in 0.5 M NaCl solution, deposited by the magnetron sputter deposition, as reported by Aperador and Delgado [80].

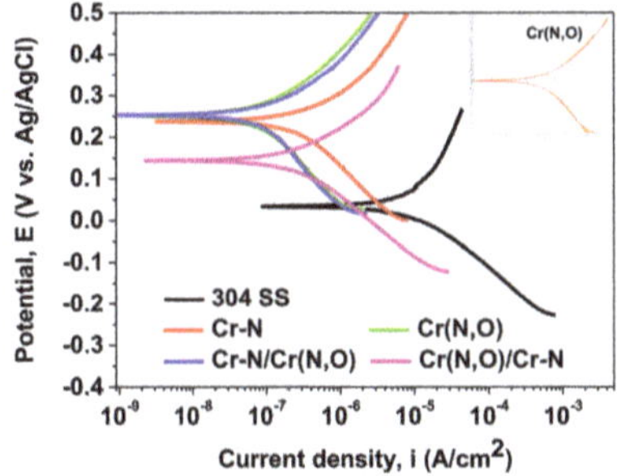

Figure 8. Tafel plots of Cr-based coatings deposited on 304 SS steel; the inset presents the fitting lines in the Tafel plot of Cr(N,O) coating.

Both porosity and protection efficiency of the investigated coatings were calculated based on Equations (2) and (3). By comparing the porosity values, one can conclude that the addition of oxygen leads to a porosity decrease in Cr-based coatings. When the Cr(N,O) layer is on top, it can be seen that the porosity is lower. The Cr(N,O)/Cr–N system shows higher porosity than that Cr–N/Cr(N,O) or Cr(N,O) coatings, its properties being more akin to that of Cr–N.

Regarding the protection efficiency, the best value was found for the Cr(N,O) monolayer, indicating better corrosion resistance. This result is in good agreement with the above-mentioned findings. We note that the Cr(N,O) coating exhibited the best protection efficiency. Moreover, its presence on the top of a bilayer also produced a high protection efficiency, superior to that observed in Cr(N,O)/Cr–N bilayer, probably as a direct result of the beneficial effect of the oxide layers surrounding the crystallites, preventing further oxidation.

Table 4. Corrosion parameters of Cr-based coatings.

Sample	E_{OC} (mV)	R_p (kΩ)	E_{corr} (mV)	i_{corr} (μA/cm^2)	P	P_e (%)
304 SS	18	2.109	34	14.689	–	–
Cr–N	263	86.136	236	0.492	0.016	96.7
Cr(N,O)	277	187.75	250	0.137	0.007	99.1
Cr(N,O)/Cr–N	117	85.599	143	0.336	0.020	97.7
Cr–N/Cr(N,O)	276	223.047	250	0.168	0.006	98.9

Figure 9a presents the evolution of corrosion current density (i_{corr}) versus the coatings' thickness. One may observe that the increase of coating thickness was accompanied, as expected, by the decrease of i_{corr} and the increase of the protection efficiency, as shown in Figure 9b. The comparison of Cr–N and Cr(N,O)/Cr–N coatings provides evidence that the superior corrosion resistance of the bilayer might be related to its higher adhesion to the substrate and with the presence at the substrate interface of the Cr oxynitride layer, conferring a higher protection efficiency to the bilayer. For the Cr–N/Cr(N,O) bilayer, a small decrease of the protection efficiency and i_{corr} values compared to Cr(N,O) monolayer was evident, despite the modest adhesion performance of the monolayer on SS. Probably the superior corrosion resistance of the Cr(N,O) monolayer might be related to the low roughness, low porosity, and low quantity of microdroplets.

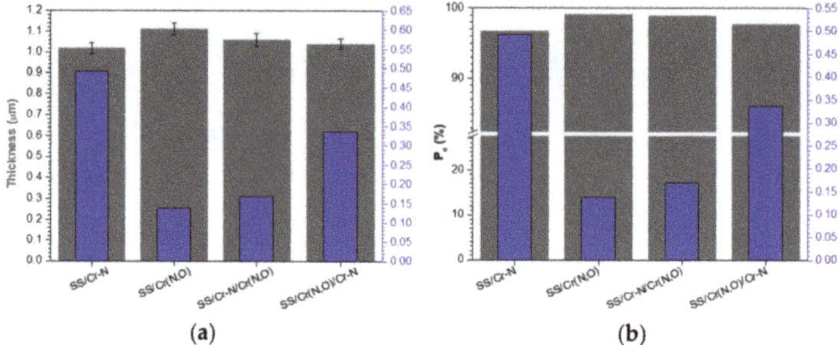

Figure 9. The variation of the corrosion current density (i_{corr}) versus: (**a**) coatings' thickness; (**b**) protection efficiency (P_e)..

3.6. Characterisation of the Coatings after the Corrosion Tests

3.6.1. Coating Roughness

The roughness of the Cr-based coatings after corrosion tests is presented in Figure 10. The most significant result is related to the roughness increase for all investigated surfaces as a result of corrosive attack, thus we could conclude that all surfaces were affected by the corrosive environment to various extents, as presented in Figure 10 and detailed below.

The roughness parameter R_a of the 304 SS substrate increased from 50 µm before corrosive attack to about 1300 µm (Figure 10a), demonstrating that a significant corrosion process affected the bare substrate. The Cr–N monolayer was also considerably affected by corrosion, the R_a increasing from 281 nm (Figure 8a) before corrosion to 549 nm (Figure 10a). This outcome is in good agreement with the electrochemical results, as the R_a value after corrosion increased from 200 nm to 316 nm. On the contrary, the bilayer with Cr(N,O) on top exhibited an almost similar R_a value before and after the corrosive attack, indicating that the roughness was not the main factor influencing the corrosion behaviour. We should underline that within the limit of experimental error, the protection efficiency P_e and the roughness parameters R_a and R_q exhibited the same trend. Moreover, the lower roughness measured on the bilayered coatings compared to the monolayered ones might be explained by the cracks and dislocation blocking at layer interfaces.

After the corrosion tests, all of the coated surfaces showed negative values of S_k compared to the surfaces before corrosion (Figure 10c), pointing to the formation of more valleys on the corroded surfaces as a consequence of significant corrosive processes taking place locally, where the electric field is more intense due to specific surface morphology. The Cr–N monolayer may well illustrate this conclusion, since compared to all the other coatings, it presented the most negative S_k value and also had the lowest corrosion resistance, as presented in Table 4.

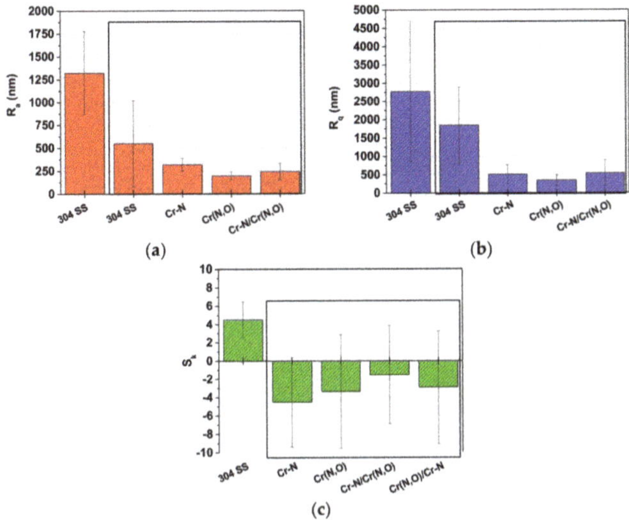

Figure 10. The roughness of Cr-based coatings after the corrosion test: (**a**) R_a parameter; (**b**) R_q parameter; (**c**) S_k parameter.

3.6.2. Surface Morphology and Elemental Composition

In order to gain a deeper insight into the corrosion mechanism in relation to the alteration of the roughness parameters after the corrosion tests, we investigated the surface morphology of corroded surfaces. SEM images of the Cr-based coatings after the corrosion test are presented in Figure 11. As can be seen, the uncoated 304 SS substrate was significantly affected by corrosion, the result being in good agreement with the electrochemical tests. Additionally, the surface of the Cr–N monolayer was deteriorated by the corrosive attack, with numerous damaged zones being observed where the coating was cracked. Some corrosion products were also found on coated surfaces. The poor corrosion resistance of Cr–N monolayer was probably due to the high density of microdroplets and high porosity, confirmed by both SEM images and electrochemical tests, allowing easy chloride penetration and reaching the coating–substrate interface, accelerating the corrosion processes.

According to the electrochemical parameters, the following evolution of corrosion resistance can be stated: Cr(N,O) > Cr–N/Cr(N,O) > Cr(N,O)/Cr–N > Cr–N > SS. By comparing the profilometry lines on selected pits (Figure 11c), it can be seen that the pits found on uncoated 304 SS substrate were deeper and larger than those found on coated surfaces. The Cr(N,O) coating exhibited the best corrosion behaviour, which also showed the small dimensions of pits that appeared during the corrosion tests. This result supports the electrochemical results.

Figure 12 shows the EDS mapping of the damaged zone found on each investigated coating after the corrosion tests. The elemental compositions of the coatings on SS substrates before and after the corrosion tests are presented in Table 5. A low content of the substrate's elements was found in coatings with high corrosion resistance, such as Cr(N,O) monolayer and Cr–N/Cr(N,O) bilayer. The highest amount of Fe originating from the substrate was detected on the Cr–N monolayer surface, confirming that this coating was significantly affected by the corrosion. In conclusion, the results obtained from SEM and EDS analyses carried out on corroded surfaces sustain the electrochemical results.

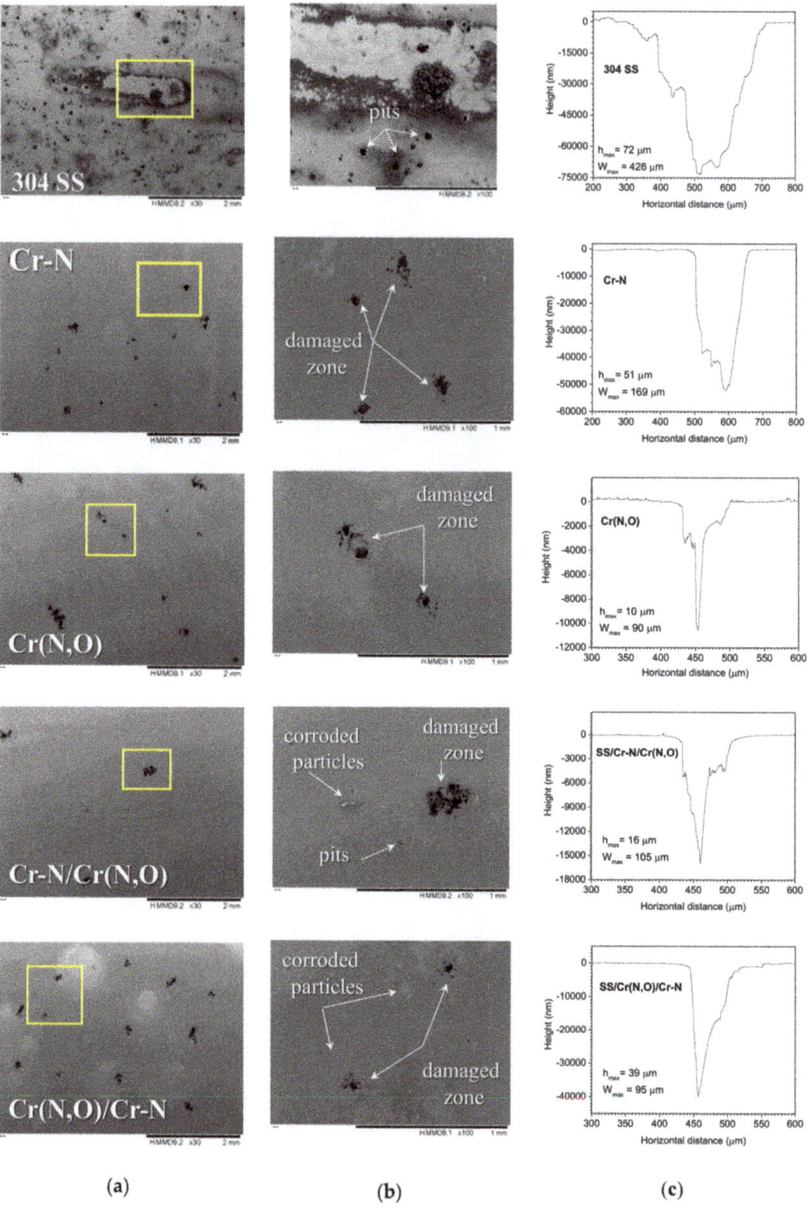

Figure 11. Investigation of the Cr-based coatings surfaces after the corrosion tests: micrographs of: (**a**) SEM micrographs at 30× magnification; (**b**) SEM micrographs at 100× magnification; (**c**) profilometry lines on selected pits.

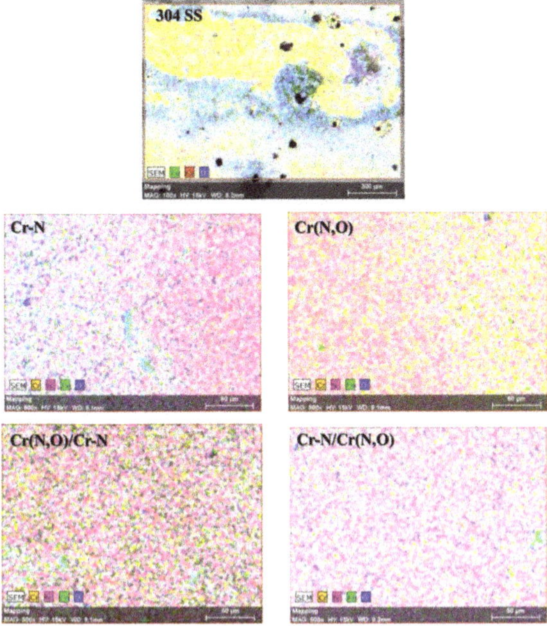

Figure 12. Energy dispersive X-ray spectrometry (EDS) mapping micrographs of the Cr-based coatings after the corrosion test.

Table 5. The elemental composition of the investigated specimens before (B) and after (A) the corrosion test in the areas shown in Figure 11a.

Substrate/Coating (Image Zone)	Elemental Composition (at.%)				
	N	O	Cr	C	Fe
304 SS (A)	–	41.0 ± 2.8	9.5 ± 0.1	–	49.5 ± 3.2
Cr–N (B)	30.5 ± 1.2	2.6 ± 0.2	63.4 ± 2.5	3.5 ± 0.2	–
Cr–N (A)	22.7 ± 0.8	17.5 ± 0.1	56.8 ± 1.8	–	3.0 ± 0.1
Cr(N,O) (B)	35.6 ± 1.7	14.9 ± 0.8	46.8 ± 2.2	2.6 ± 0.1	–
Cr(N,O) (A)	32.4 ± 1.5	20.2 ± 0.3	46.2 ± 2.1	–	1.2 ± 0.1
Cr–N/ Cr(N,O) (B)	36.9 ± 1.7	13.5 ± 0.7	49.3 ± 2.2	2.2 ± 0.1	–
Cr–N/ Cr(N,O) (A)	35.8 ± 1.5	17.4 ± 0.8	46.6 ± 1.8	–	0.5 ± 0.1
Cr(N,O)/Cr–N (B)	31.5 ± 1.3	2.8 ± 0.2	61.5 ± 2.5	4.2 ± 0.2	–
Cr(N,O)/Cr–N (A)	31.1 ± 1.3	22.9 ± 1.3	44.1 ± 2.1	–	1.9 ± 0.1

4. Conclusions

Cr-based coatings in monolayered (Cr–N and Cr(N,O)) and bilayered structures (Cr–N/Cr(N,O) and Cr(N,O)/Cr–N) were prepared by the cathodic arc method. The presence of Cr_2N, CrN, and Cr(N,O), as well as mixtures of these phases were identified and related to the composition of coatings. The oxynitrides were smooth, compact, and homogenously deposited on 304 SS, with few microdroplets. The corrosion protection performance of the developed mono- and bilayered Cr-based coatings was evaluated in 0.10 M NaCl + 1.96 M H_2O_2.

- The corrosion current densities of the coatings decreased by more than 30 times compared to the bare substrate.
- The Cr–N coating, as mono- or bilayer, had high porosity and lower protective performance.

- The protective efficiency of Cr(N,O) coating (99.1%) and porosity (0.007) were excellent when compared to other coatings, because this coating is denser, less porous, and more adherent to the substrate.

- Both bilayer coatings substantially improved the corrosion protection of 304 SS.

- The bilayer with Cr(N,O) on top possessed the best corrosion resistance behaviour, having the lowest current density corrosion and consequently the highest protective efficiency and the lowest porosity.

- The corrosion resistance can be ranked in the following order: Cr(N,O) > Cr–N/Cr(N,O) > Cr(N,O)/Cr–N > Cr–N.

Acknowledgments: The authors are grateful for the financial support of a grant of the Bilateral Cooperation South Africa–Romania Projects (no. PN-II-CT-RO-ZA-2014-1 and NRF project reference no. UID: 104018, as acknowledged by NRF/RISA Reference RJCT14042566606) signed between the Romanian National Authority for Scientific Research and Innovation, CNCS-UEFISCDI and the South Africa National Research Foundation. The work has also been funded by the Core Program PN 2018, under the support of the Romanian Ministry of Research and Innovation. The EDS, SEM, and XRD results were obtained using the systems purchased by the Romanian Sectorial Operational Programme "Increase of economic Competitiveness", ID 1799/SMIS 48589/2015.

Author Contributions: M.B. and L.F.P. designed the research and supervised the work; M.D., A.C.P. and E.S.M.M. performed the experiments; A.V. prepared the coatings. M.D. and A.V. analysed the data and provided valuable input in data interpretation; M.D., E.S.M.M. and A.V. wrote the first draft of the article; all authors reviewed and approved the work.

Conflicts of Interest: The authors declare no conflict of interest. The founding sponsors had no role in the design of the study; in the collection, analyses, or interpretation of data; in the writing of the manuscript, and in the decision to publish the results.

References

1. Baddoo, N.R. Stainless steel in construction: A review of research, applications, challenges and opportunities. *J. Constr. Steel Res.* **2008**, *64*, 1199–1206. [CrossRef]

2. Disegi, J.A.; Eschbach, L. Stainless steel in bone surgery. *Injury* **2000**, *31*, D2–D6. [CrossRef]

3. *Handbook of Stainless Steel*; Outokumpu: Helsinki, Finland, 2013.

4. Gedge, G. Structural uses of stainless steel—Buildings and civil engineering. *J. Constr. Steel Res.* **2008**, *64*, 1194–1198. [CrossRef]

5. Pocaznoi, D.; Calmet, A.; Etcheverry, L.; Erable, B.; Bergel, A. Stainless steel is a promising electrode material for anodes of microbial fuel cells. *Energy Environ. Sci.* **2012**, *5*, 9645–9652. [CrossRef]

6. Park, S.H.C.; Sato, Y.S.; Kokawa, H.; Okamoto, K.; Hirano, S.; Inagaki, M. Corrosion resistance of friction stir welded 304 stainless steel. *Scr. Mater.* **2004**, *51*, 101–105. [CrossRef]

7. Arango, S.; Peláez-Vargas, A.; García, C. Coating and surface treatments on orthodontic metallic materials. *Coatings* **2013**, *3*, 1–15. [CrossRef]

8. Hao, L.; Yoshida, H.; Itoi, T.; Lu, Y. Preparation of metal coatings on steel balls using mechanical coating technique and its process analysis. *Coatings* **2017**, *7*, 53. [CrossRef]

9. Mohamed, A.M.A.; Abdullah, A.M.; Younan, N.A. Corrosion behavior of superhydrophobic surfaces: A review. *Arab. J. Chem.* **2015**, *8*, 749–765. [CrossRef]

10. Perillo, P.M. Corrosion behavior of coatings of titanium nitride and titanium-titanium nitride on steel substrates. *Corrosion* **2006**, *62*, 182–185. [CrossRef]

11. Wicks, Z.W.; Jones, F.N.; Pappas, S.P.; Wicks, D.A. Corrosion protection by coatings. *Org. Coat.* **2007**, 137–158. [CrossRef]

12. Khelifa, F.; Habibi, Y.; Benard, F.; Dubois, P. Smart acrylic coatings containing silica particles for corrosion protection of aluminum and other metals. In *Handbook of Smart Coatings for Materials Protection*; Makhlouf, A.S.H., Ed.; Elsevier: Amsterdam, The Netherlands, 2014; pp. 423–458. [CrossRef]

13. Montemor, M.F. Functional and smart coatings for corrosion protection: A review of recent advances. *Surf. Coat. Technol.* **2014**, *258*, 17–37. [CrossRef]

14. Fenker, M.; Balzer, M.; Kappl, H. Corrosion protection with hard coatings on steel: Past approaches and current research efforts. *Surf. Coat. Technol.* **2014**, *257*, 182–205. [CrossRef]

15. Verran, J.; Packer, A.; Kelly, P.; Whitehead, K.A. Titanium-coating of stainless steel as an aid to improved cleanability. *Int. J. Food Microbiol.* **2010**, *141*, S134–S139. [CrossRef] [PubMed]

16. Mackey, E.D.; Seacord, T.F. Guidelines for using stainless steel in the water and desalination industries. *J. Am. Water Works Assoc.* **2017**, *109*, E158–E169. [CrossRef]

17. Theodore, N.D.; Holloway, B.C.; Manos, D.M.; Moore, R.; Hernandez, C.; Wang, T.; Dylla, H.F. Nitrogen-implanted silicon oxynitride: A coating for suppressing field emission from stainless steel used in high-voltage applications. *IEEE Trans. Plasma Sci.* **2006**, *34*, 1074–1079. [CrossRef]

18. Zalnezhad, E.; Hamouda, A.M.S.; Faraji, G.; Shamshirband, S. TiO$_2$ nanotube coating on stainless steel 304 for biomedical applications. *Ceram. Int.* **2015**, *41*, 2785–2793. [CrossRef]

19. Shaigan, N.; Qu, W.; Ivey, D.G.; Chen, W. A review of recent progress in coatings, surface modifications and alloy developments for solid oxide fuel cell ferritic stainless steel interconnects. *J. Power Sources* **2010**, *195*, 1529–1542. [CrossRef]

20. Pruncu, C.I.; Braic, M.; Dearn, K.D.; Farcau, C.; Watson, R.; Constantin, L.R.; Balaceanu, M.; Braic, V.; Vladescu, A. Corrosion and tribological performance of quasi-stoichiometric titanium containing carbo-nitride coatings. *Arab. J. Chem.* **2016**, *10*, 1015–1028. [CrossRef]

21. Sperko, W.J. *Rust on Stainless Steel*; Sperko Engineering Services Inc.: Greensboro, NC, USA, 2014.

22. Muslim, Z.R.; Abbas, A.A. The effect of pH and temperature on corrosion rate stainless steel 316L used as biomaterial. *Int. J. Basic Appl. Sci.* **2012**, *4*, 17–20.

23. Truman, J.E. The influence of chloride content, pH and temperature of test solution on the occurrence of stress corrosion cracking with austenitic stainless steel. *Corros. Sci.* **1977**, *17*, 737–746. [CrossRef]

24. Vitelaru, C.; Balaceanu, M.; Parau, A.; Luculescu, C.R.; Vladescu, A. Investigation of nanostructured TiSiC-Zr and TiSiC-Cr hard coatings for industrial applications. *Surf. Coat. Technol.* **2014**, *251*, 21–28. [CrossRef]

25. Yun, J.S.; Hong, Y.S.; Kim, K.H.; Kwon, S.H.; Wang, Q.M. Characteristics of Ternary Cr-O-N Coatings Synthesized by Using an Arc Ion Plating Technique. *J. Korean Phys. Soc.* **2010**, *57*, 103.

26. Chen, S.; Wu, B.H.; Xie, D.; Jiang, F.; Liu, J.; Sun, H.L.; Zhu, S.; Bai, B.; Leng, Y.X.; Huang, N.; et al. The adhesion and corrosion resistance of Ti–O films on CoCrMo alloy fabricated by high power pulsed magnetron sputtering (HPPMS). *Surf. Coat. Technol.* **2014**, *252*, 8–14. [CrossRef]

27. Ryabchikov, A.I.; Ryabchikov, I.A.; Stepanov, I.B. Development of filtered DC metal plasma ion implantation and coating deposition methods based on high-frequency short-pulsed bias voltage application. *Vacuum* **2005**, *78*, 331–336. [CrossRef]

28. Ryabchikov, A.I.; Ryabchikov, I.A.; Stepanov, I.B.; Usov, Y.P. High-frequency short-pulsed metal plasma-immersion ion implantation or deposition using filtered DC vacuum-arc plasma. *Surf. Coat. Technol.* **2007**, *201*, 6523–6525. [CrossRef]

29. Ryabchikov, A.I.; Ryabchikov, I.A.; Stepanov, I.B.; Sivin, D.O. Recent advances in surface processing with the filtered DC vacuum-arc plasma. *Vacuum* **2005**, *78*, 445–449. [CrossRef]

30. Ryabchikov, A.I.; Ryabchikov, I.A.; Stepanov, I.B.; Dektyarev, S.V. High current vacuum-arc ion source for ion implantation and coating deposition technologies. *Rev. Sci. Instrum.* **2006**, *77*, 03B516. [CrossRef]

31. Stepanov, I.B.; Ryabchikov, A.I.; Nochovnaya, N.A.; Sharkeev, Y.P.; Shulepov, I.A.; Ryabchikov, I.A.; Sivin, D.O.; Fortuna, S.V. Vacuum arc filtered metal plasma application in hybrid technologies of ion-beam and plasma material processing. *Surf. Coat. Technol.* **2007**, *201*, 8596–8600. [CrossRef]

32. Davis, J. *Corrosion: Understanding the Basics*; ASM International: Geauga County, OH, USA, 2000; p. 574.

33. Estrada-Martínez, J.; Reyes-Gasga, J.; García-García, R.; Vargas-Becerril, N.; Zapata-Torres, M.G.; Gallardo-Rivas, N.V.; Mendoza-Martínez, A.M.; Paramo-García, U. Wettability modification of the AISI 304 and 316 stainless steel and glass surfaces by titanium oxide and titanium nitride coating. *Surf. Coat. Technol.* **2017**, *330*, 61–70. [CrossRef]

34. Ali, F.; Kwak, J.S. The impact of number of interfaces on corrosion behavior of TiCrN films. *J. Nanosci. Nanotechnol.* **2017**, *17*, 4318–4321. [CrossRef]

35. Adesina, A.Y.; Gasem, Z.M.; Madhan Kumar, A. Corrosion resistance behavior of single-layer cathodic arc PVD nitride-base coatings in 1M HCl and 3.5 pct NaCl Solutions. *Metall. Mater. Trans. B Process Metall. Mater. Process. Sci.* **2017**, *48*, 1321–1332. [CrossRef]

36. Navinsek, B.; Seal, S. Transition metal nitride functional coatings. *JOM* **2001**, *53*, 51–54. [CrossRef]

37. Mayrhofer, P.H.; Rachbauer, R.; Holec, D.; Rovere, F.; Schneider, J.M. Protective transition metal nitride coatings. In *Comprehensive Materials Processing*; Hashmi, S., Ed.; Elsevier: Amsterdam, The Netherlands, 2014; Volume 4, pp. 355–388, ISBN 9780080965338.

38. Braic, L.; Vasilantonakis, N.; Mihai, A.; Garcia, I.J.V.; Fearn, S.; Zou, B.; Alford, N.M.; Doiron, B.; Oulton, R.F.; Maier, S.A.; et al. Titanium oxynitride thin films with tunable double epsilon-near-zero behavior for nanophotonic applications. *ACS Appl. Mater. Interfaces* **2017**, *9*, 29857–29862. [CrossRef] [PubMed]

39. Fenker, M.; Kappl, H.; Carvalho, P.; Vaz, F. Thermal stability, mechanical and corrosion behaviour of niobium-based coatings in the ternary system Nb–O–N. *Thin Solid Films* **2011**, *519*, 2457–2463. [CrossRef]

40. Braic, M.; Balaceanu, M.; Vladescu, A.; Kiss, A.; Braic, V.; Epurescu, G.; Dinescu, G.; Moldovan, A.; Birjega, R.; Dinescu, M. Preparation and characterization of titanium oxy-nitride thin films. *Appl. Surf. Sci.* **2007**, *253*, 8210–8214. [CrossRef]

41. Subramanian, B.; Muraleedharan, C.V.; Ananthakumar, R.; Jayachandran, M. A comparative study of titanium nitride (TiN), titanium oxy nitride (TiON) and titanium aluminum nitride (TiAlN), as surface coatings for bio implants. *Surf. Coat. Technol.* **2011**, *205*, 5014–5020. [CrossRef]

42. Rtimi, S.; Baghriche, O.; Sanjines, R.; Pulgarin, C.; Bensimon, M.; Kiwi, J. TiON and TiON-Ag sputtered surfaces leading to bacterial inactivation under indoor actinic light. *J. Photochem. Photobiol. A Chem.* **2013**, *256*, 52–63. [CrossRef]

43. Li, W.Z.; Yi, D.Q.; Li, Y.Q.; Liu, H.Q.; Sun, C. Effects of the constitution of CrON diffusion barrier on the oxidation resistance and interfacial fracture of duplex coating system. *J. Alloys Compd.* **2012**, *518*, 86–95. [CrossRef]

44. Li, W.Z.; Yao, Y.; Wang, Q.M.; Bao, Z.B.; Gong, J.; Sun, C.; Jiang, X. Improvement of oxidation-resistance of NiCrAlY coatings by application of CrN or CrON interlayer. *J. Mater. Res.* **2011**, *23*, 341–352. [CrossRef]

45. Warcholinski, B.; Gilewicz, A.; Lupicka, O.; Kuprin, A.S.; Tolmachova, G.N.; Ovcharenko, V.D.; Kolodiy, I.V.; Sawczak, M.; Kochmanska, A.E.; Kochmanski, P.; et al. Structure of CrON coatings formed in vacuum arc plasma fluxes. *Surf. Coat. Technol.* **2016**, *309*, 920–930. [CrossRef]

46. Cubillos, G.I.; Bethencourt, M.; Olaya, J.J.; Alfonso, J.E.; Marco, J.F. The influence of deposition temperature on microstructure and corrosion resistance of ZrO_xN_y/ZrO_2 coatings deposited using RF sputtering. *Appl. Surf. Sci.* **2014**, *309*, 181–187. [CrossRef]

47. Tijani, J.O.; Mouele, M.E.S.; Tottito, T.C.; Fatoba, O.O.; Petrik, L.F. Degradation of 2-nitrophenol by dielectric barrier discharge system: The Influence of carbon doped TiO_2 photocatalyst supported on stainless steel mesh. *Plasma Chem. Plasma Process.* **2017**, *37*, 1343–1373. [CrossRef]

48. Ruden, A.; Restrepo-Parra, E.; Paladines, A.U.; Sequeda, F. Corrosion resistance of CrN thin films produced by dc magnetron sputtering. *Appl. Surf. Sci.* **2013**, *270*, 150–156. [CrossRef]

49. Wang, D.; Hu, M.; Jiang, D.; Fu, Y.; Wang, Q.; Yang, J.; Sun, J.; Weng, L. The improved corrosion resistance of sputtered CrN thin films with Cr-ion bombardment layer by layer. *Vacuum* **2017**, *143*, 329–335. [CrossRef]

50. Chen, Q.; Cao, Y.; Xie, Z.; Chen, T.; Wan, Y.; Wang, H.; Gao, X.; Chen, Y.; Zhou, Y.; Guo, Y. Tribocorrosion behaviors of CrN coating in 3.5 wt.% NaCl solution. *Thin Solid Films* **2017**, *622*, 41–47. [CrossRef]

51. Shan, L.; Zhang, Y.R.; Wang, Y.X.; Li, J.L.; Jiang, X.; Chen, J.M. Corrosion and wear behaviors of PVD CrN and CrSiN coatings in seawater. *Trans. Nonferrous Met. Soc. China* **2016**, *26*, 175–184. [CrossRef]

52. Ma, F.; Li, J.; Zeng, Z.; Gao, Y. Structural, mechanical and tribocorrosion behaviour in artificial seawater of CrN/AlN nano-multilayer coatings on F690 steel substrates. *Appl. Surf. Sci.* **2018**, *428*, 404–414. [CrossRef]

53. Gulbiński, W.; Gilewicz, A.; Suszko, T.; Warcholiński, B.; Kukliński, Z. Ti–Si–C sputter deposited thin film coatings. *Surf. Coat. Technol.* **2004**, *180–181*, 341–346. [CrossRef]

54. Gilewicz, A.; Chmielewska, P.; Murzynski, D.; Dobruchowska, E.; Warcholinski, B. Corrosion resistance of CrN and CrCN/CrN coatings deposited using cathodic arc evaporation in Ringer's and Hank's solutions. *Surf. Coat. Technol.* **2016**, *299*, 7–14. [CrossRef]

55. Dinu, M.; Hauffman, T.; Cordioli, C.; Vladescu, A.; Braic, M.; Hubin, A.; Cotrut, C.M. Protective performance of Zr and Cr based silico-oxynitrides used for dental applications by means of potentiodynamic polarization and odd random phase multisine electrochemical impedance spectroscopy. *Corros. Sci.* **2017**, *115*, 118–128. [CrossRef]

56. Piedrahita, W.F.; Coy, L.E.; Amaya, C.; Llarena, I.; Caicedo, J.C.; Yate, L. Influence of the negative R.F. bias voltage on the structural, mechanical and electrical properties of Hf–C–N coatings. *Surf. Coat. Technol.* **2016**, *286*, 251–255. [CrossRef]

57. *BS EN 1071-3:2005 Advanced Technical Ceramics–Methods of Test for Ceramic Coatings*; British Standards Institution: London, UK, 2005; p. 47.

58. Oliver, W.C.; Pharr, G.M. An improved technique for determining hardness and elastic modulus using load and displacement sensing indentation experiments. *J. Mater. Res.* **1992**, *6*, 1564–1583. [CrossRef]

59. Li, M.S.; Feng, C.J.; Wang, F.H. Effect of partial pressure of reactive gas on chromium nitride and chromium oxide deposited by arc ion plating. *Trans. Nonferrous Met. Soc. China* **2006**, *16*, s276–s279. [CrossRef]

60. Münz, W.-D.; Smith, I.J.; Lewis, D.B.; Creasey, S. Droplet formation on steel substrates during cathodic steered arc metal ion etching. *Vacuum* **1997**, *48*, 473–481. [CrossRef]

61. Massalski, T.B.; Okamoto, H.; Subramanian, P.R.; Kacprzak, L. *Binary Alloy Phase Diagrams*, 2nd ed.; ASM International: Geauga County, OH, USA, 1990.

62. Ehrlich, A.; Kühn, M.; Richter, F.; Hoyer, W. Complex characterisation of vacuum arc-deposited chromium nitride thin films. *Surf. Coat. Technol.* **1995**, *76–77*, 280–286. [CrossRef]

63. Minami, T.; Nishio, S.; Murata, Y. Periodic microstructures of Cr–O–N coatings deposited by arc ion plating. *Surf. Coat. Technol.* **2014**, *254*, 402–409. [CrossRef]

64. Rebholz, C.; Ziegele, H.; Leyland, A.; Matthews, A. Structure, mechanical and tribological properties of nitrogen-containing chromium coatings prepared by reactive magnetron sputtering. *Surf. Coat. Technol.* **1999**, *115*, 222–229. [CrossRef]

65. Anna, E.M.; McMurdie, H.F.; Ondik, H.M. Borides carbides and nitrides. In *Phase Equilibria Diagrams*; McHale, A.E., Ed.; The American Ceramic Society: Westerville, OH, USA, 1994; p. 415.

66. Suzuki, K.; Endo, T.; Fukushima, T.; Sato, A.; Suzuki, T.; Nakayama, T.; Suematsu, H.; Niihara, K. Controlling oxygen content by varying oxygen partial pressure in chromium oxynitride thin films prepared by pulsed laser deposition. *Mater. Trans.* **2013**, *54*, 1140–1144. [CrossRef]

67. Suzuki, K.; Endo, T.; Sato, A.; Suzuki, T.; Nakayama, T.; Suematsu, H.; Niihara, K. Epitaxial growth of chromium oxynitride thin films on magnesium oxide (100) substrates and their oxidation behavior. *Mater. Trans.* **2013**, *54*, 1957–1961. [CrossRef]

68. Oh, U.C.; Je, J.H. Effects of strain energy on the preferred orientation of TiN thin films. *J. Appl. Phys.* **1993**, *74*, 1692–1696. [CrossRef]

69. Pelleg, J.; Zevin, L.Z.Z.; Lungo, S.; Croitoru, N. Reactive-sputter-deposited TiN films on glass substrates. *Thin Solid Films* **1991**, *197*, 117–128. [CrossRef]

70. Sebastiani, M.; Piccoli, M.; Bemporad, E. Effect of micro-droplets on the local residual stress field in CAE-PVD thin coatings. *Surf. Coat. Technol.* **2013**, *215*, 407–412. [CrossRef]

71. Steyer, P.; Mege, A.; Pech, D.; Mendibide, C.; Fontaine, J.; Pierson, J.F.; Esnouf, C.; Goudeau, P. Influence of the nanostructuration of PVD hard TiN-based films on the durability of coated steel. *Surf. Coat. Technol.* **2008**, *202*, 2268–2277. [CrossRef]

72. Musil, J.; Novák, P.; Čerstvý, R.; Soukup, Z. Tribological and mechanical properties of nanocrystalline-TiC/a-C nanocomposite thin films. *J. Vac. Sci. Technol. A Vac. Surf. Film* **2010**, *28*, 244–249. [CrossRef]

73. Yate, L.; Martínez-de-Olcoz, L.; Esteve, J.; Lousa, A. Ultra low nanowear in novel chromium/amorphous chromium carbide nanocomposite films. *Appl. Surf. Sci.* **2017**, *420*, 707–713. [CrossRef]

74. Castanho, J.M.; Vieira, M.T. Effect of ductile layers in mechanical behaviour of TiAlN thin coatings. *J. Mater. Process. Technol.* **2003**, *143–144*, 352–357. [CrossRef]

75. Mohammadpour, E.; Jiang, Z.-T.; Altarawneh, M.; Mondinos, N.; Rahman, M.M.; Lim, H.N.; Huang, N.M.; Xie, Z.; Zhou, Z.; Dlugogorski, B.Z. Experimental and predicted mechanical properties of $Cr_{1-x}Al_xN$ thin films, at high temperatures, incorporating in situ synchrotron radiation X-ray diffraction and computational modelling. *RSC Adv.* **2017**, *7*, 22094–22104. [CrossRef]

76. Jagielski, J.; Khanna, A.S.; Kucinski, J.; Mishra, D.S.; Racolta, P.; Sioshansi, P.; Tobin, E.; Thereska, J.; Uglov, V.; Vilaithong, T.; et al. Effect of chromium nitride coating on the corrosion and wear resistance of stainless steel. *Appl. Surf. Sci.* **2000**, *156*, 47–64. [CrossRef]

77. Han, S.; Lin, J.H.; Tsai, S.H.; Chung, S.C.; Wang, D.Y.; Lu, F.H.; Shih, H.C. Corrosion and tribological studies of chromium nitride coated on steel with an interlayer of electroplated chromium. *Surf. Coat. Technol.* **2000**, *133–134*, 460–465. [CrossRef]

78. Han, S.; Lin, J.H.; Wang, D.Y.; Lu, F.-H.; Shih, H.C. Corrosion resistance of chromium nitride on low alloy steels by cathodic arc deposition. *J. Vac. Sci. Technol. Part A Vac. Surf. Film* **2001**, *19*, 1442–1446. [CrossRef]

79. Yate, L.; Coy, L.E.; Aperador, W. Robust tribo-mechanical and hot corrosion resistance of ultra-refractory Ta–Hf–C ternary alloy films. *Sci. Rep.* **2017**, *7*, 3080. [CrossRef] [PubMed]
80. Aperador, W.; Duque, J.; Delgado, E. Electrochemical and tribological and mechanical performances coatings multilayer type NbC/CrN. *Int. J. Electrochem. Sci.* **2016**, *11*, 6347–6355. [CrossRef]

Article

Wear Transition of CrN Coated M50 Steel under High Temperature and Heavy Load

Chi Zhang [1], Le Gu [1,*], Guangze Tang [2] and Yuze Mao [1]

[1] Research Lab of Space & Aerospace Tribology, Harbin Institute of Technology, Harbin 150001, China; zhc_hit@163.com (C.Z.); ma0yuze@163.com (Y.M.)
[2] School of Material Science and Technology, Harbin Institute of Technology, Harbin 150001, China; tangoak@163.com
* Correspondence: gule@hit.edu.cn; Tel.: +86-451-8640-3712

Academic Editor: Shiladitya Paul
Received: 22 September 2017; Accepted: 15 November 2017; Published: 20 November 2017

Abstract: The combination of high temperature indentation and wear test provides a useful way to investigate wear of CrN coating and wear transition mechanisms. In this paper, the high temperature hardness of CrN coatings and load bearing capacity, L_b, of CrN coated M50 disks were determined from spherical indentation at temperatures up to 500 °C. Wear tests with different normal loads were carried out at the same temperatures as the indentation tests. The results show that wear mechanism of CrN coating changes with external load, P, and temperature, T. Under a tested condition of $P < L_b$ and $T < 315$ °C, abrasive is the dominant wear mechanism for CrN coating. Under a tested condition of $P < L_b$ and $T \geq 315$ °C, wear of CrN coating transitions into mild oxidation wear due to the lubricating effect of chromium oxide film. Under a tested condition of $P > L_b$ and $T < 315$ °C, wear of CrN coating was controlled by coating fracture. Under a tested condition of $P > L_b$ and $T \geq 315$ °C, wear of CrN coating transitions into the severe wear mode, due to the tensile fracture of oxidation films, thereby leading to adhesion between CrN coating and tribo-counterpart. The presented method can be helpful in predicting permissible load and working temperature in tribological applications of CrN coating.

Keywords: CrN coating; high temperature indentation; high temperature mechanical properties; wear mechanism; wear transition

---

1. Introduction

Chromium nitride (CrN) films, with good oxidation, anti-corrosive and anti-adhesive properties [1–4], are promising candidates for protection in high temperature applications, such as for tools or aerospace rolling bearings. Nevertheless, the wear of such CrN coating–steel substrate systems at elevated temperatures is still not completely understood [5–8]. This is because the material removal process for CrN coatings is dependent on operation parameters (load, temperature, etc.). Wear mechanism for CrN coatings may be changed when the external load or working temperature are changed. Lim and Ashby [9] first suggested using the terms "wear transition" for describing the changing of dominant wear mechanisms with contact pressure and velocity in a steel tribo-pair. Later, Wang et al. [10] studied the wear transition for homogenous brittle materials, such as Al_2O_3, Si_3N_4 and SiC ceramics. There is still a lack of information about the wear transition of CrN coatings. Therefore, the aim of this paper is to evaluate dependency of the wear of CrN coatings on both temperature and external load.

The wear transition from abrasive wear to oxidative wear was frequently observed for CrN coatings at high temperature. Polcar et al. [11] performed elevated temperature ball-on-disc tests on CrN coatings sliding against Al_2O_3 ball. It was found that the tribological properties of CrN coatings at high temperature were predominantly influenced by formation of a chromium oxide tribo-layer. Qi et al. [12]

studied the tribo-oxide behavior of CrN and Cr_2N coatings. The dense Cr_2O_3 scales both formed on the Cr_2N and CrN coatings after tribological tests at temperatures above 300 °C. Mandrino et al. [13] observed that the chromium-oxide films formed on CrN coatings were very thin (in 10–100 nm region). However, Scheerer et al. [14] argued that there may be critical loads, above which the lubricating of chromium oxide tribo-layer failed. He found that the high temperature wear of CrN coatings can be classified as oxidative mild wear under a low load, but different wear mechanisms under a high load, such as adhesive wear, delamination wear or the compound wear mechanisms containing oxidative wear and abrasive wear or adhesive wear. However, he did not further study how to determine the critical loads for a chosen CrN coating–substrate system.

By extension of the depth-sensing indentation techniques to elevated temperatures, the critical load for a coating–substrate system at high temperatures can be studied [15]. A comprehensive work on high-temperature sharp indentation has recently been presented by Smith et al. [16]. However, some researchers [17–19] argued that the indentation response was highly dependent on the tip geometry characteristic. As the oxidation of diamond is known to occur above 400 °C, the geometric variation of indenter's diamond tip due to the asymmetric thermal expansion or erosion by oxidation is a serious concern at high temperatures. Using a tungsten carbide (WC) ball with high hardness and oxidation resistance as the indenter tip avoids any such geometric variation or erosion.

In this paper, a joint test rig, equipped with a ϕ1.588 WC ball as the indenter tip, was developed for performing both high temperature indentation tests and ball-on-disk tribological tests. Then, the indentation and wear tests on CrN (with different thickness) coated M50 disks were performed from 25 °C to 500 °C in open atmosphere. Hardness, H, and load bearing capacity, L_b, of CrN coated M50 disks at high temperatures were predicted by load–displacement measurements. Besides, post wear test examination revealed the relationships between applied loads and material removal patterns for CrN coated M50 disks at each temperatures. Based on these results, the wear transition mechanisms for CrN coated M50 disks were explored.

2. Materials and Methods

2.1. M50 Disks and CrN Coatings

The chemical composition of the as-received M50 disks was as follows (wt %): C 0.72, Mn 0.3, Si 0.2, Cr 3.72, Mo 4.0, Ni 0.1, V 1.0, Cu 0.03 and Fe balanced. All of the M50 disks with 50 mm in diameter and 6 mm in thickness were polished to surface roughness, R_a, of approximately 0.04 μm and cleaned with alkaline in an ultrasonic bath, dried in warm air.

CrN coatings were deposited on M50 steel disks by the multi-arc ion plating technique, which were commercially available, provided by Beijing Technology Science Corp., Ltd. (Beijing, China). The M50 disks were mounted on a rotating table (2 r/min) at a distance of 180 mm from the arc source. The growth rate of was approximately 2 μm/h, using the following process parameters: target material: two Cr targets with diameter 100 mm and 99.9% in purity being used as cathodes and placed oppositely with a distance of 460 mm; atmosphere: Ar-N_2 mixture with an N_2-partial pressure of 2.4×10^{-3} mbar (0.24 Pa) and a total working pressure of 4×10^{-3} mbar (0.4 Pa); bias: -70 V; temperature: 450 °C. Coating thickness was controlled by the deposition time. The coating thickness of the two kinds of CrN coated M50 disks were around 2 ± 0.2 μm and 5 ± 0.2 μm, corresponding to processing times of 60 min and 150 min, respectively.

2.2. Test Rig

The joint test rig for high temperature micro-indentation and wear test was developed based on a Bruker CETR UMT-3 multifunction tribometer (Bruker, Billerica, MA, USA), as schematically shown in Figure 1. Balls of WC with a diameter of 1.588 mm and surface roughness, R_a, of 0.02 μm were used as the indenter tips and tribo-counterparts. The indented disk was fixed on specimen stage and heated by resistance heaters (labeled as H1) under the surface of stage. The thermocouple (labeled as T1) for controlling the sample temperature was placed 1 mm besides the disk. The thermocouple (labeled as T2) for measuring the indenter's temperature was located 2 mm from WC ball. The thermocouple (labeled as T3) for controlling chamber temperature was placed on the cover of chamber. The resistance heaters (labeled as H2 and H3) for heating chamber were placed around the wall. Besides, an air-cooling thermal shield was placed between the chamber and loading cell for minimization of any heating of electronics in the loading cell. During indentation tests or wear tests, the temperature differences among T1, T2, and T3 was ensured to be less than 5 °C.

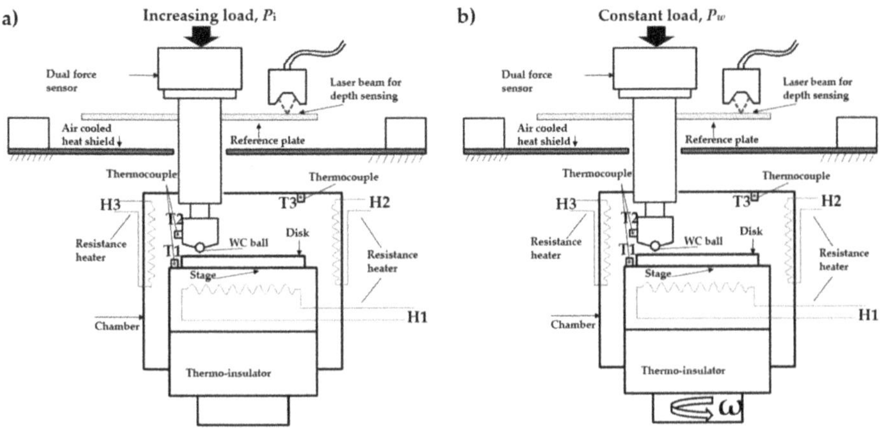

Figure 1. Schematic of the joint test rig for high temperature micro-indentation and wear test: (**a**) for indentation test; (**b**) for wear test.

In indentation tests, the stage was kept stationary. The WC ball was pressed against tested surface under an increasing load, P_i, thus displacement occurred for the reference plate which was rigidly connected to the indenter, as shown in Figure 1a. The applied force, P_i, was measured by the load sensor (UMT, DFH-100-44163, scale 1000 N, resolution 0.01%, Bruker, Billerica, MA, USA). The displacement of the reference plate was recorded by the CCD laser sensor (LK-G10, scale ± 1 mm, resolution 0.02%, Keyence, Osaka, Japan). In wear tests, the WC ball was pressed onto sample surface under a constant load, P_w, while the table was rotated at the given speed, ω.

2.3. Test Procedure

The parameters used in high temperature indentation and wear tests are shown in Table 1.

Table 1. Parameters used in high temperature indentation tests and wear tests.

Indentation Test		Wear Test	
Initial contact load	1 N	Normal load	10 N, 15 N, 25 N
Loading/unloading rate	2 N/s	Linear speed	10.5 mm/s
Peak load	300 N	Revs	900
Holding load	In unloading stage at 30 N		
Holding duration	60 s	Temperature	25 °C, 200 °C, 315 °C, 400 °C, 500 °C
Temperature	25 °C, 200 °C, 315 °C, 400 °C, 500 °C		

In the case of performing high temperature indentation tests, the WC ball was loaded against the tested surface at an initial load of 1 N. After a stabilization period about 300 s, the system was at thermal equilibrium. The loading and unloading rates were both 2 N/s. The peak load was 300 N. There was an additional holding at 10% of peak load at unloading stage for thermal drift correction. Indentation tests were performed at 25 °C, 200 °C, 315 °C, 400 °C and 500 °C for all samples. Six repetitions were carried out at each temperature to confirm the repeatability. The experimental parameters for the wear tests were as follows: the normal loads of 10 N, 15 N, and 25 N; a linear speed of 10.5 mm/s; a total of 900 revs for each tests. Wear tests were also performed at 25 °C, 200 °C, 315 °C, 400 °C and 500 °C for all samples. After the wear tests, scanning electron microscope (SEM, Quanta2000, Philips, Amsterdam, The Netherlands) and energy dispersive X-ray spectroscopy (EDS, EDAX-7760/68 ME, Ametek, Berwyn, PA, USA) were used to evaluate the wear scars.

2.4. Theoretical Background for Spherical Indentation

The measured displacement of the reference plate consisted of contact interference between WC ball and indented samples, elastic deformation of the testing frame, and thermal drift. In the present study, the linearity correction method [20] was valid for eliminating the effects of thermal drift and elastic deformation of the testing frame.

Figure 2a shows a typical indentation load–depth curve. For high alloy steels that have significant "plastic pile-up" generated in material during indention process, Figure 2b shows the typical indentation morphology by spherical indenter.

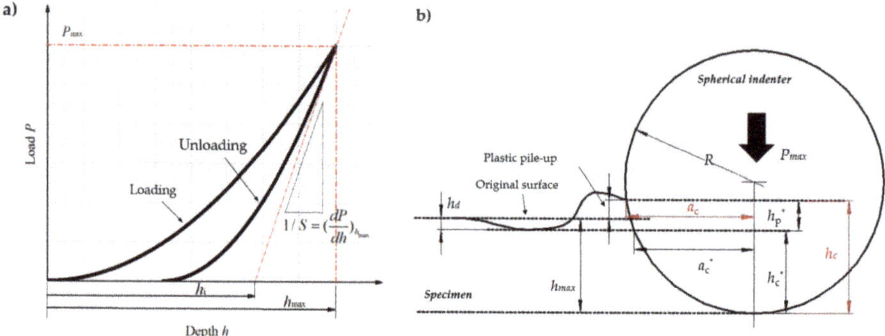

Figure 2. The typical load–depth curve and indentation morphology by spherical indenter: (**a**) the typical load–depth curve; (**b**) the typical indentation morphology by spherical indenter for steels.

As shown in Figure 2a, the maximum depth, h_{max}, was the total displacement of the indented surface and the indenter at maximum load, P_{max}. The slope of the unloading curve at P_{max} was the indentation stiffness of the specimen and indenter, noted as $1/S$. The intercept depth, h_i, was obtained by extrapolating the tangent line of the initial unloading curve to $P = 0$. In addition, some indentation depths were defined from indentation morphology, as shown in Figure 2b. h_c was the actual contact depth considering "plastic pile-up". a_c was the actual contact radius. The deflection depth, h_d, was the depth to which the maximum indentation depth, h_{max}, was reduced by elastic deflection of the indented material. The difference of h_{max} and h_d was the elastic contact depth, h_c^*. The increase in depth from h_c^* to h_c by the "plastic pile-up" phenomenon was defined as h_p^*. Finally, the actual contact depth, h_c, can be expressed as:

$$h_c = h_c^* + h_p^* = h_{max} - h_d + h_p^* \tag{1}$$

Kim et al. [21] studied the actual contact depth during spherical indentation of many kinds of tool steels. According to his study, the elastic contact depth, h_c^*, and the pile-up depth, h_p^*, could be expressed as:

$$h_c^* = h_{max} - h_d \tag{2}$$

$$h_d = \chi(h_{max} - h_i) \tag{3}$$

$$h_p^*/h_c^* = 0.131(1 - 3.423n + 0.079n^2) \times [1 + 6.258(h_{max}/R) - 8.072(h_{max}/R)^2] \tag{4}$$

where χ is a constant related to the indenter shape; $\chi = 0.75$ for a spherical indenter. n is the working-hardening exponent for indented material, and n was set as 0.25 in the present study. R was the radius of the ball. Then, the composite hardness, H_{sys}, and composite elastic modulus, E_r, of the CrN coating–M50 substrate system were determined as follows:

$$H_{sys} = \frac{P_{max}}{A}, A = \pi(2Rh_c - h_c^2) \tag{5}$$

The composite hardness of coating–substrate system was thus the result of the two, coating and substrate, contributed. In order to determine the true hardness of the coating, it was necessary to separate these contributions. According to the model advanced by Puchi-Cabrera [22], the relationship between the composite hardness of coating–substrate system and the independent hardness of coating and substrate can be expressed as:

$$H_{sys} = H_s + (H_c - H_s)\exp(-kZ_r^n) \tag{6}$$

where H_s is the substrate hardness. H_c is the coatings hardness. Z_r is the relative indentation depth, and $Z_r = 2a_c/7t_c$. t_c is the coating thickness. k and n are the constants characteristic of the coating–substrate system. Puchi-Cabrera [22] studied the indentation response of the CrN coated tool steel. According to his study, k and n were determined to be 2.3 and 0.25, respectively.

3. Results

3.1. Indentation Tests at Elevated Temperatures

Elevated temperature indentation test results of uncoated M50 disks, 2 µm CrN coated M50 disks and 5 µm CrN coated ones are shown in Figure 3.

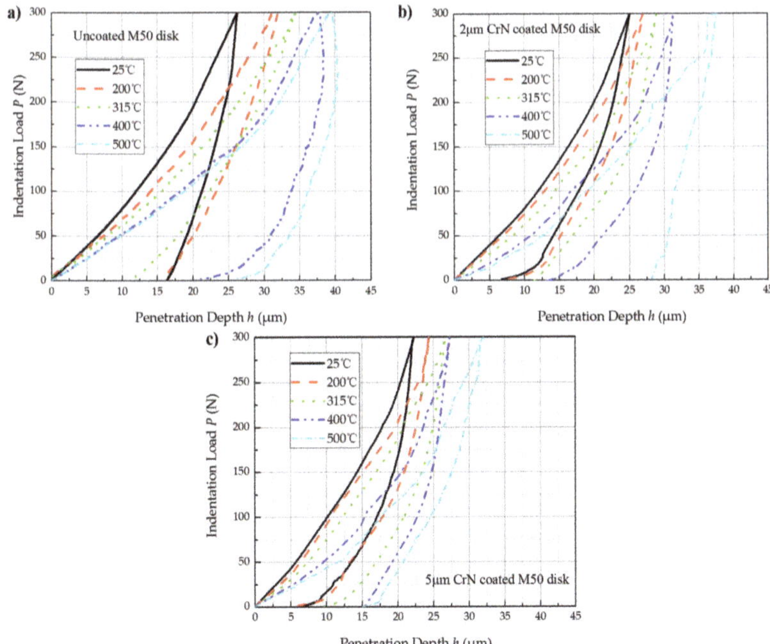

Figure 3. Load–depth curves of the samples as a function of temperatures: (**a**) uncoated M50 disk; (**b**) 2 µm CrN coated M50 disk; and (**c**) 5 µm CrN coated M50 disk.

Using Equations (1)–(5), hardness of M50 substrate, H_s, and composite hardness of CrN coated M50 disks, H_{sys}, were calculated from the load–depth curve shown Figure 3. Then, using Equation (6), the absolute hardness of 2 μm CrN coating and 5 μm CrN coating were calculated, as presented in Figure 4. In general, CrN coatings with all thicknesses showed a decreasing of hardness with temperatures. When the tested temperature was below 315 °C, the hardness of 2 μm thick CrN coating was higher than 5 μm thick CrN coating. However, when the tested temperature was above 400 °C, the 5 μm thick CrN coating had higher values in hardness. The 2 μm thick CrN coating had reduction of 49% in hardness values when the temperature increased from 25 to 500 °C. By comparison, the 5 μm thick CrN coating only had a reduction of 26%. It can be concluded that a thicker CrN coating would be helpful in maintaining the stability of surface hardness in high temperatures.

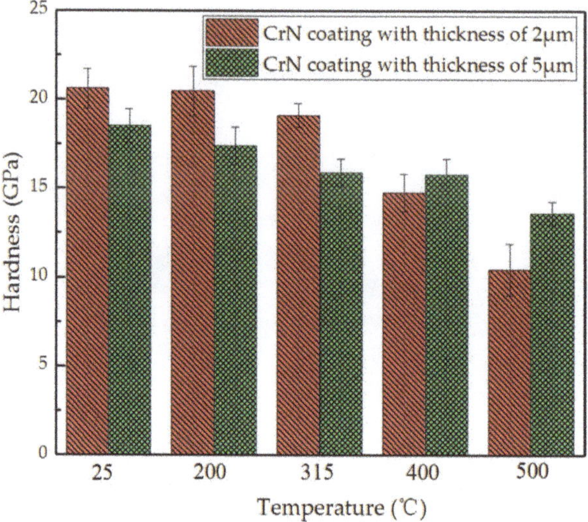

Figure 4. Hardness of CrN coating as a function of temperatures.

Figure 5 presents the loading curves and corresponding contact stress curves of 2 μm CrN coated M50 disks and 5 μm CrN coated ones in the load range of 1–50 N. The contact stress during indentation tests can be calculated using the following equation [23]:

$$\sigma = \frac{P}{\pi a_c^2} \tag{7}$$

where σ is mean contact stress, and a_c is the contact radius during indentation.

In Figure 5a,c, the loading curves were continuous up to a certain load, at which a sudden displacement burst occurred, indicating the occurrence of "pop-in" behavior. According to Shao [24], the indentation stress-induced phase transition with material volume reduction (i.e., coating cracking) was the prerequisite to "pop-in" behavior. Hence, the indentation load, corresponding to the appearance of a "pop-in" event, could possibly be considered as an evaluation tool of load bearing capacity for the coating–substrate system. In another aspect, as shown in Figure 5b,d, it was found that the contact stress reached its maximum values at this indentation load. The external load corresponding to the "pop-in" behavior was named as the load bearing capacity, L_b, for a coating–substrate system.

It can be concluded that the load bearing capacities of CrN coated M50 disks decreased with temperatures. As illustrated in Figure 5a,b, for the 2 μm CrN coated M50 disks, a normal load of 10 N was always less than its load bearing load bearing capacity, L_b, in temperature range of 25–500 °C. However, an external load of 15 N was higher than its load bearing capacity when the temperature was elevated to 500 °C. An external load of 25 N was always higher than its load bearing capacity at any temperature. As illustrated in Figure 5c,d, for the 5 μm CrN coated M50 disks, an external load of 10 N, 15 N or 25 N was always less than its load bearing capacity in temperature range of 25–500 °C.

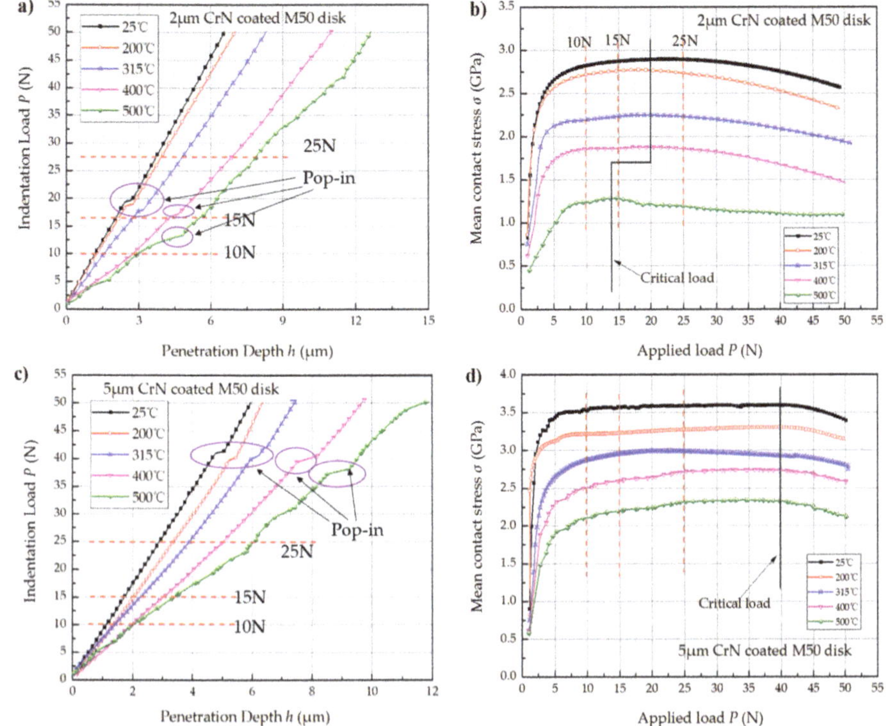

Figure 5. Loading curves and corresponding contact stress curves of the samples as a function of temperatures in the load range of 1–50 N: (**a**) loading curves of 2 μm CrN coated M50 disk; (**b**) contact stress curves of 2 μm CrN coated M50 disk; (**c**) loading curves of 5 μm CrN coated M50 disk; and (**d**) contact stress curves of 5 μm CrN coated M50 disk.

3.2. Tribological Responses

Figure 6a–f shows the friction coefficient curves and penetration depth during wear test of 5 μm CrN coated M50 disks as a function of temperatures under the tested loads of 10 N, 15 N, and 25 N. The variation of width of wear scars with temperatures is depicted in Figure 6g.

As illustrated in Figure 6a,c,e, running-in was presented by an increase of the friction coefficient during first several dozens to hundreds of cycles for all measured temperatures and loads. In the running-in stage, the friction force is mainly contributed to deform and fracture the contact asperities [25]. Due to the lower hardness of the coating material at higher tested temperature, it would take less number of cycles for the friction coefficient to reach the steady stage.

For 5 μm CrN coated M50 disks tested under each load, the stable values of friction coefficient were all found to be decreased with temperatures. At the temperatures of 25 °C and 200 °C, the friction coefficient was determined to be about 0.6. However, it decreased to 0.4 at 315 °C, then gradually decreased to the lowest value of 0.3 at the highest temperature of 500 °C.

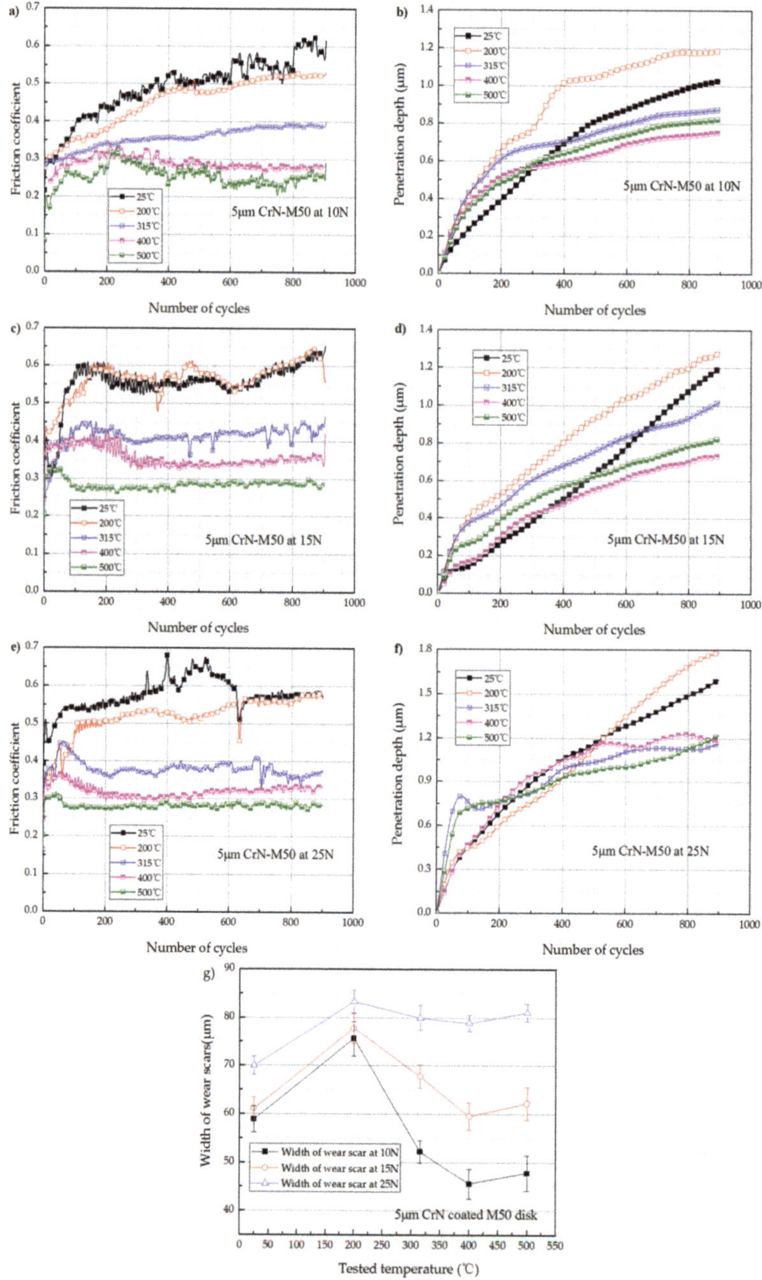

Figure 6. Friction coefficient curves and penetration depth during wear test of the 5 μm CrN coated M50 disks: (**a**) variation of friction coefficient with temperatures under 10 N; (**b**) variation of penetration depth with temperatures under 10 N; (**c**) variation of friction coefficient with temperatures under 15 N; (**d**) variation of penetration depth with temperatures under 15 N; (**e**) variation of friction coefficient with temperatures under 25 N; (**f**) variation of penetration depth with temperatures under 25 N; and (**g**) variation of width of wear scars with temperatures.

As shown in Figure 6b,d,f, the variation trends of penetration depth at the temperatures above 315 °C were quite different from those at the temperatures below 200 °C. Due to the lower hardness of CrN coating at high temperatures, the penetrate depth increased more sharply at the temperatures above 315 °C than at the temperature below 200 °C in first 300 cycles. After 300 cycles of relative sliding, the penetration depth went on increasing linearly with distance at the temperatures below 200 °C, however, was much slowed down the temperatures above 315 °C. Meanwhile, the fluctuation in friction coefficient with sliding distance was also weakened when the tested temperatures were equal or above 315 °C. According to Polcar [11], the reduction of penetration depth at the temperatures above 315 °C might contributed to the self-lubricant function of the formed chromium oxide tribo-layer.

The high temperature tribological response of 2 μm CrN coated M50 disks under each tested loads were quite different from the 5 μm CrN coated M50 disks. Figure 7a–f shows the friction coefficient curves and penetration depth during wear test of 2 μm CrN coated M50 disks as a function of temperatures under the tested loads of 10 N, 15 N and 25 N. The variation of width of wear scars with temperatures is depicted in Figure 7g.

As shown in Figures 6a and 7a, under the tested load of 10 N, the stable values of friction coefficients of 2 μm CrN coated M50 samples were similar to those of the 5 μm CrN coated M50 samples at each tested temperatures. As shown in Figure 7c, under the applied load of 15 N, the friction coefficients of 2 μm CrN coated M50 samples were still similar to those for the 5 μm CrN coated M50 samples when the tested temperatures were equal or below 400 °C. However, at the tested temperature of 500 °C, the friction coefficient substantially increased from 0.25 to 0.55 during the wear test. As shown in Figure 7e, under the applied load of 25 N, the values of friction coefficients at each temperatures were all above 0.5. As shown in Figure 7g, the width of wear scars was found to be always increased with the tested temperatures under the tested load of 25 N.

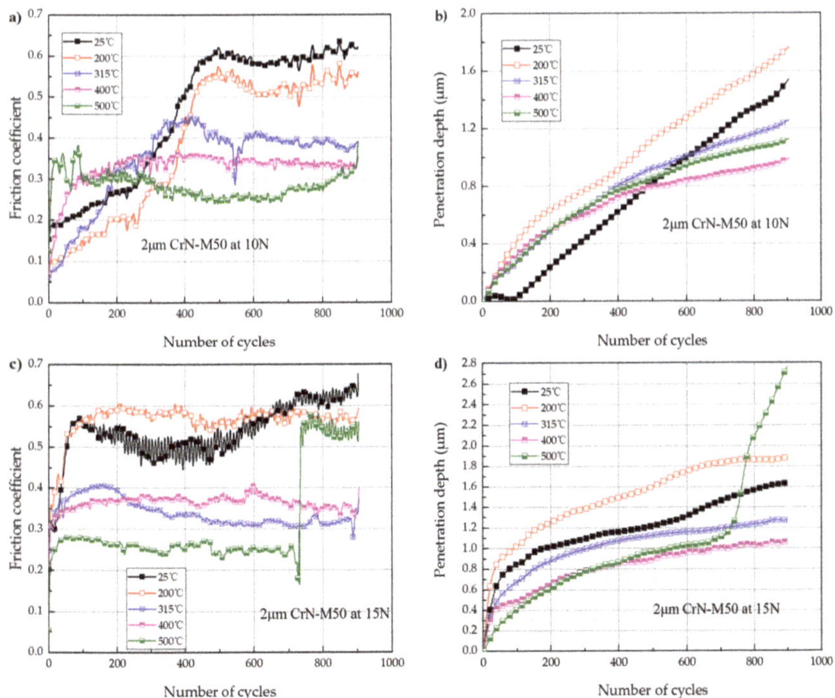

Figure 7. *Cont.*

self-lubricant function. Thus, the friction coefficient and width of wear scars decreased at 315 °C. It is clear that wear of CrN coating at temperatures equal or above 315 °C was mild oxidative wear.

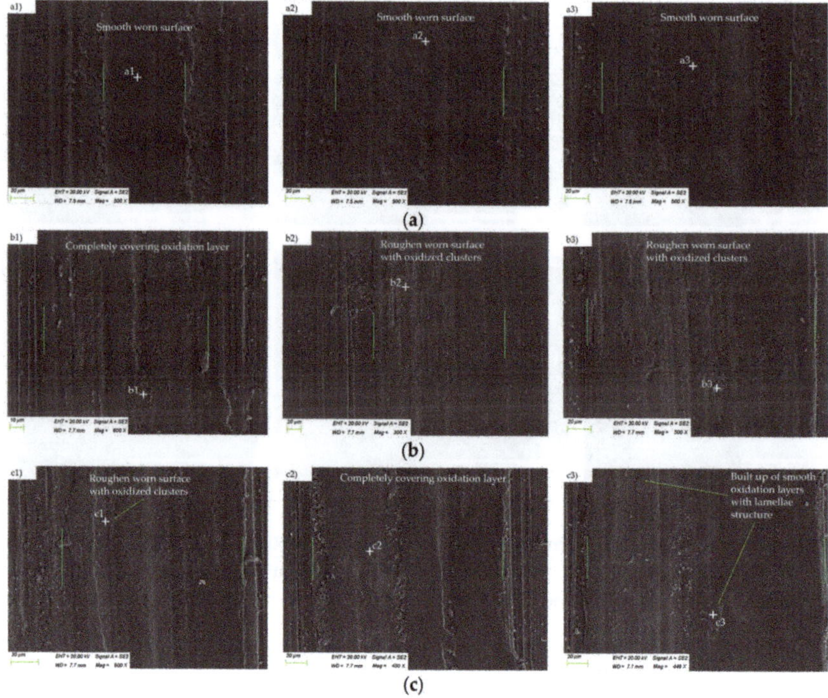

Figure 9. Morphology of the worn tracks on 5 μm CrN coated M50 disks: (**a**) at the temperature of 315 °C under the load of (**a1**) 10 N, (**a2**) 15 N, and (**a3**) 25 N; (**b**) at the temperature of 400 °C under the load of (**b1**) 10 N, (**b2**) 15 N, and (**b3**) 25 N; and (**c**) at the temperature of 500 °C under the load of (**c1**) 10 N, (**c2**) 15 N, and (**c3**) 25 N.

Table 3. The EDS analysis results corresponding to the positions in Figure 9.

Element, at.%	Positions								
	a1	a2	a3	b1	b2	b3	c1	c2	c3
Cr	46.39	46.23	36.6	30.19	36.2	37.99	38.36	35.08	32.35
N	23.72	21.95	25.67	17.75	13.61	21.28	13.65	11.61	7.88
O	22.43	25.61	30.58	37.97	44.96	38.38	41.51	44.98	51.35
Fe	4.19	5.39	6.34	12.43	3.39	2.02	3.87	6.47	5.81
W	0.07	0.04	0.05	–	–	–	–	–	–

It can be concluded that, for 5 μm CrN coated M50 disks, the dominant wear mechanism transitioned from mild abrasive wear to mild oxidation wear when the temperatures were above 315 °C.

Figure 10 presents the morphology of worn tracks on 2 μm CrN coated M50 disks under various loads at the temperatures of 25 °C and 200 °C. As shown in Figure 10a1,a2,b1,b2, under the applied load of 10 N or 15 N, the worn tracks on 2 μm CrN coated M50 disks and the worn tracks on 5 μm CrN coated M50 disks presented similar morphology. The dominant wear mechanism for the 2 μm CrN coated M50 samples under the applied load of 10 N or 15 N was mild abrasive wear, corresponding to smooth worn surfaces with wear debris and narrow grooves presented in the worn area. However, when the load was increased to 25 N, local coating fracture can be observed on the edge of the wear scars for the 2 μm CrN coated M50 disks, as illustrated in Figure 10a3,b3. EDS analysis results are listed in Table 4. EDS analysis of the coating delamination area (Points a3, b3) confirmed the significant increasing of Fe element, indicating the partial detachment of coating material. The results above indicated that, at lower tested temperatures, the wear of CrN coatings transitioned from mild abrasive wear to the coating fracture controlled wear when the applied load exceeded its the load carrying capacity.

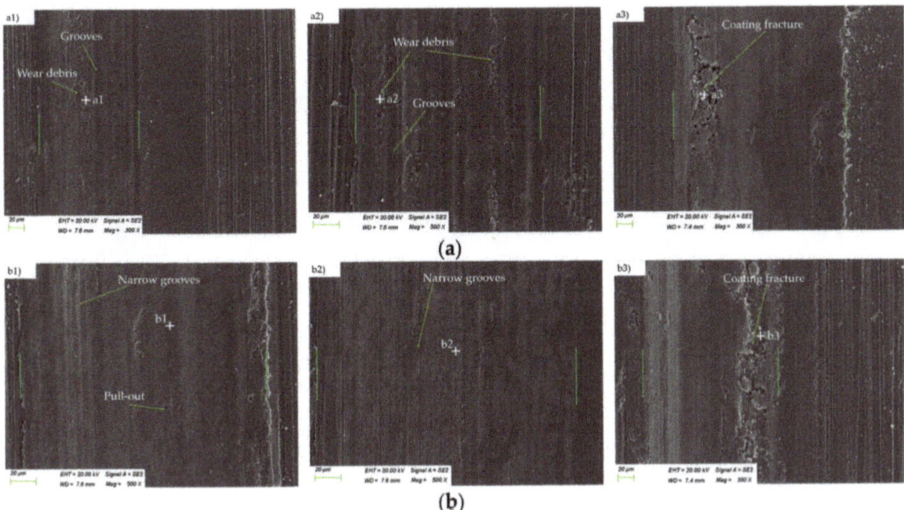

Figure 10. Morphology of the worn tracks on 2 μm CrN coated M50 disks: (**a**) at the temperature of 25 °C under the load of (**a1**) 10 N, (**a2**) 15 N, and (**a3**) 25 N; and (**b**) at the temperature of 200 °C under the load of (**b1**) 10 N, (**b2**) 15 N, and (**b3**) 25 N.

Table 4. The EDS analysis results corresponding to the positions in Figure 10.

Element, at.%	Positions					
	a1	a2	a3	b1	b2	b3
Cr	48.32	42.71	34.96	31.25	36.7	12.7
N	40.27	36.84	5.49	37.53	37.93	24.65
O	3.41	9.28	7.77	10.4	13.76	14.26
Fe	8.32	9.43	48.05	15.94	9.2	46.96
W	4.27	2.95	0.71	3.07	2.93	–

Figure 11 shows the morphology of worn tracks on 2 μm CrN coated M50 disks when the tested temperatures were equal or above 315 °C. As shown in Figure 11a1,b1,c1, under a tested load of 10 N, the dominant wear mechanism for the 2 μm CrN coated M50 samples was mild-oxidation wear when the tested temperatures were above 315 °C. In additional, under a tested load of 15 N, the wear of 2 μm CrN coated M50 samples could still be characterized as mild-oxidation wear at the temperature of 315 °C and 400 °C, as illustrated in Figure 11a2,b2. However, when the tested temperature was elevated to 500 °C, the oxidation layers were stripped from the surface of coating under the applied load of 15 N, as illustrated in Figure 11c2. It should be pointed out that the external load of 15 N just exceeded the load-bearing limitation of 2 μm CrN coated M50 disk at the temperature of 500 °C, as shown in Figure 5c,d. While the tested load was increased to 25 N, no obvious oxidation layers were detected after wear tests at temperatures of 315 °C, 400 °C and 500 °C, as illustrated in Figure 11c3,d3,e3.

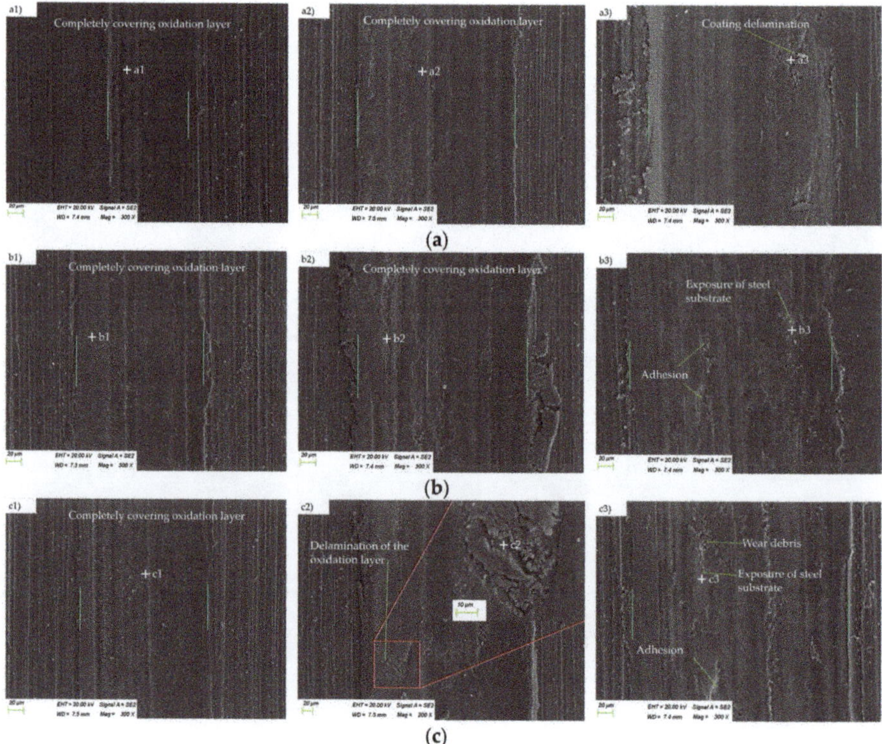

Figure 11. Morphology of the worn surfaces of 2 μm CrN coated M50 disks: (**a**) at the temperature of 315 °C under the load of (**a1**) 10 N, (**a2**) 15 N, and (**a3**) 25 N; (**b**) at the temperature of 400 °C under the load of (**b1**) 10 N, (**b2**) 15 N, and (**b3**) 25 N; and (**c**) at the temperature of 500 °C under the load of (**c1**) 10 N, (**c2**) 15 N, and (**c3**) 25 N.

EDS analysis results are listed in Table 5. EDS analysis of Points a3, b3, and c3 revealed that the contents of Fe, O significantly surpassed that of Cr, N in the worn area after 900 cycles of relative sliding under the load of 25 N at 315 °C, 400 °C or 500 °C. This indicated the severe materials removal occurred for CrN coatings, thereby leading to exposure of M50 substrate. Besides, it can be noticed that there were delaminated regions and oxidation clusters in a widespread dispersion on the worn tracks. It can be conclude that, at higher tested temperatures, the wear of CrN coated M50 steel transitioned from mild oxidation wear to the compound wear mechanisms containing oxidative wear, coating delamination or adhesive wear when the external load exceeded its load bearing capacities. This was also the key reason why the high temperature friction coefficient of the 2 μm CrN coated M50 samples suddenly increased under the tested load of 25 N.

Table 5. The EDS analysis results corresponding to the positions in Figure 11.

Element, at.%	Positions								
	a1	a2	a3	b1	b2	b3	c1	c2	c3
Cr	37.48	32.19	17.92	32.65	21.61	4.9	23.58	18.3	4.0
N	21.93	18.67	10.45	20.97	13.72	4.53	22.68	13.85	3.47
O	34.39	35.16	36.24	39.64	44.06	35.14	44.59	40.78	31.0
Fe	5.52	6.22	32.76	6.61	18.75	52.79	8.48	24.87	54.79
W	–	–	1.9	–	–	2.15	–	–	3.21

4. Discussion

The lubricating function of chromium oxide film were further utilized by many researchers to develop the so-called "chameleon" coating for high temperature application, such as CrN–Ag [26]. For those CrN based hard composited coatings, the soft metal possessed solid lubricating function at lower temperatures, while the chromium oxide tribo-layer possessed lubricating function at higher temperatures. It is provn that formed chromium oxide film was found to provide a long endurance and friction coefficients within 0.3–0.4 at 400–500 °C in air [27]. This was coincidence with the present study. However, Voevodin et al. [28] argued that, to develop the "chameleon" coating with better temperature-adaptive property, the functional temperature range of soft metal lubricating and chromium oxide film lubricating should be with certain degrees of overlapping. In this study, it was found that the chromium oxide film lubricating started at temperature about 300 °C. Furthermore, we believe that this conclusion would be useful in the CrN based hard composited coating design.

Many researchers also concerned on the critical load for high temperature application of CrN coating or CrN based temperature-adaptive coating. In this study, the high temperature indentation test was employed to study the load bearing capacity of CrN coating–substrate system, and it was found that the obtained load bearing capacity can be cited as the critical load for high temperature application of CrN coatings. The failure mechanism of the chromium oxide film is summarized as follows.

Taking the 2 µm CrN coated M50 disks as an example, under an external load of 15 N, which just exceeded its load bearing capacity at 500 °C, the stripping and piling up of the oxide film was observed, as illustrated in Figure 11c2. As pointed by Holmberg [29], this kind of thin film failure is related to the tensile fracture and subsequently detachment from the interface. Cracks on the interface were an essential factor triggering this kind of failure. SEM observation verified that radical cracks were generated on the coating surface when the 2 µm CrN coated M50 disk was indented under a load of 15 N at 500 °C (see Figure 12).

Figure 12. SEM observation of the indented surface of 2 µm CrN coated M50 disks under a load of 15 N and temperature of 500 °C.

The chromium oxide film would act as a cantilever beam when cracks were on the surface of CrN coating. The stress concentration would be generated in the center of the cantilever beam, as shown in Figure 13a, thereby resulting in the tensile fracture of the oxide film, as depicted in Figure 13b. Then, shear stress concentration would be generated on the interface between the chromium oxide film and CrN coating surface. The shear stress concentration finally resulted in the separation of the oxide film from the coating surface, as illustrated in Figure 13c. As the failure of oxidation film occurred, the adhesion between coating material and tribo-counterpart occurred, resulting in the wear transition for CrN coated M50 steel, as illustrated in Figure 13d. The failure mechanisms for the oxide film was also confirmed by the morphology of wear tracks on 2 µm CrN coated M50 disks at the load of 15 N and temperature of 500 °C, as shown in Figure 13e.

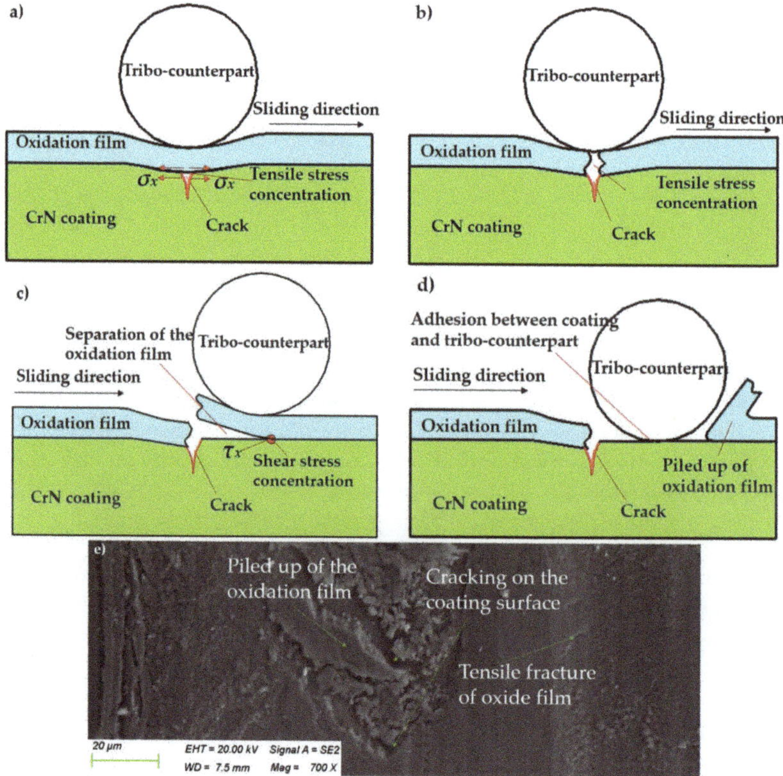

Figure 13. The schematic drawing of the failure of oxidation film: (**a**) tensile stress concentration; (**b**) tensile fracture of oxidation film; (**c**) shear stress concentration; (**d**) adhesion occurred after the failure of oxidation film; and (**e**) verification of the failure mechanisms from the wear scar morphologies of the 2 µm CrN coated M50 disks under the load of 15 N and temperature of 500 °C.

In the current study, four types of wear were detected for the CrN coated M50 disks under various of loads and temperatures: abrasive wear; mild oxidation wear; coating fracture controlled wear; and the compound wear mechanisms containing oxidative wear, coating delamination or adhesive wear. According to the above discussion, the characteristics of wear transition for CrN coated M50 steel at elevated temperatures are summarized in Table 6.

Table 6. Wear transition for CrN coated M50 steel at elevated temperatures and its characteristics.

Temperatures	Load	
	Below the Load Bearing Limit	**Above the Load Bearing Limit**
Below 315 °C	Mild-abrasive wear	Coating fracture controlled wear
Equal or above 315 °C	Mild-oxidation wear	Compound wear mechanisms containing oxidative wear, coating delamination and adhesive wear

5. Conclusions

The combination of high temperature spherical indentation and wear test provides a useful and unique way to study the relationships between operating conditions and wear mechanisms of CrN coating at elevated temperatures. The main conclusions are summarized as follows:

- The hardness of CrN coating decrease with temperatures: When the tested temperature was below 315 °C, the hardness of 2 μm thick CrN coating was higher than 5 μm thick CrN coating. However, when the tested temperature was above 400 °C, the 5 μm thick CrN coating had higher values in hardness. A thicker CrN coating would be helpful in maintaining the stability of surface hardness in high temperatures.

- Wear of CrN coating changes with external load, P, and temperature, T: Under the tested condition of $P < L_b$ and $T < 315$ °C, abrasive is the dominant wear mechanism for CrN coating. With a tested condition of $P < L_b$ and $T \geq 315$ °C, wear of CrN coating transitions into mild oxidation wear due to the lubrication effect of oxidation layers. Under the tested condition of $P > L_b$ and $T < 315$ °C, wear of CrN coating was controlled by coating fracture. Under the tested condition of $P > L_b$ and $T \geq 315$ °C, wear of CrN coating transitions into the severe wear mode with a combination of detachment, adhesion and oxidation, due to the tensile fracture of oxidation films, thereby leading adhesion between CrN coating and tribo-counterpart.

- The presented analysis method can be helpful in predicting the permissible loads for a CrN coating–M50 substrate system at the given temperature. To conclude, it can be helpful in the tribological design for CrN coatings and allow the rational selection of coating thickness for a particular high temperature application.

Acknowledgments: This work was supported by the National Natural Science Foundation of China (Grant No. 51675120).

Author Contributions: Guangze Tang contributed the deposition of CrN coatings. Le Gu contributed to the discussion of failure mechanism of chromium oxide film. Yuze Mao contributed to the SEM observation and EDS analysis. Chi Zhang analyzed the test results and wrote the paper.

Conflicts of Interest: The authors declare no conflicts of interest.

References

1. Gilewicz, A.; Chmielewska, P.; Murzynski, D.; Dobruchowska, E.; Warcholinski, B. Corrosion resistance of CrN and CrCN/CrN coatings deposited using catholic arc evaporation in Ringer's and Hank's solutions. *Surf. Coat. Technol.* **2016**, *299*, 7–14. [CrossRef]

2. Podgornik, B.; Sedlaček, M.; Mandrino, D. Performance of CrN coatings under boundary lubrication. *Tribol. Int.* **2016**, *96*, 247–257. [CrossRef]

3. Ding, J.; Zhang, T.; Yun, J.M.; Kang, M.C.; Wang, Q.; Kim, K.H. Microstructure, mechanical, oxidation and corrosion properties of the Cr-Al-Si-N Coatings deposited by a hybrid sputtering system. *Coatings* **2017**, *7*, 119. [CrossRef]

4. Khanna, R.; Ong, J.L.; Oral, E.; Narayan, R.J. Progress in wear resistant materials for total hip arthroplasty. *Coatings* **2017**, *7*, 99. [CrossRef]

5. Ramadoss, R.; Kumar, N.; Dash, S.; Arivuoli, D.; Tyagi, A.K. Wear mechanism of CrN/NbN super lattice coating sliding against various counter-bodies. *Int. J. Refract. Met. Hard Mater.* **2013**, *41*, 547–552. [CrossRef]

6. Silva, F.; Martinho, R.; Andrade, M.; Baptista, A.; Alexandre, R. Improving the wear resistance of moulds for the injection of glass fibre-reinforced plastics using PVD coatings: A comparative study. *Coatings* **2017**, *7*, 28. [CrossRef]

7. Gouveia, R.M.; Silva, F.J.G.; Reis, P.; Baptista, A.P.M. Machining duplex stainless steel: Comparative study regarding end mill coated tools. *Coatings* **2016**, *6*, 51. [CrossRef]

8. Kawahara, Y. An overview on corrosion-resistant coating technologies in biomass/waste-to-energy plants in recent decades. *Coatings* **2016**, *6*, 34. [CrossRef]

9. Lim, S.C.; Ashby, M.F. Overview No. 55 wear mechanism maps. *Acta Metall.* **1987**, *35*, 1–24. [CrossRef]

10. Wang, Y.; Stephen, M.H. Wear and wear transition mechanisms of ceramics. *Wear* **1996**, *195*, 112–122. [CrossRef]

11. Polcar, T.; Parreira, N.M.G.; Novák, R. Friction and wear behaviour of CrN coating at temperatures up to 500 °C. *Surf. Coat. Technol.* **2007**, *201*, 5228–5235. [CrossRef]

12. Qi, Z.; Liu, B.; Wu, Z.; Zhu, F.; Wang, Z.; Wu, C. A comparative study of the oxidation behavior of Cr_2N and CrN coatings. *Thin Solid Films* **2013**, *544*, 515–520. [CrossRef]

13. Mandrino, D.; Podgornik, B. XPS investigations of tribofilms formed on CrN coatings. *Appl. Surf. Sci.* **2017**, *396*, 554–559. [CrossRef]

14. Scheerer, H.; Hoche, H.; Broszeit, E.; Berger, C. Tribological properties of sputtered CrN coatings under dry sliding oscillation motion at elevated temperatures. *Surf. Coat. Technol.* **2001**, *142–144*, 1017–1022. [CrossRef]

15. Beake, B.D.; Fox-Rabinovich, G.D.; Veldhuis, S.C.; Goodes, S.R. Coating optimization for high speed machining with advanced nanomechanical test methods. *Surf. Coat. Technol.* **2009**, *203*, 1919–1925. [CrossRef]

16. Smith, J.F.; Zheng, S. High temperature nanoscale mechanical property measurements. *Surf. Eng.* **2000**, *16*, 143–146. [CrossRef]

17. Everitt, N.M.; Davies, M.I.; Smith, J.F. High temperature nanoindentation—The importance of isothermal contact. *Philos. Mag.* **2011**, *91*, 1221–1244. [CrossRef]

18. Sander, T.; Tremmel, S.; Wartzack, S. A modified scratch test for the mechanical characterization of scratch resistance and adhesion of thin hard coatings on soft substrates. *Surf. Coat. Technol.* **2011**, *206*, 1873–1878. [CrossRef]

19. Ghosh, S.; Yadav, S.; Das, G. Ball Indentation Technique: A currently developed tool to study the effect of various heat treatments on the mechanical properties of En24 steel. *Mater. Lett.* **2008**, *62*, 3966–3968. [CrossRef]

20. Wheeler, J.M.; Armstrong, D.E.J.; Heinz, W.; Schwaiger, R. High temperature nanoindentation: The state of the art and future challenges. *Curr. Opin. Solid State Mater. Sci.* **2015**, *19*, 354–366. [CrossRef]

21. Kim, S.H.; Lee, B.W.; Choi, Y.; Kwon, D. Quantitative determination of contact depth during spherical indentation of metallic materials—A FEM study. *Mater. Sci. Eng. A* **2006**, *415*, 59–65. [CrossRef]

22. Puchi-Cabrera, E.S. A new model for the computation of the composite hardness of coated systems. *Surf. Coat. Technol.* **2002**, *160*, 177–186. [CrossRef]

23. Kot, M.; Rakowski, W.; Lackner, J.M.; Major, L. Analysis of spherical indentations of coating-substrate systems: Experiments and finite element modeling. *Mater. Des.* **2013**, *43*, 99–111. [CrossRef]

24. Huang, X.; Etsion, I.; Shao, T.M. Indentation pop-in as a potential characterization of weakening effect in coating/substrate systems. *Wear* **2015**, *338–339*, 325–331. [CrossRef]

25. Mo, J.L.; Zhu, M.H. Sliding tribological behaviors of PVD CrN and AlCrN coatings against Si_3N_4 ceramic and pure titanium. *Wear* **2009**, *267*, 874–881. [CrossRef]

26. Mulligan, C.P.; Papi, P.A.; Gall, D. Ag transport in CrN–Ag nanocomposite coatings. *Thin Solid Films* **2012**, *520*, 6774–6779. [CrossRef]

27. Mulligan, C.P.; Blanchet, T.A.; Gall, D. CrN–Ag nanocomposite coatings: High-temperature tribological response. *Wear* **2010**, *269*, 125–131. [CrossRef]

28. Voevodin, A.A.; Muratore, C.; Aouadi, S.M. Hard coatings with high temperature adaptive lubrication and contact thermal management: Review. *Surf. Coat. Technol.* **2014**, *257*, 247–265. [CrossRef]

29. Matthews, A.; Franklin, S.; Holmberg, K. Tribological coatings: Contact mechanisms and selection. *J. Phys. D Appl. Phys.* **2007**, *40*, 5463–5475. [CrossRef]

Article

Corrosion Resistance of Boronized, Aluminized, and Chromized Thermal Diffusion-Coated Steels in Simulated High-Temperature Recovery Boiler Conditions

Amirhossein Mahdavi [1], Eugene Medvedovski [2], Gerardo Leal Mendoza [2] and André McDonald [1,*]

[1] 10-203 Donadeo Innovation Center for Engineering, Department of Mechanical Engineering, University of Alberta, Edmonton, AB T6G 1H9, Canada; amahdavi@ualberta.ca

[2] Endurance Technologies Inc., 7940–56 St. SE, Calgary, AB T2C 4S9, Canada; eugenem@endurancetechnologies.com (E.M.); gerardom@endurancetechnologies.com (G.L.M.)

* Correspondence: andre.mcdonald@ualberta.ca; Tel.: +1-780-492-2675

Received: 30 May 2018; Accepted: 22 July 2018; Published: 24 July 2018

Abstract: In this study, the high-temperature molten salt corrosion resistance of bare steels and steels with protective coatings, fabricated by thermal diffusion processes (boronizing, aluminizing and chromizing), were investigated and compared. Surface engineering through thermal diffusion can be used to fabricate protective coatings against corrosion, while alleviating issues around possible cracking and spallation that is typical for conventional thermal-sprayed coatings. In this regard, samples of low carbon steel and 316 stainless steel substrates were boronized, chromized, and aluminized through a proprietary thermal diffusion process, while some of the samples were further coated with additional thin oxide and non-oxide layers to create new surface architectures. In order to simulate the actual corrosion conditions in recovery boilers (e.g., from black liquor combustion), the surfaces of the samples sprayed with a modeling salt solution, were exposed to low-temperature (220 °C) and high-temperature (600 °C) environments. According to microstructural and X-ray diffraction (XRD) studies and results of hardness determination, the coatings with multilayered architectures, with and without additional oxide layers, showed successful resistance to corrosive attack over bare steels. In particular, the samples with boronized and chromized coatings successfully withstood low-temperature corrosive attack, and the samples with aluminized coatings successfully resisted both low- and high-temperature molten salt corrosive attacks. The results of this study conducted for the first time for the thermal diffusion coatings suggest that these coatings with the obtained architectures may be suitable for surface engineering of large-sized steel components and tubing required for recovery boilers and other production units for pulp and paper processing and power generation.

Keywords: corrosion; microstructural analysis; recovery boilers; thermal diffusion coatings

---

1. Introduction

Recovery boilers are widely used in pulp and paper mills to produce process steam and electrical power by burning black liquor fuel which is an aqueous solution of lignin hemicellulose and inorganic chemicals [1–3]. The primary function of recovery boilers is to recycle inorganic cooking chemicals (black liquor) and to provide the major volume of the process steam used in the Kraft process, as well as auxiliary electricity for the production plant [2–4]. As a result of combustion of black liquor with solid contents varying from 20% to 80% of inorganic solids, depending on the used starting materials and process features, low-melting temperature ash or smelt is produced. The produced smelt mainly consists of different salts, e.g., sodium and potassium chlorides, sulfates, and carbonates. At elevated temperatures, mostly greater than 500 °C, the aforementioned salts in certain combination likely become molten. In the presence of oxygen, the molten salts can be highly corrosive [2,3]. Combustion of the black liquor fuel also results in the formation of hot corrosive gases, such as SO_2, CO_2, Cl_2, and some others [5].

Generally, high-temperature black liquor corrosion occurs according to two main mechanisms, namely (i) high-temperature active oxidation and (ii) corrosion due to the formation of sulfidic and chlorine gasses and residual deposits of molten salts and their interaction with the steel surface [6]. In the former mechanism, which generally occurs in high temperatures (above 450 °C), the anions of the molten salt, such as chlorides (Cl^-), sulfates (SO_4^{2-}) and sulfides (S^{2-}), continuously diffuse into the oxide-metal interface and actively sustain the oxidation. On the other hand, in the latter mechanism, the presence of the anions of the molten salt at elevated temperatures forms low melting point eutectics, which gradually dissolves the protective oxide layer from the metal surface and facilitates the oxidation of the metal surface [6]. During the high-temperature interaction of the molten salts and formed iron and chromium salts and oxides in the ash and scale layers, the corrosive gases (Cl_2, CO_2, HCl, and some others) are produced, which can expedite the corrosion of steels. The occurring "self-catalytic active oxidation" corrosion depends on the Kraft process parameters, temperature, formed molten salt and gaseous phase compositions, and many other factors. The features of the corrosion process and the chemical reactions occurring among the steels and formed molten salts and corrosive gases at high temperatures are described in details elsewhere [7–12].

Stress corrosion cracking and/or thermal fatigue cracking also occur at specific locations within the boiler system [13–15]. In summary, inherent in the boiler system are various species of salts that become molten at the high operating temperatures of the boiler, increasing corrosion of the boiler surfaces and tubes. Flue gases in recovery boilers used in pulp and paper mills pass over the super-heater tubes at temperatures of 700 °C or even greater, and they contain large concentrations of corrosive gases and particulates with melting points as low as 500 °C [14]. Relative to utility boilers, this hot, corrosive gas environment results in significant corrosion of the steels used to fabricate the boiler walls and tubes. Similar problems also occur in the evaporators, superheater systems, fireside, fans, and exhausting systems in the pulp and paper industry. In many cases, the metallic components used in industry are made from carbon steels and low-alloy steels, and, in some case, from expensive stainless steels [2,3,12]. Although stainless steels perform better than low-alloy steels, they also experience sufficient degradation at high temperature molten salt corrosion environments. Anyway, serious corrosion problems with steel components and deposit accumulation reducing equipment efficiency lead to unscheduled shutdown of the boilers and to necessity of expensive maintenance. With regards to the briefly described problems occurring with steels, the reliable protection of steel components, including of long size tubing, is highly important.

Different surface engineering methods, particularly those based on coatings, are used in industry to protect steels and alloys against corrosion. They include painting, dipping, sol-gel processes, cold and thermal spraying, chemical and electrochemical methods (anodizing, electroplating, electroless plating, and electrophoretic deposition), plasma-assisted technologies, physical vapor deposition (PVD), and chemical vapor deposition (CVD), among others [16–25]. However, most of these methods cannot be widely utilized in industry, especially for processing of large complex shape components and long tubing when the inner surface needs to be protected. Moreover, the application of some of the aforementioned coating processes is limited, since they might experience spallation or delamination of the coating layers due to poor adhesion to the substrate and weak bonding or thermal expansion mismatch at the coating–substrate interface. Coatings with low thickness (sub-micron and micron length scales) and low hardness often experience rapid degradation, especially due to inadvertent mechanical action during service or installation. Organic-based coatings should be excluded because of decomposition or degradation at elevated temperatures.

Typically, thermal-sprayed coatings are used to provide protection against high-temperature corrosion [26–28]. Thus, high-velocity oxy-fuel (HVOF) thermal-sprayed coatings of stainless steel or Inconel alloy 625 provide high resistance against corrosion in the pulp and paper processing [26]. Some studies have even explored the use of nanostructured yttria-stabilized zirconia as a top coat for a multi-layered environmental barrier coating system for the boiler tubes [26]. However, due to the high operating temperatures (3,500–10,000 °C) of the thermal spray processes [29–31], and possible formation of detrimental cracks and delamination of the final coating, these coatings may not be suitable for all substrate materials and practical conditions. In addition, there is a real challenge to apply thermal spraying techniques for protection of the inner surface of long tubing, as well as the components with complex shapes. To the contrary, surface engineering through the Chemical Vapor Deposition (CVD) technology may be an efficient and reliable process for protection of industrial components with different sizes and shapes [32–37]. Thus, thermal diffusion technology, which is based on the CVD principles, is an advanced and novel surface engineering process that can alleviate the issues around cracking and delamination of the surface treatment [32,33]. Therefore, they can be applicable for large size and long components, including for tubing with a high length-to-diameter ratio. This technology and its variations are successfully used to produce thin, hard, and corrosion/wear resistant layers of carbides, borides, aluminides, chromides, and some other compounds on the surface of steel substrates [38,39]. In the light of growing attention towards this technology in recent years, investigation on the corrosion/wear resistance of thermal diffusion coatings is of interest.

The thermal diffusion process allows the formation of the coatings on the steel substrates via deposition of vapors of certain atoms and species occurred through high temperature reactions originated from specially formulated powder mixes, diffusion of these atoms into the metal substrate and consequent high-temperature interaction of these diffused atoms with the metal substrate resulted in the formation of new phases and their consolidation [33–35]. In this process, the steel substrates are usually immersed into the special powder mixes which all together are placed into retorts, then heated to selected temperatures. Consequently, new compounds, such as borides or intermetallides (e.g., aluminides and chromides) depending on the selected process, are formed on the surfaces of steel substrates. Since the aforementioned technology is related to diffusion processes, the diffusion laws are applicable to the coating formation. Due to inward diffusion of the deposited atoms (e.g., B, Al, and Cr) and outward diffusion of the metallic constituents from the substrates (e.g., Fe, Cr, Ni, and others), the coatings with a few layers consisted of borides, aluminides, or chromides of certain compositions are formed simultaneously. Their structures, thicknesses, and compositions are defined by the initial (starting) mix compositions, chemistry, and other features of the substrate materials and the process parameters such as temperature, exposure time, gas pressure, and others. As an example of the thermal diffusion process and coating formation, the boronizing process employed by Endurance Technologies Inc. (ETI) has been described in detail elsewhere [40,41]. This process significantly distinguishes from the thermal reactive carbide-based coatings that utilize a salt bath process. The later process has been described in detail by Liu et al. elsewhere [38].

Numerous studies on the corrosion resistance of boronizing thermal diffusion coatings on steels have been conducted with respect to a variety of practical applications, particularly for oil and gas production, refinery, and power generation applications [40–42]. It was shown that the boronizing process provides enhanced corrosion resistance of steel components, especially for applications dealt with large-size complex shapes and long tubular components [40]. These coatings with thicknesses of 50–400 µm and uniform, dense "saw-tooth" structures (for carbon steels) demonstrated high wear and corrosion resistance in steam with hydrocarbons, carbonaceous gases (CO + CO$_2$), hydrogen sulfide (H$_2$S), and chloride salts (NaCl and CaCl$_2$) at elevated pressures (14–28 MPa) and temperatures (200–300 °C) for long time exposure, as well as in the actual oil well corrosive conditions [41]. These boronized thermal diffusion coatings provided a noticeable extension of the service life, on average 3–10 times greater longevity compared to uncoated steels, in harsh corrosive environments of oil well production conditions and power generation. However, due to their oxidation at higher temperatures, boronized coatings can be suitable in environment with temperatures only up to 500 °C [41,42].

It has been shown that the aluminizing and chromizing processes enhanced corrosion resistance of steels and alloys exposed to gaseous environments at elevated temperatures up to 800–950 °C [43–47]. The diffusion aluminized coatings usually have two distinctive layers, an outer aluminide layer and an inner inter-diffusion layer, that are formed on top of the original stainless steel substrate [43]. In a detailed study, Wang [45] investigated the corrosion resistance of both the aluminized and chromized thermal diffusion coatings on steel substrates at elevated temperatures (600 °C) and proposed that chromized coating with more than 30 wt.% Cr and a thickness of 120 μm provided suitable corrosion resistance in sulfidation-oxidation conditions. According to Bai et. al. [46], the chromized thermal diffusion coating on low-carbon AISI 1020 steel showed improved resistance to corrosion compared to the original substrate. While these studies have placed emphasis on the corrosion behavior of the boronized, aluminized, and chromized thermal diffusion coatings on steel substrates, no comparative research work has been conducted to explore and develop thermal diffusion coatings to provide protection against high temperature molten salt corrosive attack. Moreover, coatings with more complex architectures, where a few potential protective layers may be involved, have never been considered for application in severely corrosive environmental conditions.

The main principle of the protection of the steels through the formation of the boride or intermetallide coatings is based on the fact that these coating materials have crystalline structures and strong diffusion bonding to the substrates. These coating compounds (intermetallides or borides) have high values of thermodynamic properties (e.g., high crystalline lattice energy and enthalpies) and strong, short covalent bonds of Fe–B, Fe–Al, Cr–Al, among others. As known, high lattice energies provide greater stability of crystal structures and higher degree of covalence associated with chemical bonds in solids, and therefore it has a direct correlation with chemical stability and integrity of the materials in harsh environments [48,49]. Therefore, these compounds have high corrosion resistance in general [50,51]. These coatings do not contain "free" Fe and other elements, like in steels or alloys, and the "introduced" elements (B, Al, and Cr) are not in their "free" forms either. In other words, the possibility of interaction of these metals with the anions of the corrosive environment (e.g., of the salts) is inhibited sufficiently. The interaction with the corrosive medium is possible only when the covalent bonds Fe–B, Fe–Al, Fe–Cr, and other related compounds are "broken".

The objective of this comparative study was to investigate the corrosion resistance of low carbon steel and 316 stainless steel with boronized, aluminized, or chromized surfaces obtained by using a thermal diffusion process and to compare them with bare steels. In order to simulate a corrosive environment to the typical of actual recovery boilers, the materials were exposed to modeling molten salts in low-temperature (220 °C) and high-temperature (60′0 °C) environments. In this regard, through a comparative study, the microstructure, composition, and corrosion behaviors of the modified surfaces were analyzed.

2. Materials and Methods

2.1. Sample Preparation

In this study, the target substrate steel materials were A36/44W low carbon steel (hereafter "CS"), which has similar composition and properties as steels J55 and K55 widely used in industry for tubing manufacturing, and 316 stainless steel (hereafter "316"). Steel substrates with dimensions of 25 mm × 25 mm × 6.3 mm (1″ × 1″ × $\frac{1}{4}$″) were coated according to a proprietary ETI thermal diffusion process. In this regard, steel substrates were boronized (–B), aluminized (–A), or chromized (–Cr). To that end, special powder compositions formulated for boronizing, aluminizing, and chromizing, which are the proprietary of ETI, have been prepared. The steel substrates have been immersed into the prepared powder mixes and placed into the retorts. The sealed retorts contained powders and substrates have been heated to certain temperatures provided the gas formation from the selected mixes. As mentioned above, the required atoms deposited onto the substrates diffused into the steel structures with consequent new phase formation, growth, and consolidation of the coating structures. According to the selected process, borides, aluminides, and chromides have been formed on the steels. Some of the obtained thermal diffusion coatings were further coated with thin oxide or non-oxide layers in combinations of boronized + boron nitride (–B–BN), boronized + yttria partially stabilized zirconia (–B–Z), aluminized + tin oxide (–A–Sn), and aluminized + partially stabilized zirconia (–A–Z), creating new composite "more complex" coating architectures. These materials have been selected for the additional layer due to their well-known chemical inertness and high temperature stability (up to 900 °C for BN and greater than 1200 °C for the oxides) [52,53]. The formation of the top thin layer was accomplished by immersion of the diffusion coated samples, which were first cleaned by brushing and washing with acetone, into the selected ceramic suspension, containing nano-size ceramic particles, with subsequent heat treatment at elevated temperatures. This heat treatment allowed consolidation of the top layer and adequate bonding to the boride or aluminide coating. The surface preparation and the use of the suspensions with nanoparticles promoted the adhesion of the coating materials. The obtained top layers had thicknesses varying of below 1 µm (for BN) to several microns for (zirconia and tin oxide). The top coating thickness depended on the concentration of the ceramic suspensions, particle size of the ceramic ingredient, and immersion time.

Five samples of each coating options, along with ten uncoated samples (control samples) were prepared. All the samples were weighed before exposure to corrosive attack by using a high-precision scale (Adventurer Pro AV313, OHAUS Corporation, Parsippany, NJ, USA). The maximum capacity and the tolerance of the scale were 310 g and 1 mg, respectively. The masses of the samples before the corrosive attack were used in the determination of the mass change after corrosive attack.

2.2. Corrosion Testing

For the corrosion study, the coated and uncoated (control) samples were first sprayed with a prepared dissolved salt solution. The simulated salt was a combination of four reagent grade chemicals selected to simulate the deposits on upper boiler tubes [27]. The chemicals in the salt mixture and their weight percentages are listed in Table 1. The simulated salt mixture was dissolved in distilled water to produce an unsaturated solution with a concentration of 192 g/L. According to the results of Differential Thermal Analysis (DTA) conducted by Rao et. al. [27], this salt mixture had a single first melting point of approximately 520 °C. To ensure that the amount of salt sprayed on all samples was approximately equal, the samples were weighed before and after application of the salt.

The corrosion testing was performed in two conditions, namely low temperature (220 °C, i.e., below the melting point of the modeling salt) and high temperature (600 °C, i.e., above the melting point of the modeling salt), both in air, which simulated two different conditions of the actual recovery boiler applications. For this purpose, the low-temperature furnace (ISOTEMP, Fisher Scientific, Ottawa, ON, Canada) and the high-temperature furnace (Wheelco Instruments, Chicago, IL, USA) were used. Accordingly, a list of the samples exposed to the low-temperature and the high-temperature corrosive environments are presented in Table 2. For each experimental series (low- and high-temperature corrosive attack), one sample of each coating was exposed in the corrosive oxidation environments for 24, 48, 96, and 168 h. The samples designated for each experimental series were placed all together to the furnace and exposed simultaneously due to the capacity of the employed furnaces, that is, these samples were tested at the same temperature and time conditions.

Table 1. Composition of the simulated salt mixture.

Chemical	Weight Percent (%)
KCl	10.2
Na_2CO_3	11.5
Na_2SO_4	73.9
K_2SO_4	4.4

Table 2. List of the samples exposed to low-temperature and high-temperature corrosive attack.

Test Type	1	2	3	4	5	6	7	8	9	10	11
High Temperature Corrosion Test	Uncoated Samples (Control Sample)				Aluminized Samples				Boronized Samples		
	CS-Bare	316-Bare	CS-A	CS-A-Sn	CS-A-Z	316-A	316-A-Sn	316-A-Z	CS-B	CS-B–BN	CS-B-Z
High Temperature Corrosion Test	Uncoated Samples (Control Sample)				Aluminized Samples				Chromized Samples		
	CS-Bare	316-Bare	CS-A	CS-A-Sn	CS-A-Z	316-A	316-A-Sn	316-A-Z	CS-Cr	316-Cr	–

2.3. Materials Examination

After the corrosion tests, the samples were rinsed with warm water and 95 vol.% ethanol to clean their surfaces from residual salt, which may remain from the test. Afterward, the samples were kept at room temperature to dry them completely. All the samples were weighed before and after washing and rinsing for the appropriate record of their masses, and the mass change was calculated. A so-called "relative mass change" was presented as a percentage of the mass of the samples before corrosive attack.

$$\text{Relative Mass Change } (\%) = \frac{(\text{Mass Before Corrosive Attack}) - (\text{Mass After Corrosive Attack})}{(\text{Mass Before Corrosive Attack})} \times 100 \tag{1}$$

The samples also were visually examined for the color, presence or absence of the flakes, delamination, macro-cracks, and pitting. After that, the samples were cut in half, mounted in epoxy, and then ground and polished according to the ETI established procedure. The examination of the microstructures of the cross-sections was conducted using a light optical microscope (IM7200, MEIJI Techno, San Jose, CA, USA). For the high quality images, acidic-based etching (e.g., Nital consisted of the nitric acid solution in ethanol) was employed. X-ray Diffraction (XRD) analysis was conducted by using an X-ray diffractometer (Ultima IV, Rigaku Corporation, The Woodlands, TX, USA) to elucidate an evidence of possible phase change within the coated samples after extended time exposure to a high-temperature corrosive environment. A cobalt tube was used to characterize the surface treatments. The tube was used at 38 kV and 38 mA with a curved graphite monochromator. Continuous XRD mode, where the two-theta diffraction angle was changed from 10° to 120° at a rate of 2°/min, was used. Moreover, a portable surface roughness gage (Pocket Surf. III, Mahr Federal Inc., Providence, RI, USA) was used to measure the roughness of the substrates after the corrosive attack. The nominal traverse length of the gage was 5 mm and the traverse speed was 5.08 mm/s.

Two methods of hardness determination were used in this study. Micro-indentation via Knoop hardness testing (Clark Micro-hardness Tester CM-400AT, CM Series, MetLab Corp., Niagara Falls, NY, USA) was employed by applying a diamond indenter with a 100-g load (HK 0.1) onto the coating cross-section under a microscope for a dwell time of 10–15 s. The Rockwell hardness test (Starrett Rockwell Hardness Tester, The L.S. Starrett Company, Athol, MA, USA) was also used and a load of 100 kgf (HRB) was applied onto the substrate material for a dwell time of 10–15 s. Two methods of hardness determination have been selected because, considered together, they provide better evaluation of the integrity of the materials. Rockwell hardness can reflect the property of the steel substrates at high indentation load; while, Knoop micro-hardness shows how the coating resists the action of sharp indentation applied exact to the coating. The measurements of the Knoop micro-hardness for the coating and for the substrate allow for comparison of the two different materials under the same testing conditions. The comparison of the hardness values was carried out for the original samples and for the samples exposed for 168 h; in the corrosion environments at low- and high-temperature conditions. Since the action of corrosion and temperature can create stress and micro-crack formation, phase changes, microstructure modification, etc., which can weaken the structure, the hardness comparison using both methods would provide better understanding of the corrosion resistance and integrity of the materials.

3. Results and Discussion

3.1. Microstructural Analysis

3.1.1. Coating Microstructure before Corrosive Attack

The obtained coatings had multi-layered architectures that were formed simultaneously during the thermal diffusion fabrication process (i.e., a few coating layers were formed through one heat treatment cycle). Besides, for some coatings, a thin top layer of either tin dioxide (SnO_2) or partially stabilized zirconia (ZrO_2) or boron nitride (BN) was applied as an additional step. Thus, the case depth consisted predominantly of a few layers, and the coating architecture depended on the type of the thermal diffusion process option (i.e., boronizing, aluminizing, or chromizing), substrate material, and whether the additional top layer was applied or not.

Figure 1 shows microstructure images of the aluminized low carbon steel and 316 stainless steel. For the aluminized samples, the top aluminum-rich aluminides layer contained 40–50 wt.% of Al. The subsequent inner aluminide layer (main layer) consisted of aluminides ($Fe(Ni,Cr)_xAl_y$) with Al contents ranging 30–38 wt.%. In many cases, the area between the main aluminized layer and the Al-rich layer consisted of a mixture of the aluminides with different compositions. In some cases, as shown on Figure 1, a third layer with ~20–28 wt.% of Al could be observed. Finally, the transition layer with Al contents of ~3–8 wt.% was observed between the substrate and the actual protective layers contained the aluminides with sufficient Al contents. With regards to the transition zone, its composition and thickness varied depending on the steel composition. Thus, the transition layer for the aluminized carbon steel was very thin (only a few microns), while this layer for the aluminized stainless steel was significantly greater (approximately 60–70 μm). The difference between the thickness of the transition layer could be explained by the percentage of the contents of the alloying elements, such as Cr, Ni, and some others, in the steel, and their outward diffusion during the aluminizing process and limited inward diffusion of Al. Basically, the structure, thickness of the layers, and Al contents in each layer strongly depended on the substrate material, the features of the inward diffusion of Al and outward diffusion of Fe, Ni, and Cr, which resulted in the formation of aluminides with different compositions, i.e., with different Fe, Cr, Ni, and Al contents. Further discussion of the formation of the coating structure and composition is beyond the scope of the current study. The images of the structures of the aluminized samples with the additional top oxide layers were not shown since these top layers are very thin (a few microns) and could not be well visible at the selected (200×) magnification (they were analyzed in the unpublished ETI work).

Figure 2 shows the microstructure of chromized carbon steel and 316 stainless steel. While the chromized carbon steel consisted of two well-visible layers with different compositions of the formed chromides (the top layer had Cr-rich chromides), the chromized 316 stainless steel had a significantly larger protective layer (greater than 100 μm) with rather homogeneous composition. The latter can be explained by the sufficient content of Cr in the stainless steel. In this case, Cr from the gas phase diffused into the substrate, while Cr from the substrate diffused from the opposite side, and, as a result of the inward-outward diffusion of Cr, two chromide layers with rather close contents of Cr may be observed.

Figure 3 shows the microstructure images of the boronized carbon steel (CS) substrates. This coating consisted of two phases, Fe_2B (near the substrate) and FeB as the outer layer, and it had a "saw-tooth" morphology. The features of the boronized coating formation, its composition, and structure have been described elsewhere [41]. For the samples with an additional thin coating layer (boron nitride and zirconia) applied on the top of the boronized coating, this top layer was not fully visible on the cross-section image at the selected magnification (200×) from the light optical microscope; thus, the microstructure images of these samples were not shown.

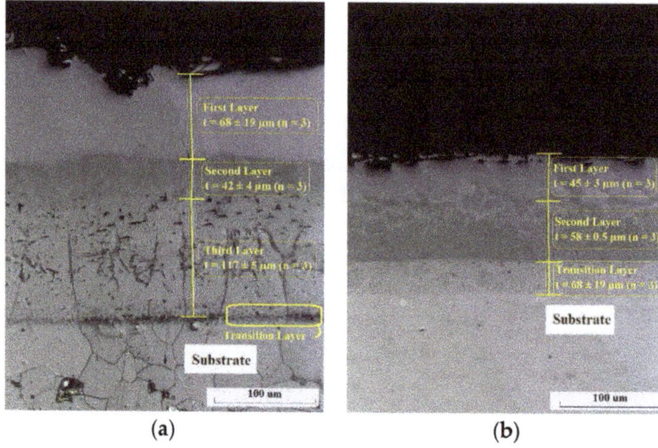

Figure 1. Microstructure image of aluminized (a) low carbon steel, and (b) 316 stainless steel.

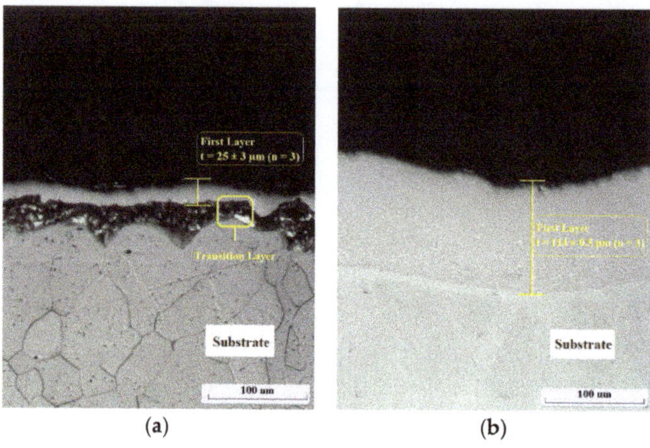

Figure 2. Microstructure image of chromized (a) low carbon steel, and (b) 316 stainless steel.

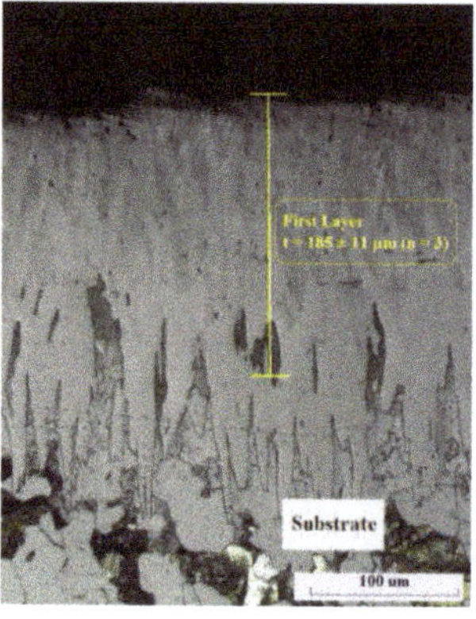

Figure 3. Microstructure image of boronized low carbon steel.

3.1.2. Qualitative Comparison of Top Surface of Samples (Appearance) after Corrosive Attack

In order to provide a qualitative comparison of the samples after the corrosive attack, macroscopic imaging of the top surface of the samples was conducted before mounting them in epoxy. Therefore, macroscopic degradation, if any, due to corrosive attack could be observed. The images of the samples before and after exposure in the corrosion environment conducted according to the tabulated test method (see Table 2) are presented in Figures A1 and A2 in Appendix A. The black mark on the top right- or left-hand side of the samples was painted to indicate the surface that was exposed to the hot corrosive environment. In the figure, the changes of the samples as represented (a) before the corrosion tests, (b) after 24 h, (c) after 48 h, (d) after 96 h, and (e) after 168 h of exposure to the simulated corrosive environment are shown.

Most of the samples changed their colors after the exposure that was related to the surface oxidation. Greater discoloration was observed after high-temperature exposure. However, no blistering, pitting, surface cracking, delamination, and peeling-off was observed for the samples with the protective coatings, even after high-temperature corrosive attack. Considering the aluminized and chromized coatings, it was unlikely that these intermetallides would oxidize at the low-temperature (220 °C) test conditions with formation of "measurable" oxide films, since these intermetallides are rather stable and inert. Therefore, their oxidation had only a superficial effect. On the contrary, the boronized coatings oxidized to a greater extent at 220 °C with formation of the iron-boron oxide film that may be of the "glassy" nature (boron oxide is known as a glassy-forming oxide). However, this oxide film was not easy visible yet even at high magnifications under optical microscope. The simulated salt, as expected, did not melt at 220 °C, and its attack was negligible. At the high-temperature (600 °C) testing conditions, slight oxidation of the aluminized and chromized coatings occurred resulting in the formation of the thin Al_2O_3 and Cr_2O_3 films. These oxides had, as is well known, high thermal stability and chemical inertness [52,53]. Therefore, these films, which were well-adhered to the diffusion coatings, protecting these coatings from further oxidation and chemical attack of the molten salt. The additional thin layers of SnO_2, ZrO_2 and BN applied onto some coatings also protected the boronized and aluminized coatings against oxidation and interaction with the simulated salt, regardless of the low-temperature (unmolten salt) or the high-temperature (molten salt) corrosive attacks. Although, as expected, the adhesion of the molten salt (at 600 °C) to the samples was significantly higher than the adhesion of the unmolten salt (at 220 °C), the coated samples could be easy washed from the residue of the molten salt without visible damage of the coating that may be considered as a positive point.

As opposed to the coated steel samples, a thin scale of the rust could be observed on bare carbon steel samples even after exposure at 220 °C, and this rusting was significantly larger after the exposure at 600 °C. While the surface of stainless steel samples did not have visible rusting after low-temperature testing, the rusting issue was clearly observed after the high-temperature exposure, however, to a significantly less extent compared to low carbon steel. The formed rusting layer was rather soft and porous, loosely adhered to the original steel, and it could be detached from the surface, especially in the case of carbon steel and after longer exposures. The thicknesses of the soft rusting layers were uneven. This matter will be further discussed in the sections related to the materials' microstructure examination with demonstration of the thickness of the rusting for the bare steel samples. The spallation issue for the steels at high-temperature oxidation or corrosion-oxidation is generally obvious and described in numerous literature sources [54,55]. Because of the delamination issue with the uncoated materials, including stainless steels, it can be expected that the scale formation may grow with extension of the exposure time.

3.1.3. Microstructural Analysis after Corrosive Attack

Low-Temperature Corrosive Attack

The images of the coatings' microstructures (cross-sections) of the samples examined before and after 168 h of exposure at the low-temperature corrosive attack are presented in Appendix B.

It could be seen that the coated samples were not considerably affected by the corrosive testing condition (see Figures A3–A13 in Appendix B). Thus, no micro-cracks, delamination, or coating structure were observed. As mentioned earlier, since the simulated salt did not melt at the selected temperature (220 °C), the effectiveness of the salt action to induce corrosion of the surface of all samples was limited. Under this low-temperature condition, no difference in the case depths was observed for the coated samples examined before and after the testing (see Figures A3–A13 in Appendix B). Even for the boronized steel, no noticeable oxidation of iron borides occurred. Moreover, based on the microstructure images of the coatings with additional thin oxide or non-oxide layers, it was likely that the presence of the chemically inert top coatings (e.g., SnO_2 and ZrO_2, as well as BN) also prevented oxidation of iron borides and aluminides.

As opposed to the coated steel samples, the corrosion-oxidation degradation was observed on the surface of bare carbon steel. A scale with a thickness of several microns could be observed. This scale was soft and poorly bonded to the steel, and it was easily detached at the cross-section preparation. Also pits and cavities with the depths of 20–35 μm were observed (Figure A3), and their presence could facilitate further corrosion in molten salts at elevated temperatures. Stainless steel 316 samples did not experience degradation under the low-temperature unmolten salt testing condition.

High-Temperature Corrosive Attack

The coating microstructure images of the samples examined before and after corrosion testing at 600 °C are presented in Appendix C, while Figures 4 and 5 show the microstructure (cross-sections) of low-carbon steel and 316 stainless steel also before and after high-temperature testing at the same conditions. It was clearly observed that the uncoated steel samples experienced corrosion from their surface. Thus, carbon steel had the uneven rusting scale with a thickness from 40 to 70 μm, and this scale contains pores and even large voids. This scale formation could be detected at short times of molten salt-oxidation exposure (a thin scale was already detected for the sample exposed for 24 h). The uncoated 316 stainless steel sample had, as expected, a significantly smaller scale (from 9 to 15 μm) that was partially flaked-off at the cutting-polishing of the cross-section preparation. The corrosion scale formation can be explained by the joint interaction of the molten salt with the steel surface and oxidation, e.g., by the formation of iron oxides (Fe_2O_3, Fe_3O_4, Cr_2O_3, iron sulfates, and chlorides). The formation of these new compounds was confirmed by the XRD data, which will be discussed later.

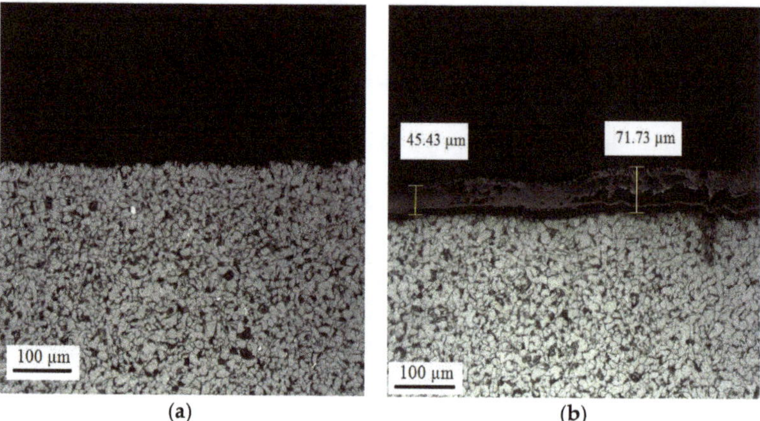

Figure 4. Microstructure images of the cross-section of CS-Bare (**a**) before, and (**b**) after the high-temperature corrosive attack after 168 h.

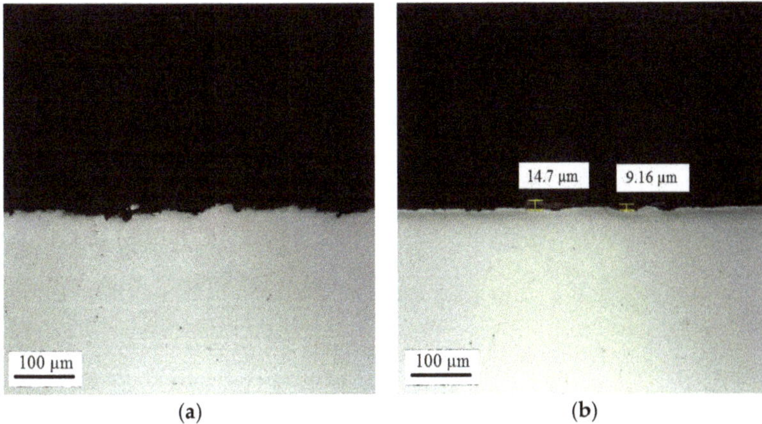

Figure 5. Microstructure images of the cross-section of 316-Bare (**a**) before, and (**b**) after the high-temperature corrosive attack after 168 h.

In contrary to bare steels, the structures of the aluminized coatings did not experience sufficient changes, and no micro-cracks and delamination were observed (see Figures A14–A16 and Figures A18–A20 in Appendix C). However, some changes of the thicknesses of the coatings' layers could be detected, but without changes of the total case depths. This small change of the thicknesses of the layers can be explained by the continuation of the diffusion processes occurred at the continual action of high temperature (600 °C); particularly, by the inward diffusion of Al and outward diffusion of Fe, Cr, and Ni from the substrate, but not due to the corrosion. As mentioned above, the aluminides of the Al-rich layer started oxidizing with formation of the thin Al_2O_3 film with high inertness that inhibited the coating surface from further oxidation and from contact with the molten salt ingredients, i.e., this Al_2O_3 film also enhanced corrosion resistance of the materials [53]. A similar effect can be inherent to the samples with additional top oxide layers (SnO_2 and ZrO_2).

In order to analyze the surface topology of the materials, the values of their average roughness after high-temperature corrosive attack were measured and presented in Table 3. In Table 3, n represents the number of measurements. In the case of the aluminized coating, the roughness of the coating on the 316 stainless steel was significantly less than that of the coating on the low carbon steel substrate. The lower value of the average roughness of the uncoated low-carbon steel is probably due to the delamination of the corroded parts from top surface; otherwise, a greater value for the average roughness of the uncoated low-carbon steel was expected. Low values of roughness should be more favorable for corrosion protection due to the significantly lower potential contacts with the corrosive media and the exposed material and due to the reduced possibility of fracture and removal of the asperities during occasional mechanical loading of the surface. In the case of chromized samples, insignificant changes in the chromide layers were detected, but these changes were more noticeable in the case of the carbon steel substrate compare to the 316 stainless steel (see Figures A17 and A21 in Appendix C). These changes with the chromized samples may also be related to the continuation of the diffusion processes, but to a significantly smaller extent compared to the aluminized coatings. Since the chromized samples had a smooth top layer, the measurements were unaffected by the topology of the surface (see Table 3).

Table 3. Average roughness of the substrates after the high-temperature corrosive attack.

Coating Information	CS-Bare	CS-A	CS-A-Sn	CS-A-Z	CS-Cr	316-Bare	316-A	316-A-Sn	316-A-Z	316-Cr
Average roughness (R_a) (µm)	2.62	6.4	6.22	6.32	3.05	1.41	1.26	1.37	1.36	3
Standard deviation (n = 3)	±0.2	±0.1	±0.28	±0.06	±0.06	±0.32	±0.07	±0.24	±0.18	±0.44

The difference in the diffusion rate among various metal atoms in the layers within the coating might have caused the mass transfer at the interface during the thermal diffusion process occurred during the high-temperature corrosion testing. This has been characterized as the Kirkendall effect [56]. After the low- and high-temperature corrosive attack, no porosity (voids), as a result of the Kirkendall effect, was observed. Moreover, no evidence of delamination or cracking was found for the coated samples. Since the diffusion process is highly dependent on temperature and exposure time, it was expected that extended exposure to the high-temperature environment might cause minimal phase changes in the coating. This is related to the fact that the diffusion process exhibits a parabolic behavior, in which after a short period of time, any "change" in the coating during the thermal diffusion process will approach a "plateau". Similarly, with oxidation, the Al_2O_3 scale formed was chemically inert and protected the inert aluminides against further oxidation. Therefore, it was found that the coated samples, particularly aluminized, showed good resistance against the high-temperature corrosive environment.

3.2. Qualitative XRD Analysis

XRD analysis was performed in order to qualify possible phase changes on the surface of the coated and bare steel samples associated with corrosion-oxidation. In this regard, the top surface of the samples was analyzed before and after the low- and high-temperature corrosive attack, and the positions of the diffraction peaks and their intensities were compared. This approach is widely used for the corrosion evaluation, and it was successfully utilized by Melendez and McDonald for the studies of some metal–matrix composite coatings [57]. All the samples with coatings did not experience the formation of new phases after low-temperature exposure in the contact with the unmolten salt. The XRD profiles of the coated samples confirmed that since the location of the peaks associated with the two-theta coordinates remained unchanged for the samples examined before and after low-temperature corrosive attack (see Figures A22–A27 and Figures A29–A31 in Appendix D).

After the high-temperature exposure in the molten salt, the aluminized samples with top SnO_2 and ZrO_2-based layers also did not experience the phase formation; their XRD profiles were similar for the samples examined before and after corrosion testing, although the intensity of some peaks changed (See Figure A23, Figure A24, Figure A30, and Figure A31 in Appendix D). It may be related to the slight surface oxidation of the aluminides with formation of the Al_2O_3 film. The XRD peaks that were due to the very thin Al_2O_3 films likely overlapped with other peaks that were due to the materials under this film; therefore, the XRD results may not be very clear or conclusive as it relates to the thin Al_2O_3 films. Generally, the aluminized coatings experience oxidation at temperatures of 600–1000 °C with formation of very thin Al_2O_3-based films. Higher temperatures will lead to greater oxidation of the aluminized coatings. These were observed by, and confirmed in, studies by other investigators [58–60]. In addition, due to the relatively low temperature (600 °C) for the formation, crystallization, and consolidation of thick layers of Al_2O_3, it was suspected that the Al_2O_3 film that was formed consisted of transition phases, which would not be clearly detected by XRD. Therefore, the XRD patterns, with the detailed indication of the phases, were not included in this manuscript. However, the results of the XRD analyses showed that the chromized low-carbon steel and 316 steel samples experienced phase changes at their surfaces after high-temperature corrosive attack. As shown in Figures A28 and A32 in Appendix D, the predominant phase in the coatings after the high-temperature corrosive attack for both samples was chromium oxide Cr_2O_3 (which is also known as Eskolaite), and its formation was associated with the remarkable oxidation of the chromides in the coatings. As explained earlier, the presence of ions originated from the molten salt under the elevated temperature conditions (600 °C) stimulated the oxidation of chromium, which resulted in the observed phase change in the coating, which in turn, may be considered as a degradation of the thermal diffusion chromized coatings. However, the XRD data did not show the presence of the chromium salts, which may be associated with interaction of the chromide coatings with molten salts.

Regarding the uncoated steels (low-carbon steel and 316 stainless steel), new peaks corresponded to the formation of new phases can be observed on the diffractograms. Also the noticeable increase in the intensity of some peaks after high-temperature corrosive attack may be related to the formation of iron oxides (Fe_2O_3 and Fe_3O_4), and, probably, to the formation of iron chlorides and sulfates (see Figures A22 and A29 in Appendix D). The suggested formation of new phases is supported by the data obtained by other researchers [7,10,12]. The obtained results of the XRD analysis correspond well to the data obtained from the microstructural examination of the studied materials. Since the intensity of the peaks was only the representative of the percentage of a particular phase in the coating, the intensity differed after long time exposure to the high-temperature environment. The XRD analysis of the coatings with particular identification of certain phases requires an additional work.

3.3. Mass Change after Corrosive Attack

The relative mass change after 168 h of exposure to the low- and high-temperature environments is shown on Figure 6. Each sample, on average, had about 225.7 $\pm$ 81.8 mg (n = 11) of the salt solution on their surfaces before the corrosive attack. The relative mass change for all samples, both for the control samples (bare steels) and the aluminized and boronized substrates, was negligible (less than 0.02%) in the case of the low-temperature corrosive conditions (Figure 6a). However, the relative mass change increased after the high-temperature corrosive attack. As shown in Figure 6b, the maximum mass change was recorded for the uncoated low carbon steel (approximately 1%). The mass change gradually increased at the exposure time increase. In addition to the uncoated low carbon steel, Figure 6b confirmed that the mass change for all other samples, as a result of the high-temperature corrosive attack, was low (significantly less than 1%). The top surface of the uncoated steels (control samples) experienced delamination of the rusting scale after the high-temperature corrosive attack (see Figures 4 and 5 demonstrating the microstructures of the steels exposed and Figure A2 in Appendix A) that is especially related to the carbon steel sample.

The weight change in the samples typically corresponds to the mass loss due to corrosion ("dissolution") of the material exposed to the salt at elevated temperatures and mass gain due to oxidation of the surface. The formation of the corrosion products with lower density may also affect the mass reduction. Moreover, the weight change may depend on whether or not the formed oxide or corrosion layers have a strong adhesion to the surface of samples. If the adhesion is not sufficient, delamination of the oxide layer will take place. In the case of the coated samples in this study (boronized, aluminized, and chromized samples), no delamination occurred. In the case of bare steels, especially for carbon steel exposed to the high-temperature corrosion, the scale layer flaked-off (delaminated) from the surface. Because of these two "concurrent" cases (mass loss due to corrosion and mass gain due to oxidation at elevated temperature), the values of the mass change for the uncoated samples cannot be considered as very accurate and may vary at different situations.

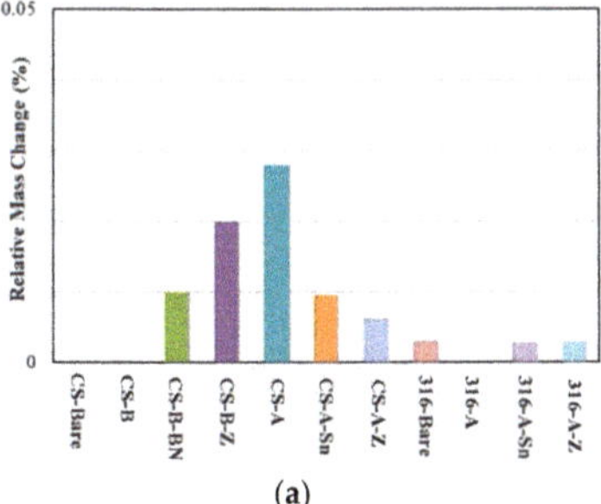
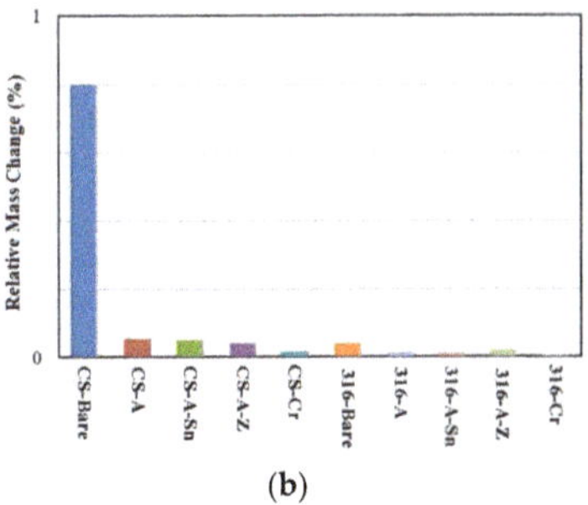

(**a**)   (**b**)

Figure 6. Relative mass change of samples after (**a**) low-temperature, and (**b**) high-temperature corrosive attack after 168 h.

3.4. Hardness Evaluation

3.4.1. Hardness of the Substrates

The hardness of the samples was examined before and after the corrosion tests by employing two different methods of measuring the hardness, namely, the micro-indentation hardness test (Knoop test) and Rockwell hardness test. These tests were performed both on the cross-section of the original substrate, and at a specific location in the middle area of the case depth of the sample. Figure 7 shows the location of the indentation of the Knoop and Rockwell hardness tests on the original substrate.

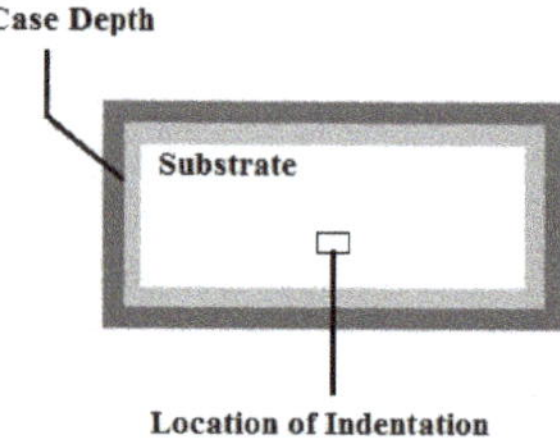

Figure 7. Location of the indentation on the original substrate.

The hardness determination results, e.g., Knoop micro-hardness and Rockwell B hardness, for the samples examined before and after corrosion-oxidation testing for 168 h are presented in Figures 8–11. From a comparison of the hardness values based on HK 0.1 and HRB for the substrates, it was observed that the values did not change noticeably after low-temperature corrosive attack (Figures 8 and 9). The difference between the values was within the standard deviation and standard error of the mean. However, in the case of high-temperature testing, some decrease of the hardness values could be observed for low carbon steel substrates (Figures 10 and 11). The 316 stainless steel substrates did not experience significant reduction of hardness after 168 h of exposure. The decrease in the hardness values of the substrates, particularly low carbon steel, could be related to the grain growth after long exposure at high temperature. In fact, some grain growth in steels occurs both at thermal diffusion process and high-temperature corrosion testing, and this effect is especially observed in the substrate area close to the coating. As a sequence of the grain growth and related dislocations within the grains, the associated stresses result in the decrease of hardness and strength of the material [52]. The dependence of the mechanical properties reduction with the grain size increase is described by the Hall-Petch equation [52]:

$$\sigma_y = \sigma_0 + \frac{k}{\sqrt{D}} \tag{2}$$

where σ_y and σ_0 represent the yield stress and the intrinsic stress of the material, respectively, k is the strengthening coefficient, and D denotes the grain size. The intrinsic stress is defined as the initial stress for dislocation movements. According to the Hall-Petch equation, the grain size increase leads to decrease of the yield stress, and as a result, to decrease of hardness of the material. The experimental studies conducted by Busby et al. [61] confirmed the relation between yield stress and hardness of austenitic and ferritic steels. Although the stainless steel substrate did not experience the grain growth and related reduction of hardness after rather short (168 h) high temperature exposure, some grain growth and related hardness reduction of steels may be expected in the long term exposures adhered to the industrial applications. Moreover, the coarsening and possible phase transformation may occur both during the thermal diffusion process and at long exposure at elevated temperatures, and they would affect the hardness of the area close to the coating. However, the determination of the phase transformation within the carbon steel substrates requires additional study with detailed XRD analyses.

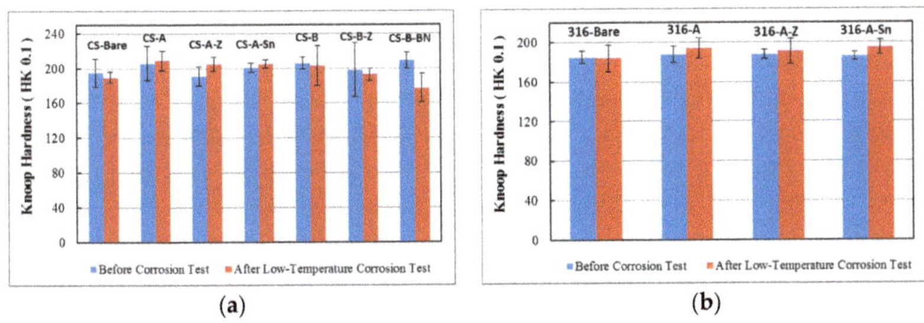

(**a**)　　　　　(**b**)

Figure 8. Comparison of Knoop hardness of the original substrate before and after the low-temperature corrosive attack for (**a**) low carbon steel, and (**b**) 316 stainless steel substrates.

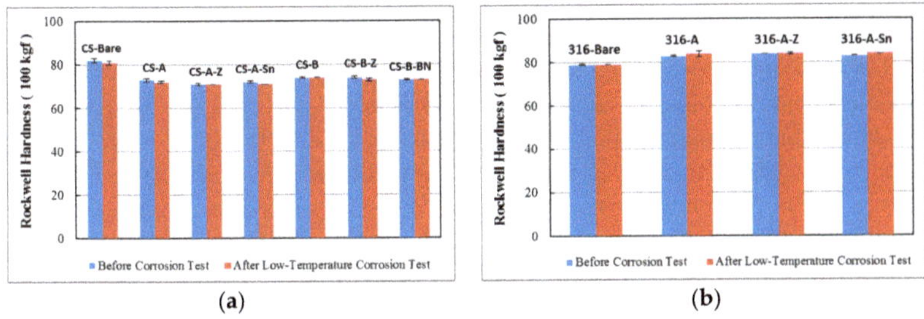

(**a**)　　　　　(**b**)

Figure 9. Comparison of Rockwell hardness of the original substrate before and after the low-temperature corrosive attack for (**a**) low carbon steel, and (**b**) 316 stainless steel substrates.

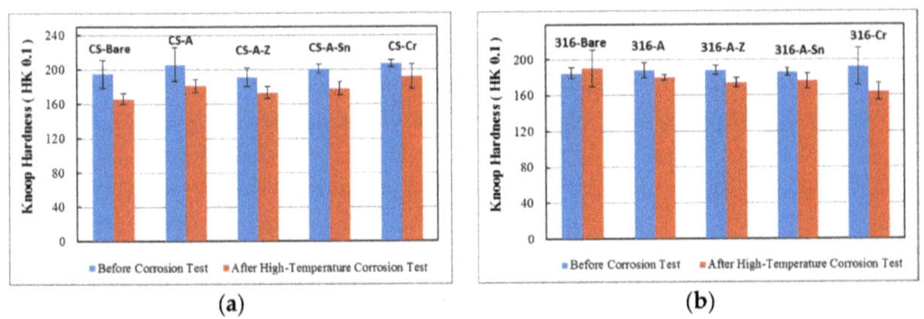

(**a**)　　　　　(**b**)

Figure 10. Comparison of Knoop hardness of the original substrate before and after the high-temperature corrosive attack for (**a**) low carbon steel, and (**b**) 316 stainless steel substrates.

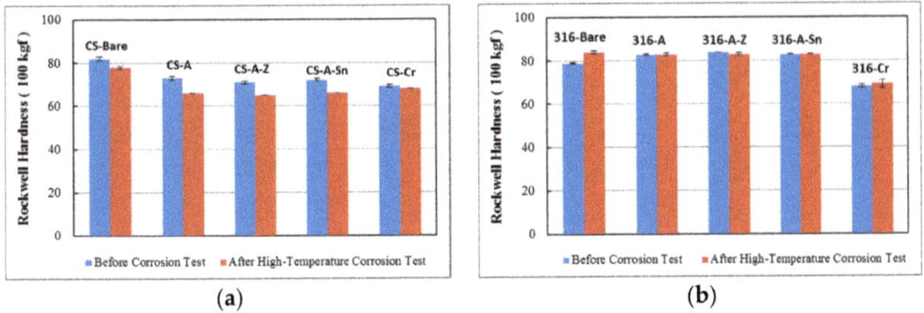

(**a**)　　　　　(**b**)

Figure 11. Comparison of Rockwell hardness of the original substrate before and after the high-temperature corrosive attack for (**a**) low carbon steel, and (**b**) 316 stainless steel substrates.

3.4.2. Hardness of the Case Depth

The hardness of the case depth of the samples was examined by employing a micro-indentation hardness test (Knoop test). This method of determining the hardness is recognized as the optimal test for measuring the coating characterization, according to extensive studies on aluminized, boronized, and other coatings by thermal diffusion process [41,62]. Since the coatings studied in this work had multi-layered architectures consisting of layers with different phase compositions and structures, and therefore with different hardness, it is important to indicate the location of where the indenter was applied at the hardness determination. Because one of the aims of this study was to evaluate whether the coating hardness decreases or not after the corrosive attack, i.e., to evaluate the coating integrity as the capability to maintain the structure and properties during corrosion at elevated temperatures, it was decided to determine a "general hardness" of the coatings, but not to determine a hardness for each individual layer. In this regard, the Knoop diamond indenter was applied in the middle of the coatings. Figures 12–14 demonstrate the locations of where the indenters were applied depending on the type of the coatings. For the aluminized coatings, the indenters were applied onto the junction of the Al-rich and the main aluminized layers (see Figure 12). Figure 13 showed the locations of the applied indenters for the chromized coatings; e.g., the indenter was applied onto the middle of the top Cr-rich layer in the case of the low carbon steel substrate and onto the middle of the chromide coating (onto the junction of two layers) in the case of the 316 stainless steel substrate. In the case of the boronized samples, the indenter was also applied in the middle of the coatings where both iron boride phases, FeB and Fe_2B, coexist (see Figure 14). The hardness data for the middle of the coating for the considered materials was important, not only to compare the coatings' integrity in general, but it is also related to the fact that this area, where two crystalline phases are in "contact", was the weakest area with an elevated stress condition.

Comparing micro-hardness data for the coatings and for the substrates (see Figures 8, 10, 15 and 16), it is clear that the coatings had significantly greater hardness values than the steels (both carbon steel and stainless steel). The hardness of the case depth for all the studied materials determined through the Knoop indentation remained on the original level after 168 h of low-temperature exposure being in contact with the unmolten salt (Figure 15). The hardness determination for the aluminized coatings examined after the high-temperature molten salt corrosion also demonstrated that these coatings had a high integrity and could successfully withstand the selected harsh conditions; their micro-hardness remained on the original level (Figure 16). Similar results were obtained for the chromized coating applied onto carbon steel. However, the results obtained for the chromized stainless steel 316 showed that the coating hardness decreased noticeably (Figure 16). The reduction of hardness of the chromized coating was likely related to the lower stability of the chromides under the corrosive conditions considered in this study. It is likely that the stress generated in the chromized coating structure at oxidation at elevated temperatures (600 °C and higher) with a rather thick Cr_2O_3 layer formation (see the data of the XRD analysis) led to the hardness reduction. This may correspond to the fact that the phase transformation and new phase formation may further affect the hardness of the materials [52]. Overall, the obtained results related to the hardness data and their comparison for the samples examined before and after corrosion testing suggested that the aluminized samples have a higher potential withstanding against the corrosive conditions (e.g., in high-temperature molten salt), which are expected in recovery boilers.

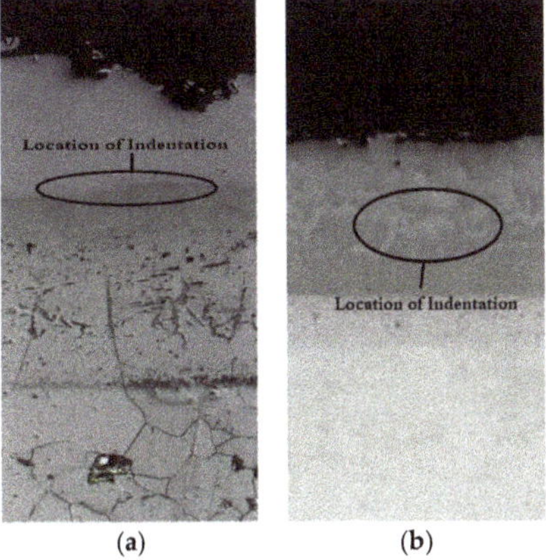

(**a**) (**b**)

Figure 12. Location of the area where the micro-indentation hardness test was conducted for the aluminized coatings (**a**) low carbon steel, and (**b**) 316 stainless steel substrates.

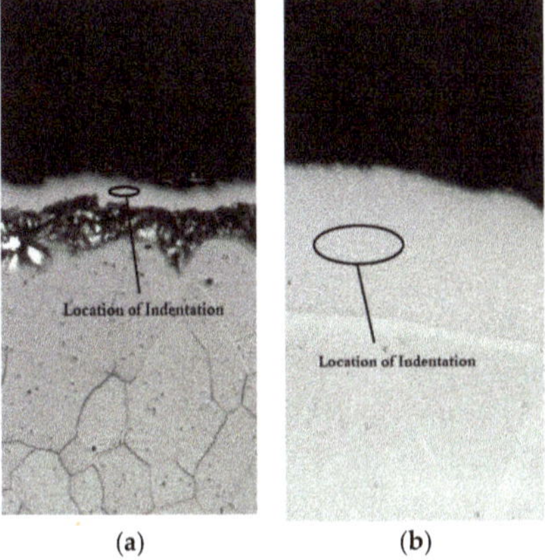

(**a**) (**b**)

Figure 13. Location of the area where the micro-indentation hardness test was conducted for the chromized coatings (**a**) low carbon steel, and (**b**) 316 stainless steel substrates.

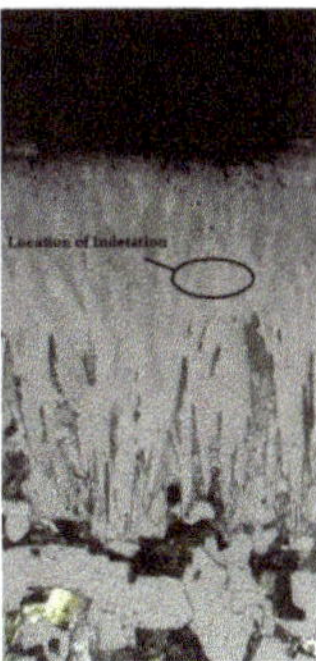

Figure 14. Location of the area where the micro-indentation hardness test was conducted for the boronized low carbon steel.

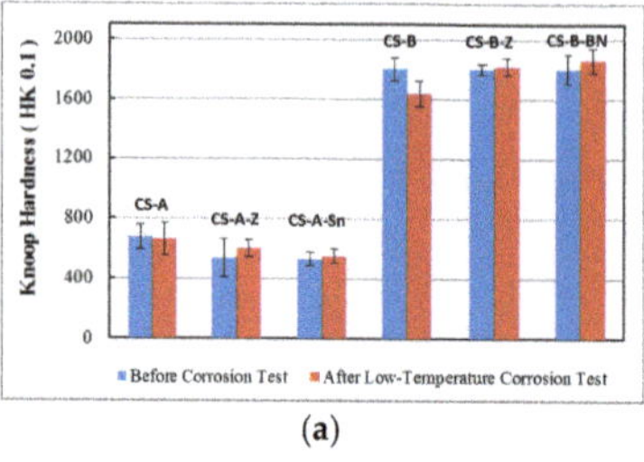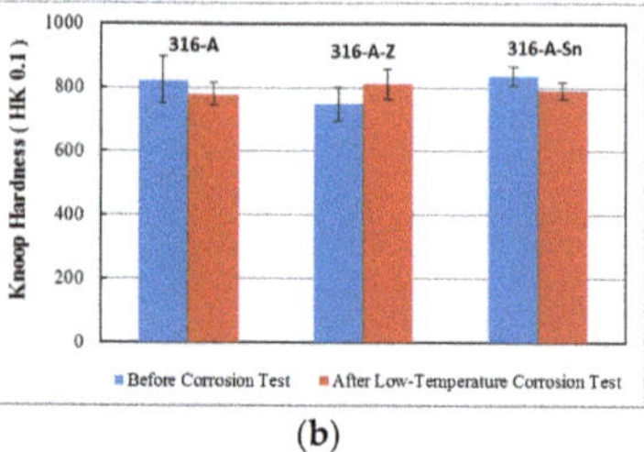

Figure 15. Comparison of Knoop hardness of the case depth before and after the low-temperature corrosive attack for (**a**) coated low carbon steel, and (**b**) coated 316 stainless steel substrates.

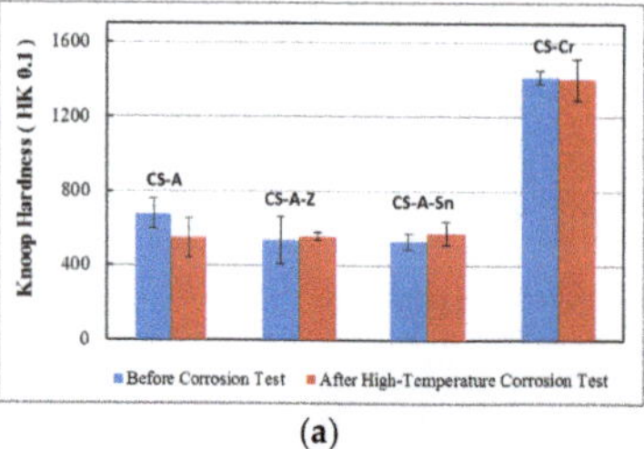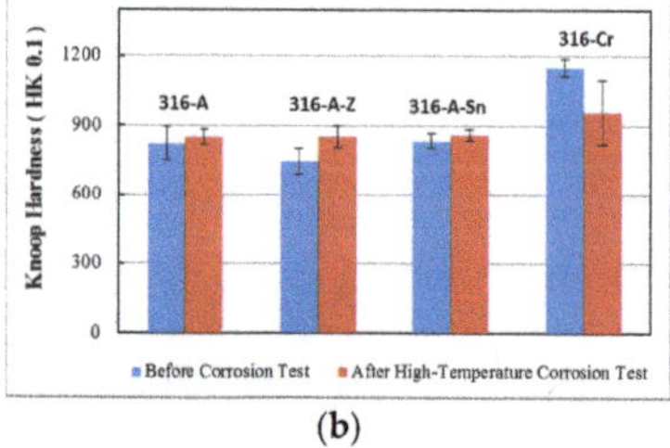

Figure 16. Comparison of Knoop hardness of the case depth before and after the high-temperature corrosive attack for (**a**) coated low carbon steel, and (**b**) coated 316 stainless steel substrates.

4. Conclusions

Corrosion resistance of low carbon steel and 316 stainless steel with thermal diffusion coatings (boronized, aluminized, and chromized), including the coatings with additional layers from inert ceramics, such as ZrO_2, SnO_2, and BN, have been studied, for the first time, in the molten salt oxidation conditions simulated industrial corrosive conditions in recovery boilers and some other corrosive environments in pulp and paper processing. The systematic studies included comparison of the appearance, mass change, microstructure examination (cross-sections), phase analysis (XRD of the surface), and hardness of the coated steels vs. uncoated (bare) steels have been conducted. Based on the results, the most important findings of the current study are highlighted below:

The samples from the coated steels did not have any blistering, surface cracks, coating delamination, and peeling after low- and high-temperature exposures in corrosive-oxidation conditions, although surface discoloration could be observed for most of samples. In contrary, the top surface of the bare steels experienced more visible oxidation with partial delamination after the corrosive attack, particularly, low carbon steel substrates after high-temperature exposure in the molten salt. The coated samples had a negligible weight change (significantly less than 1%) after the high-temperature molten salt corrosive attack.

The results of the microstructural and XRD analyses suggested that the aluminized samples had high corrosion resistance in both low- and high-temperature molten salt oxidation environments. The high-temperature environment seemed to stimulate the continuation of the thermal diffusion process and resulted in some changes in the thickness of the layers within the case depth but without total thickness reduction and without micro-cracks and delamination in the aluminide layers. The surface oxidation of the aluminized coatings with the formation of a very thin chemically inert Al_2O_3 skin further promoted corrosion resistance of these coatings. Additional top layers that were applied (e.g., SnO_2, ZrO_2, and BN) inhibited the interaction of the coatings with the corrosive salts, especially at elevated temperature. The boronized and chromized samples were not affected by the low-temperature test conditions; however, the chromide coatings experienced some phase changes after the high-temperature exposure that is confirmed by the data of the comparative XRD analysis. Although bare 316 stainless steel resisted the low-temperature contact with the salt, it experienced interaction with molten salt at high temperature that was confirmed by the microstructural and XRD analyses (visible formation of thin layer of the products of corrosion). Moreover, low carbon steel samples had visible corrosion even at low-temperature, and the formation of the thick (more than 50 μm) corrosion-oxidation layer with subsequent delamination was clearly observed.

The hardness of the case depth of the aluminized and boronized samples remained on the original level (as before testing), which confirmed the coatings' integrity under the selected simulated testing conditions. However, the hardness of the case depth of the chromized samples decreased, which likely occurred due to possible phase change of the chromides after high temperature exposure. The hardness of the substrate decreased after the high-temperature corrosive attack for almost all samples that can be explained by the grain growth in the steels.

The studied coatings, particularly, aluminized coatings, demonstrated high integrity in the simulating unmolten salt and molten salt oxidation corrosive conditions. The protection of steels employing the aluminized coatings is defined by the inertness of aluminides, by the inert Al_2O_3-oxidation film formation at elevated temperatures, and by the diffusion-related bonding between the protective layers and the substrate materials. A lower-roughness aluminized coating on stainless steel is preferable for corrosion protection. Multi-layer coating architectures are favorable for corrosion protection.

The tested materials, especially aluminized steels, can be recommended for application in high-temperature molten salt corrosion environments in the pulp and paper processing. The thermal diffusion technology, when the multi-layered coating structure is formed through one processing "step", can be successfully employed for manufacturing of the industrial components, including long recovery boiler tubing.

Author Contributions: Conceptualization, Amirhossein M., André M. and E.M.; Methodology, Amirhossein M., André M., E.M. and G.L.M.; Formal Analysis, Amirhossein M., André M., E.M. and G.L.M.; Investigation, Amirhossein M., André M. and E.M.; Writing-Original Draft Preparation, Amirhossein M. and E.M.; Writing-Review & Editing, André M. and E.M.; Supervision, André M.

Funding: This research received no external funding.

Appendix A.

Figures A1 and A2 shows the evolution of the top surfaces of the samples after low-temperature (220 °C) and high-temperature (600 °C) corrosive attack. Each image is labeled based on the coating material. In this figure, the changes of the samples as (a) represented before the corrosion tests, (b) after 24 h, (c) after 48 h, (d) after 96 h, and (e) after 168 h of exposure to the simulated corrosive environment are shown. Even though the samples were washed after the test, some of the sample surfaces were likely still tainted with residual salt material. The black mark on the top right- or left-hand side of the samples was painted to distinguish the surface that was exposed to the hot corrosive environment.

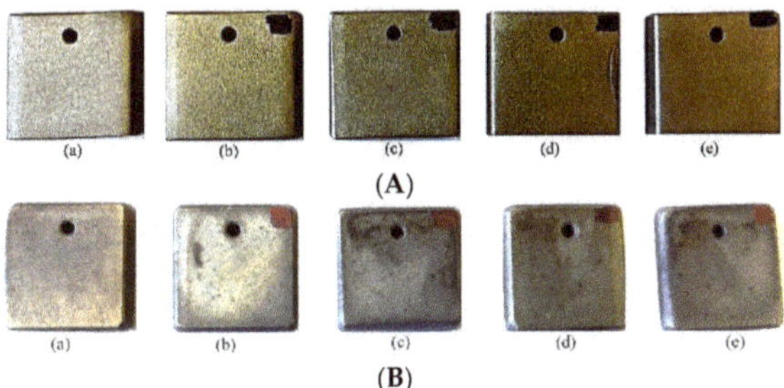

Figure A1. *Cont.*

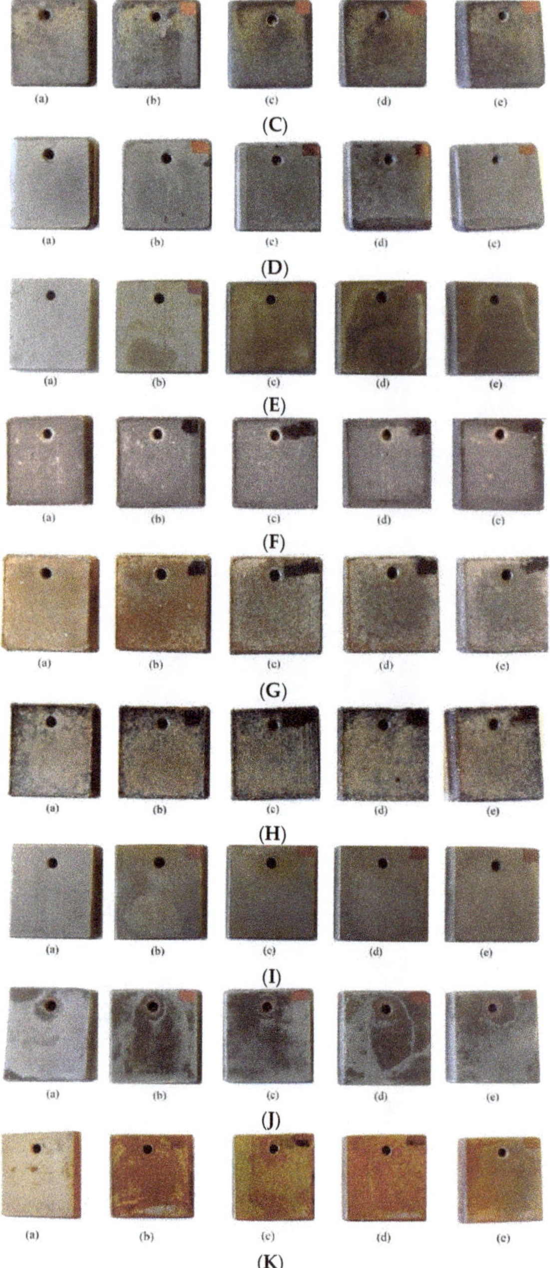

Figure A1. Photographs of the samples (**A**) 316-Bare, (**B**) 316-A, (**C**) 316-A-Sn, (**D**) 316-A-Z, (**E**) CS-Bare, (**F**) CS-A, (**G**) CS-A-Sn, (**H**) CS-A-Z, (**I**) CS-B, (**J**) CS-B-BN, and (**K**) CS-B-Z: (**a**) before salt corrosion testing; (**b**) after 24 h; (**c**) after 48 h; (**d**) after 96 h; and (**e**) after 168 h of exposure in low-temperature (at 220 °C) corrosive environment. We moved the words in the original column (left) and re-edited the title, pls confirm.

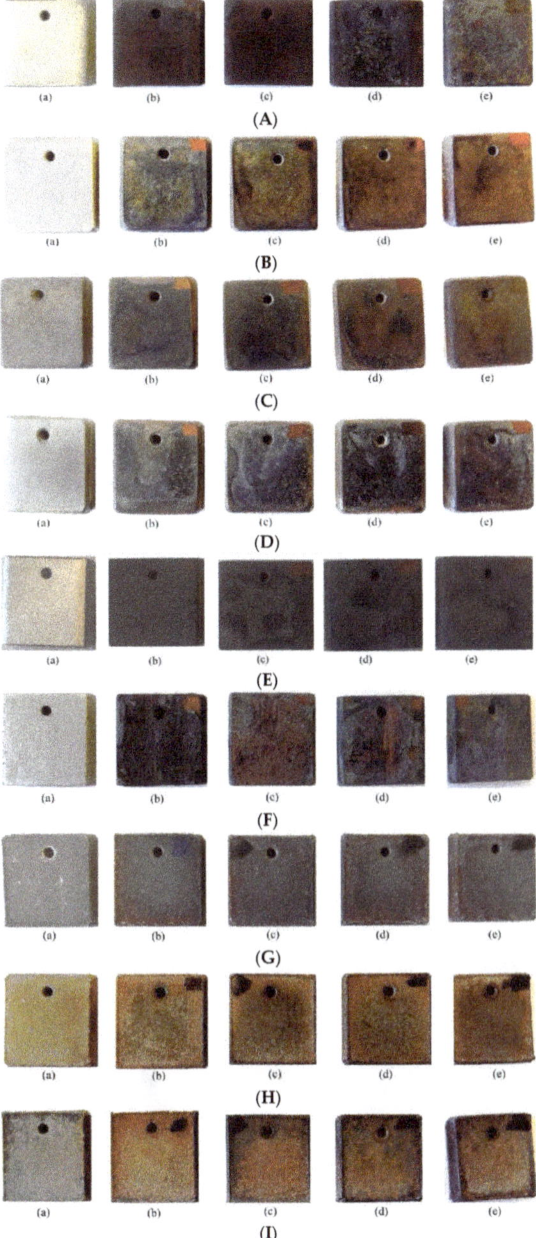

Figure A2. *Cont.*

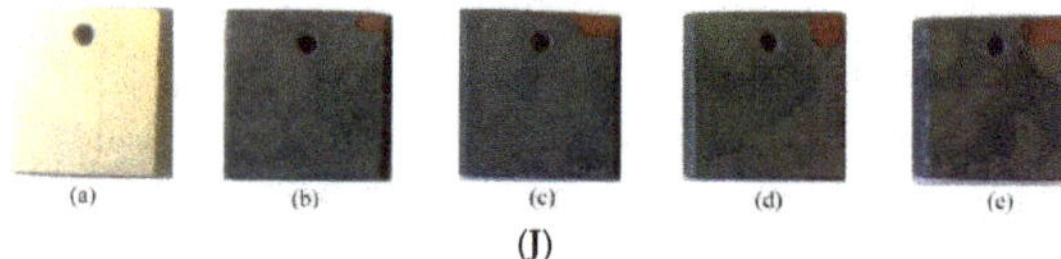

Figure A2. Photographs of the samples (**A**) 316-Bare, (**B**) 316-A, (**C**) 316-A-Sn, (**D**) 316-A-Z, (**E**) 316-Cr, (**F**) CS-Bare, (**G**) CS-A, (**H**) CS-A-Sn, (**I**) CS-A-Z, and (**J**) CS-Cr: (**a**) before corrosion testing; (**b**) after 24 h; (**c**) after 48 h; (**d**) after 96 h; and (**e**) after 168 h of exposure in high-temperature (at 600 °C) corrosive environment.

Appendix B.

The microstructure images of the samples before and after 168 h of exposure to the low-temperature corrosive attack are presented here.

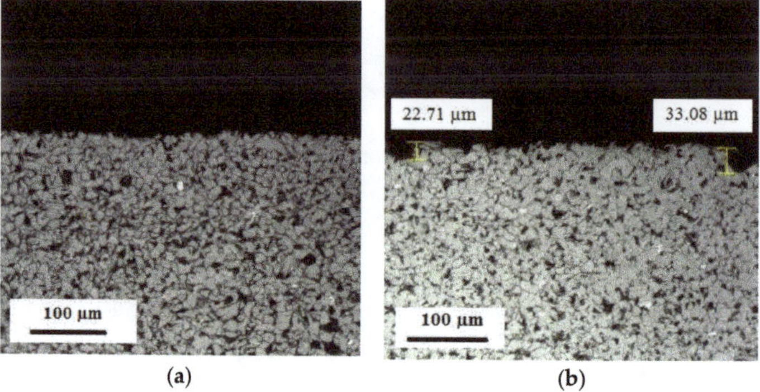

Figure A3. Microscopic images of the cross-section of CS-Bare (**a**) before and (**b**) after the low-temperature corrosive attack after 168 h.

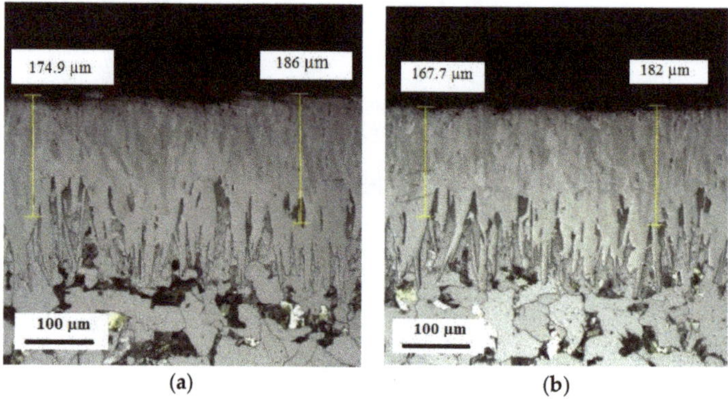

Figure A4. Microscopic images of the cross-section of CS-B (**a**) before and (**b**) after the low-temperature corrosive attack after 168 h.

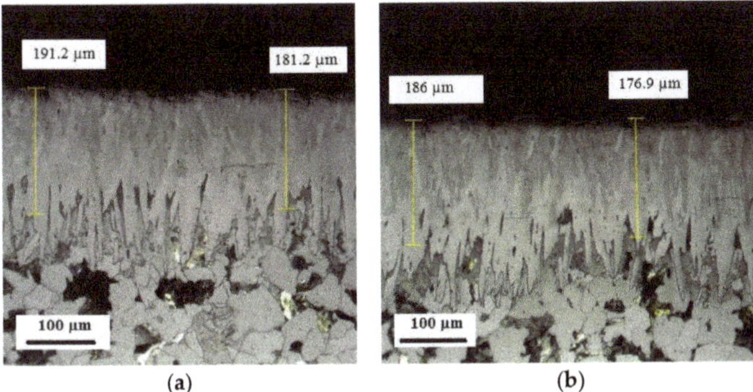

(a) (b)

Figure A5. Microscopic images of the cross-section of CS-B-BN (**a**) before and (**b**) after the low-temperature corrosive attack after 168 h.

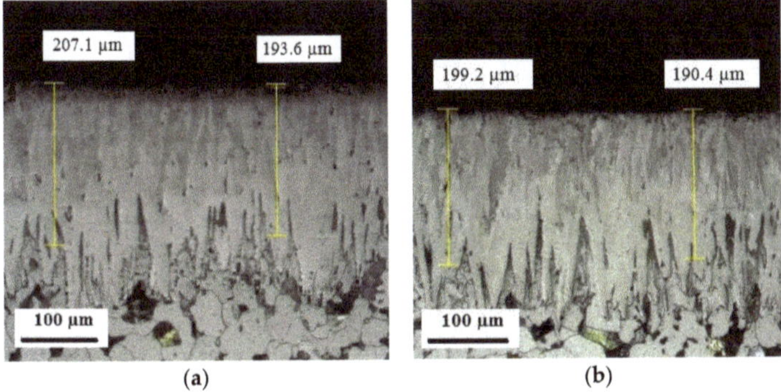

(a) (b)

Figure A6. Microscopic images of the cross-section of CS-B-Z (**a**) before and (**b**) after the low-temperature corrosive attack after 168 h.

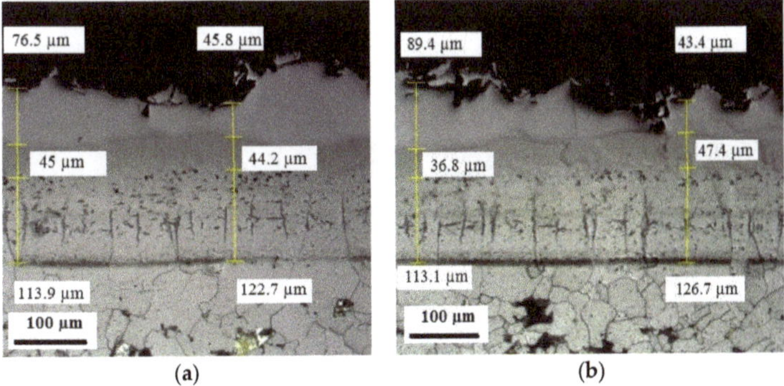

(a) (b)

Figure A7. Microscopic images of the cross-section of CS-A (**a**) before and (**b**) after the low-temperature corrosive attack after 168 h.

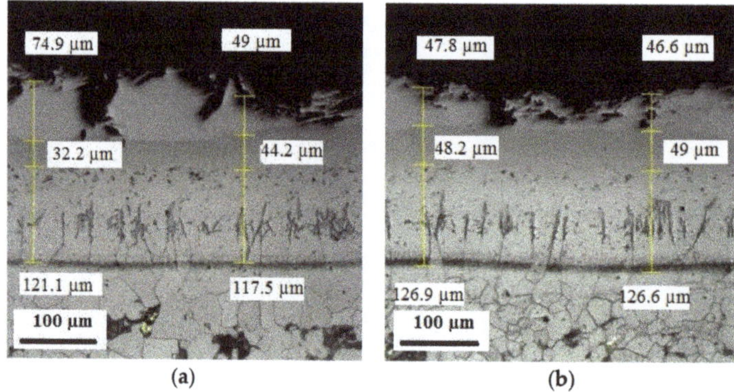

(a)   (b)

Figure A8. Microscopic images of the cross-section of CS-A-Sn (**a**) before and (**b**) after the low-temperature corrosive attack after 168 h.

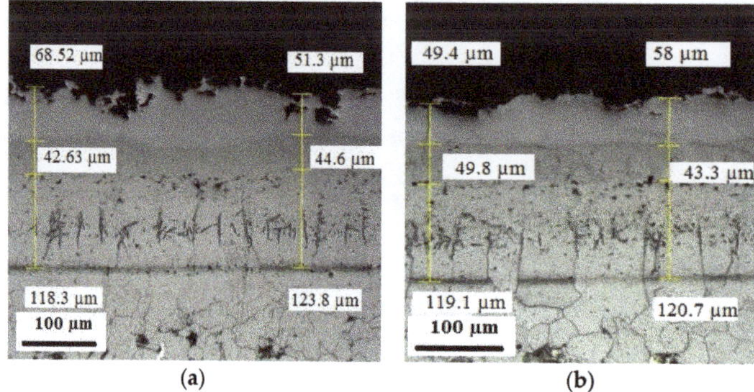

(a)   (b)

Figure A9. Microscopic images of the cross-section of CS-A-Z (**a**) before and (**b**) after the low-temperature corrosive attack after 168 h.

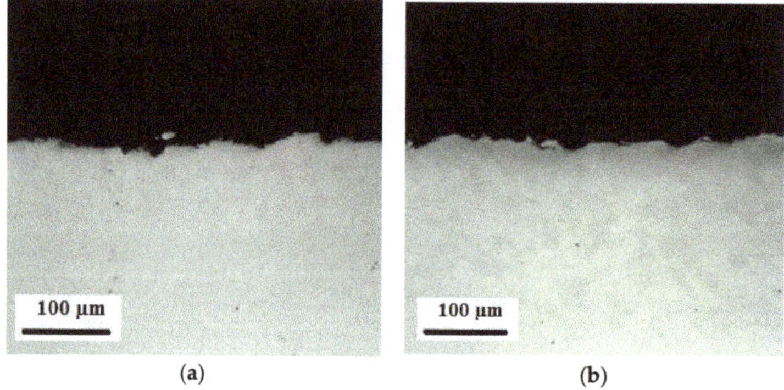

(a)   (b)

Figure A10. Microscopic images of the cross-section of 316-Bare (**a**) before and (**b**) after the low-temperature corrosive attack after 168 h.

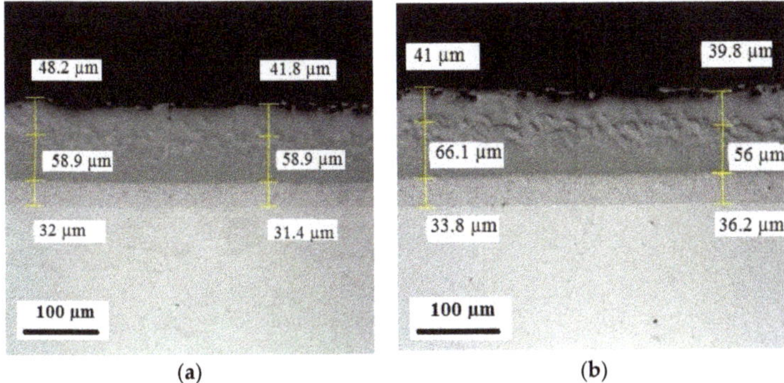

(a)   (b)

Figure A11. Microscopic images of the cross-section of 316-A (**a**) before and (**b**) after the low-temperature corrosive attack after 168 h.

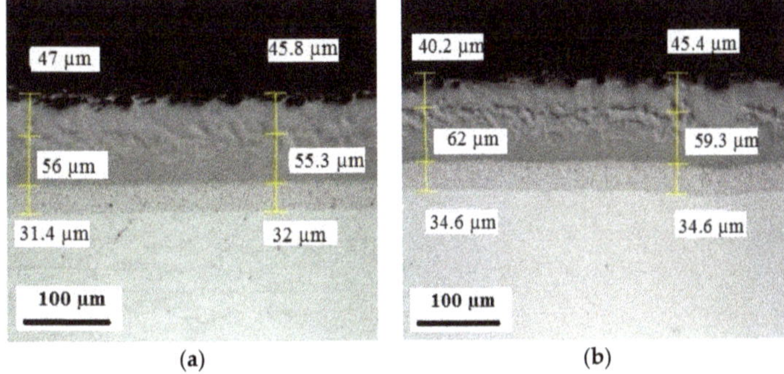

(a)   (b)

Figure A12. Microscopic images of the cross-section of 316-A-Sn (**a**) before and (**b**) after the low-temperature corrosive attack after 168 h.

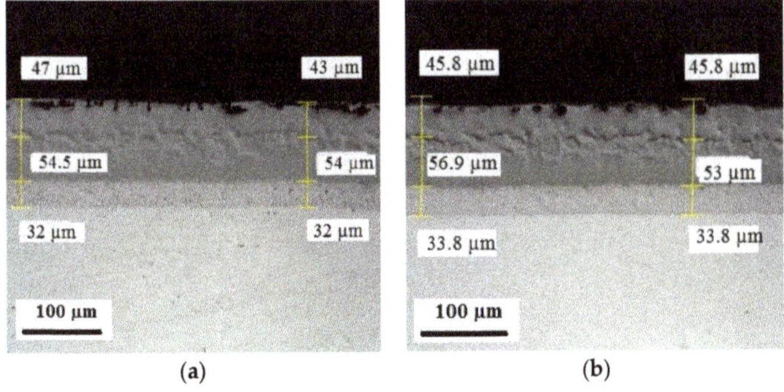

(a)   (b)

Figure A13. Microscopic images of the cross-section of 316-A-Z (**a**) before and (**b**) after the low-temperature corrosive attack after 168 h.

Appendix C.

The microstructure images of the samples before and after 168 h of exposure to the high-temperature corrosive attack are presented here.

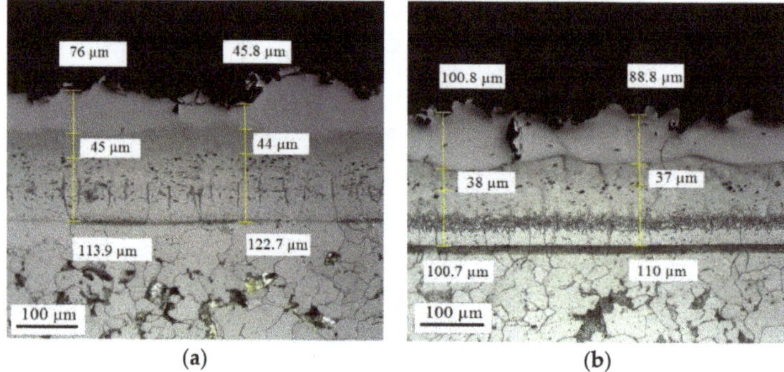

(a)   (b)

Figure A14. Microscopic images of the cross-section of CS-A (**a**) before and (**b**) after the high-temperature corrosive attack after 168 h.

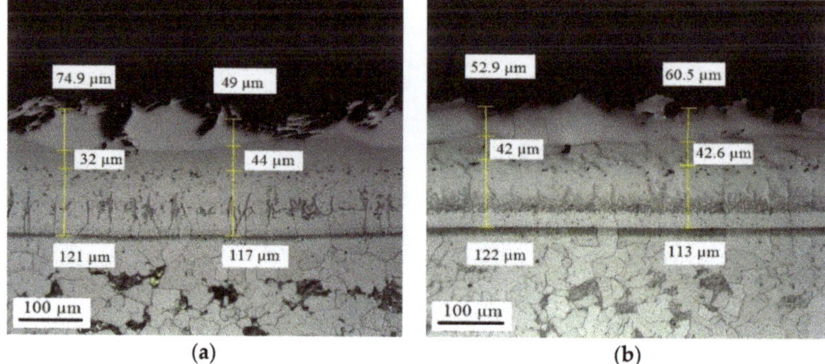

(a)   (b)

Figure A15. Microscopic images of the cross-section of CS-A-Sn (**a**) before and (**b**) after the high-temperature corrosive attack after 168 h.

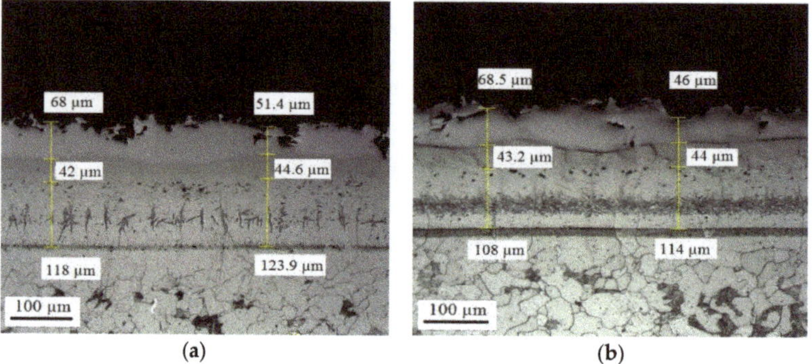

(a)   (b)

Figure A16. Microscopic images of the cross-section of CS-A-Z (**a**) before and (**b**) after the high-temperature corrosive attack after 168 h.

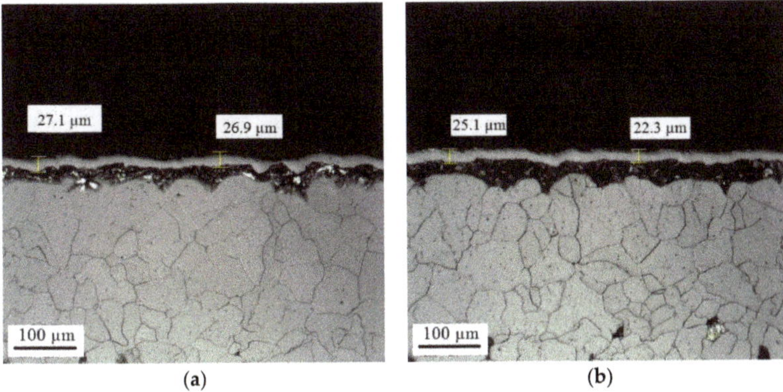

Figure A17. Microscopic images of the cross-section of CS-Cr (**a**) before and (**b**) after the high-temperature corrosive attack after 168 h.

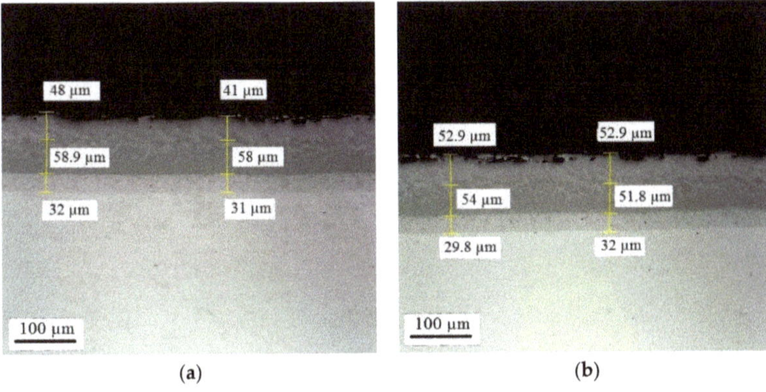

Figure A18. Microscopic images of the cross-section of 316-A (**a**) before and (**b**) after the high-temperature corrosive attack after 168 h.

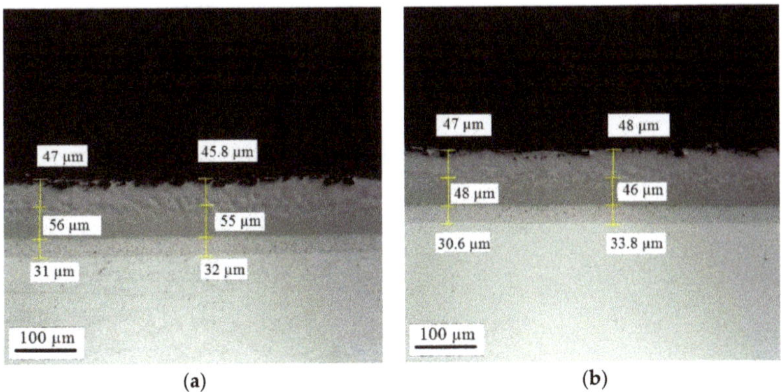

Figure A19. Microscopic images of the cross-section of 316-A-Sn (**a**) before and (**b**) after the high-temperature corrosive attack after 168 h.

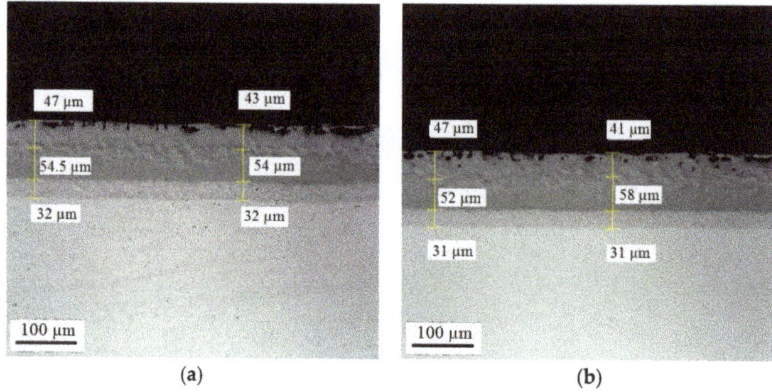

(a)   (b)

Figure A20. Microscopic images of the cross-section of 316-A-Z (**a**) before and (**b**) after the high-temperature corrosive attack after 168 h.

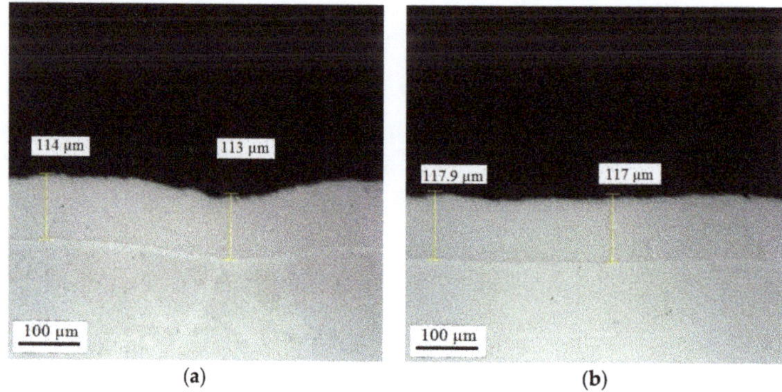

(a)   (b)

Figure A21. Microscopic images of the cross-section of 316-Cr (**a**) before and (**b**) after the high-temperature corrosive attack after 168 h.

Appendix D.

XRD profiles of the top surface of the samples before and after low- and high-temperature corrosive attack.

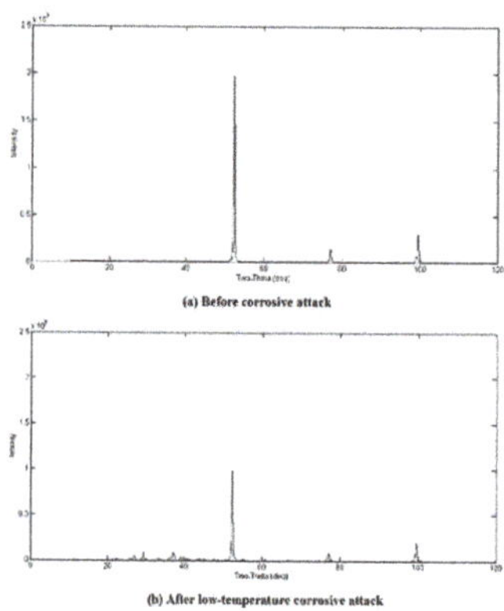

Figure A22. *Cont.*

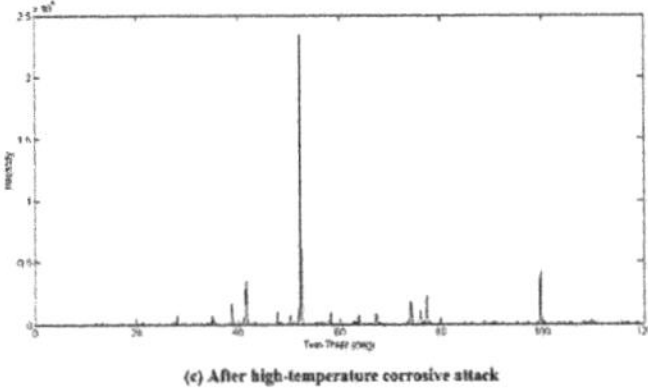

Figure A22. Qualitative XRD analysis of the surface of CS-Bare (**a**) before corrosive attack, (**b**) after low-, and (**c**) high-temperature corrosive attack.

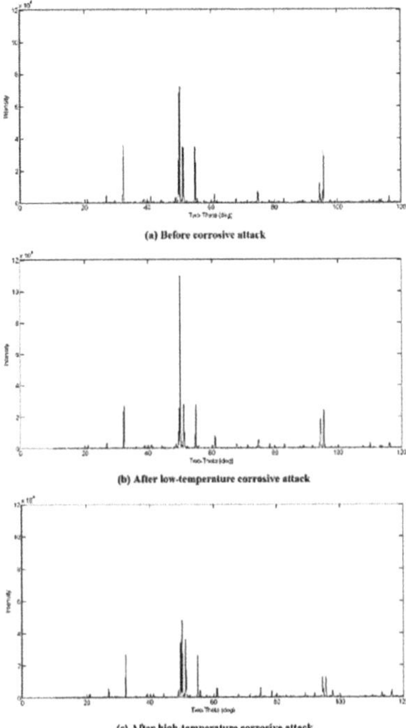

Figure A23. Qualitative XRD analysis of the surface of CS-A-Sn (**a**) before corrosive attack, (**b**) after low-, and (**c**) high-temperature corrosive attack.

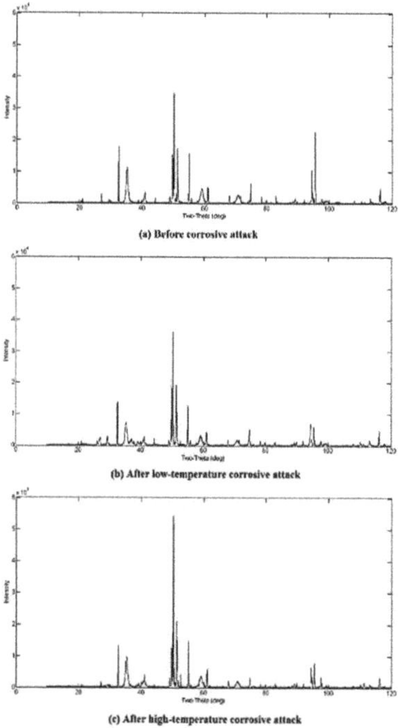

Figure A24. Qualitative XRD analysis of the surface of CS-A-Z (a) before corrosive attack, (b) after low-, and (c) high-temperature corrosive attack.

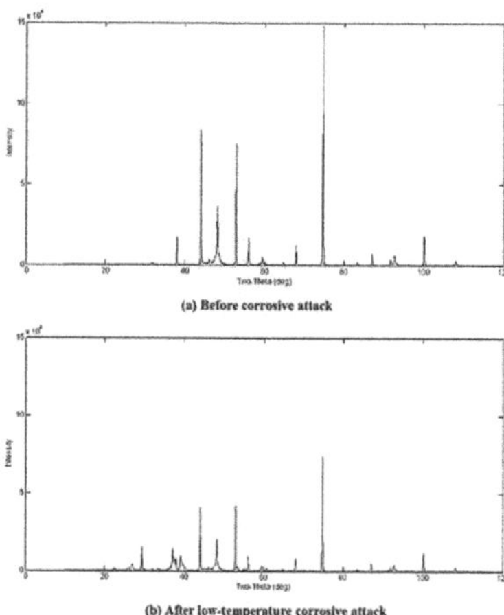

Figure A25. Qualitative XRD analysis of the surface of CS-B (a) before corrosive attack, (b) after low-temperature corrosive attack.

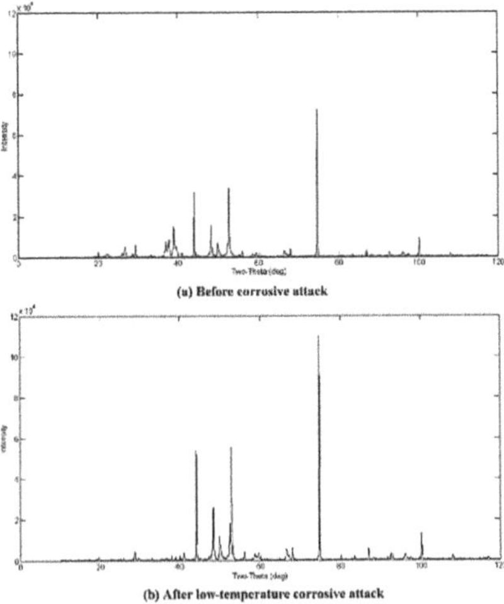

Figure A26. Qualitative XRD analysis of the surface of CS-B-BN (**a**) before corrosive attack, (**b**) after low-temperature corrosive attack.

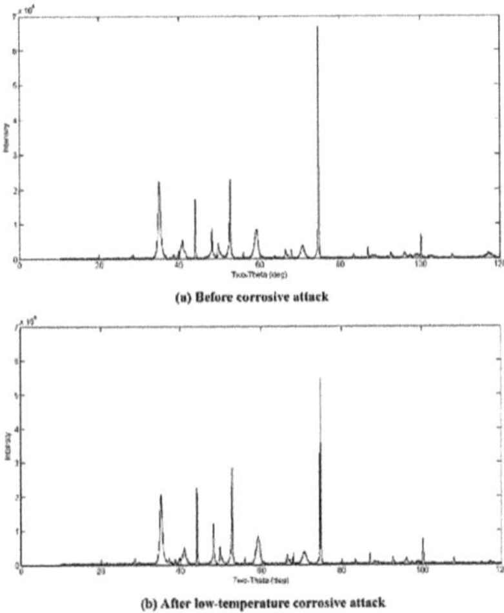

Figure A27. Qualitative XRD analysis of the surface of CS-B-Z (**a**) before corrosive attack, (**b**) after low-temperature corrosive attack.

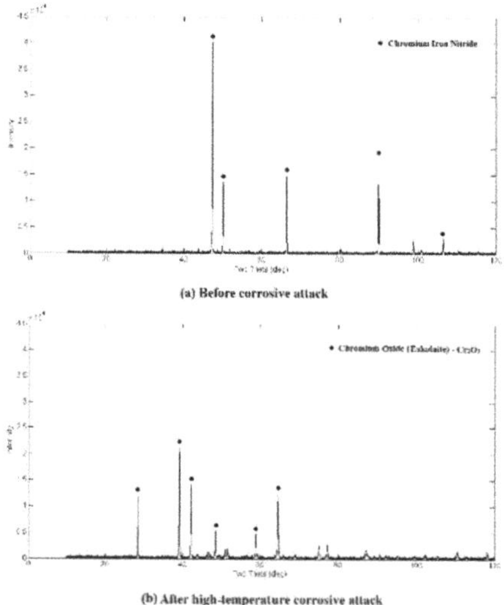

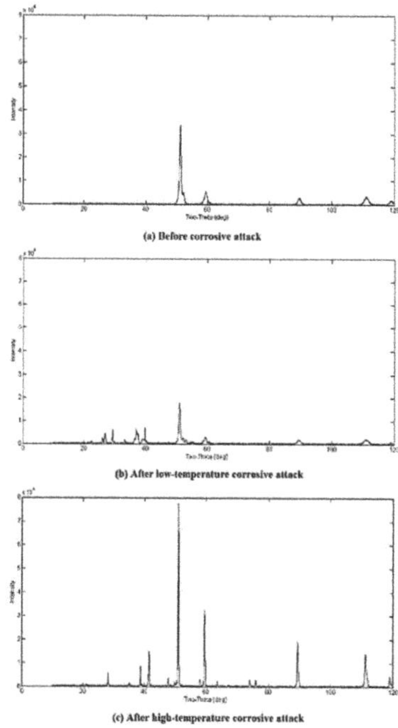

Figure A28. Qualitative XRD analysis of the surface of CS-Cr (**a**) before corrosive attack, (**b**) after high-temperature corrosive attack.

Figure A29. Qualitative XRD analysis of the surface of 316-Bare (**a**) before corrosive attack, (**b**) after low-, and (**c**) high-temperature corrosive attack.

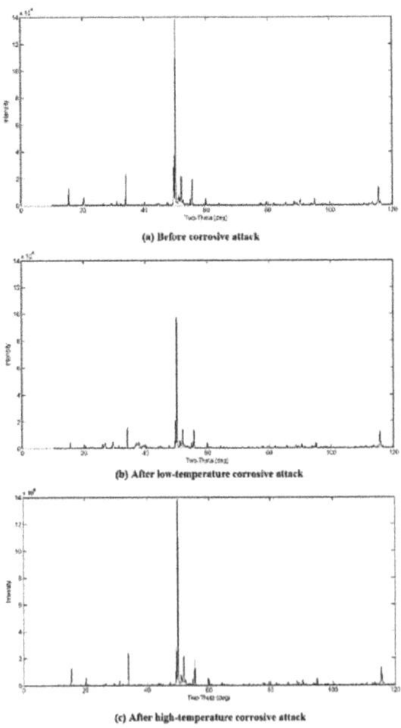

Figure A30. Qualitative XRD analysis of the surface of 316-A-Sn (**a**) before corrosive attack, (**b**) after low-, and (**c**) high-temperature corrosive attack.

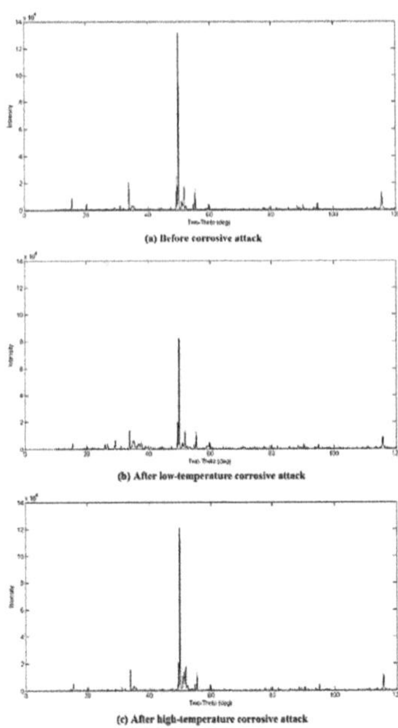

Figure A31. Qualitative XRD analysis of the surface of 316-A-Z (**a**) before corrosive attack, (**b**) after low-, and (**c**) high-temperature corrosive attack.

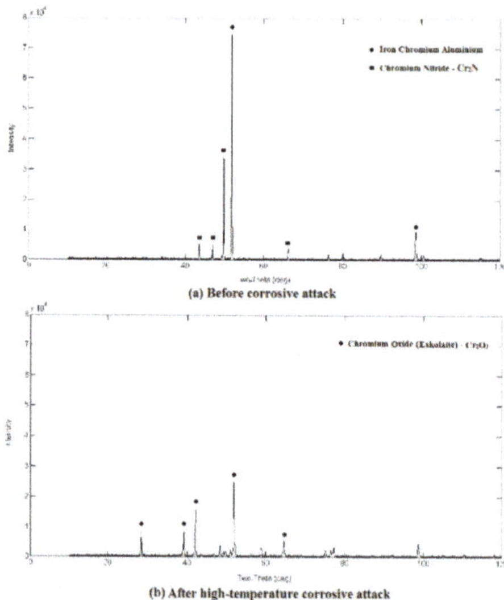

Figure A32. Qualitative XRD analysis of the surface of 316-Cr (**a**) before corrosive attack, (**b**) high-temperature corrosive attack.

References

1. Darmawan, A.; Hardi, F.; Yoshikawa, K.; Aziz, M.; Tokimatsu, K. Enhanced process integration of black liquor evaporation, gasification, and combined cycle. *Appl. Energy* **2017**, *204*, 1035–1042. [CrossRef]
2. Tran, H.N. Upper furnace deposition and plugging. In *Kraft Recovery Boilers*; Adams, T.N., Ed.; Tappi Press: Atlanta, GA, USA, 1997; pp. 245–282.
3. Tran, H.N.; Barham, D.; Hupa, M. Fireside corrosion in Kraft recovery boilers—An overview. *Mater. Preform.* **1988**, *27*, 40–45.
4. Naqvi, M.; Yan, J.; Dahlquist, E. Black liquor gasification integrated in pulp and paper mills: A critical review. *Bioresour. Technol.* **2010**, *101*, 8001–8015. [CrossRef] [PubMed]
5. Ji, X.; Bie, H.; Zhang, Y.; Chen, P.; Fang, W.; Bie, R. Release of K and Cl and emissions of NO_x and SO_2 during reed black liquor combustion in a fluidized bed. *Energy Fuels* **2017**, *31*, 1631–1637. [CrossRef]
6. Li, S.; Themelis, N.; Castaldi, M. High-temperature corrosion in waste-to-energy boilers. *J. Therm. Spray Technol.* **2007**, *16*, 104–110.
7. Riedl, R.; Dahl, J.; Obernberger, I.; Narodoslawsky, M. Corrosion in fire tube boilers of biomass combustion plants. In Proceedings of the China International Corrosion Control Conference, Beijing, China, 26–28 October 1999; p. 90129.
8. Grabke, H.J.; Spigel, M.; Zahs, A. Role of alloying elements and carbides in the chlorine-induced corrosion of steels and alloys. *Mater. Res.* **2004**, *7*, 89–95. [CrossRef]
9. Grabke, H.J.; Reese, E.; Spigel, M. The effects of chlorides, hydrogen chloride, and sulfur dioxide in the oxidation of steels below deposits. *Corros. Sci.* **1995**, *37*, 1023–1043. [CrossRef]
10. Sorell, G. The role of chlorine in high temperature corrosion in waste-to-energy plants. *Mater. High Temp.* **1997**, *14*, 207–220. [CrossRef]
11. Buscaglia, V.; Nanni, P.; Bottino, C. The mechanism of sodium sulphate-induced low temperature hot corrosion of pure iron. *Corros. Sci.* **1990**, *311*, 327–349. [CrossRef]
12. He, J.; Xiong, W.; Zhang, W.; Li, W.; Long, K. Study on the high-temperature corrosion behavior of superheater steels of biomass-fired boiler in molten alkali salts' mixtures. *Adv. Mech. Eng.* **2016**, *8*, 1687814016678163. [CrossRef]
13. Xu, S.; Wang, C.; Wang, W. Failure analysis of stress corrosion cracking in heat exchanger tubes during start-up operation. *Eng. Fail. Anal.* **2015**, *51*, 1–8. [CrossRef]

14. Barbosa, C.; De Barros, S.K.; De Cerqueira Abud, I.; Do Nascimento, J.L.; De Carvalho, S.S. Failure analysis of an aqua tubular boiler tube. *J. Fail. Anal. Prev.* **2012**, *12*, 654–659. [CrossRef]

15. Ahmad, J.; Purbolaksono, J.; Beng, L.C. Thermal fatigue and corrosion fatigue in heat recovery area wall side tubes. *Eng. Fail. Anal.* **2010**, *17*, 334–343. [CrossRef]

16. Roberge, P.R. *Handbook of Corrosion Engineering*, 2nd ed.; McGraw-Hill Education: New York, NY, USA, 2012; ISBN 9780071750370.

17. Levy, A.V. The erosion-corrosion behavior of protective coatings. *Surf. Coat. Technol.* **1988**, *36*, 387–406. [CrossRef]

18. Wang, D.; Bierwagen, G.P. Sol-gel coatings on metals for corrosion protection. *Prog. Org. Coat.* **2009**, *64*, 327–338. [CrossRef]

19. Cha, S.C.; Gudenau, H.W.; Bayer, G.T. Comparison of corrosion behaviour of thermal sprayed and diffusion-coated materials. *Mater. Corros.* **2002**, *53*, 195–205. [CrossRef]

20. Günthner, M.; Kraus, T.; Krenkel, W.; Motz, G.; Dierdorf, A.; Decker, D. Particle-filled PHPS silazane-based coatings on steel. *Appl. Ceram. Technol.* **2009**, *6*, 373–380. [CrossRef]

21. Movchan, B.A.; Yu, Y.K. High-temperature protective coatings produced by EB-PVD. *J. Coat. Sci. Technol.* **2014**, *1*, 96–110.

22. Drozdz, M.; Kyziol, K.; Grzesik, Z. Chromium-based oxidation-resistant coatings for the protection of engine valves in automotive vehicles. *Mater. Technol.* **2017**, *51*, 603–607.

23. Panjan, P.; Drnovšek, A.; Kovač, J. Tribological aspects related to the morphology of PVD hard coatings. *Surf. Coat. Technol.* **2018**, *343*, 138–147. [CrossRef]

24. Chicatun, F.; Cho, J.; Schaab, S.; Brusatin, G.; Colombo, P.; Roather, J.A.; Boccaccini, A.R. Carbon nanotube deposits and CNT/SiO_2 composite coatings by electrophoretic deposition. *Adv. Appl. Ceram.* **2007**, *106*, 186–195. [CrossRef]

25. Musil, J.; Vlcek, J.; Zeman, P. Hard amorphous nanocomposite coatings with oxidation resistance above 1000 °C. *Adv. Appl. Ceram.* **2008**, *107*, 149–154. [CrossRef]

26. Tuthill, A.H. *Stainless Steels and Specialty Alloys for Modern Pulp and Paper Mills*; Nickel Development Institute (NiDI): Toronto, ON, Canada, 2002; pp. 47–50.

27. Rao, S.; Frederick, L.; McDonald, A. Resistance of nanostructured environmental barrier coatings to the movement of molten salts. *J. Therm. Spray Technol.* **2012**, *21*, 887–899. [CrossRef]

28. Heath, G.R.; Heimgartner, P.; Irons, G.; Miller, R.; Gustafsson, S. An assessment of thermal spray coating technologies for high temperature corrosion protection. *Mater. Sci. Forum* **1997**, *251–254*, 809–816. [CrossRef]

29. Fauchais, P.; Montavon, G. Thermal and cold spray: Recent developments. *Key Eng. Mater.* **2008**, *384*, 1–59. [CrossRef]

30. Davis, J. *Handbook of Thermal Spray Technology*; ASM International: Materials Park, OH, USA, 2004; pp. 1–36.

31. Nelson, G.; Nychka, J.; McDonald, A.G. Flame spray deposition of titanium alloy-bio-active glass composite coatings. *J. Therm. Spray Technol.* **2011**, *20*, 1339–1351. [CrossRef]

32. Davis, J. *Surface Engineering for Corrosion and Wear Resistance*, 1st ed.; ASM International: Materials Park, OH, USA, 2001; pp. 1–43.

33. Dearnley, P. Surface engineering with diffusion technologies. In *Introduction to Surface Engineering*; Cambridge University Press: Cambridge, UK, 2017; pp. 35–115.

34. Mittemeijer, E.J.; Somers, M.A.J. *Thermochemical Surface Engineering of Steels*, 1st ed.; Elsevier-Woodhead Publishing: Cambridge, UK, 2014.

35. Choy, K.L. Chemical vapor deposition of coatings. *Prog. Mater. Sci.* **2003**, *48*, 57–170. [CrossRef]

36. Budinski, K.G. *Surface Engineering for Wear Resistance*; Prentice Hall: Englewood Cliffs, NJ, USA, 1988.

37. Takadoum, J. *Materials and Surface Engineering in Tribology*; Wiley: London, UK, 2010.

38. Liu, X.; Wang, H.; Li, D.; Wu, Y. Study on kinetics of carbide growth by thermal diffusion Process. *Surf. Coat. Technol.* **2006**, *201*, 2414–2418. [CrossRef]

39. Nicholls, J.R. Designing oxidation-resistant coatings. *JOM* **2000**, *52*, 28–35. [CrossRef]

40. Medvedovski, E. Formation of corrosion-resistant thermal diffusion boride coatings. *Adv. Eng. Mater.* **2016**, *18*, 11–33. [CrossRef]

41. Medvedovski, E.; Chinski, F.; Stewart, J. Wear- and corrosion-resistant boride-based coatings obtained through thermal diffusion CVD processing. *Adv. Eng. Mater.* **2014**, *16*, 713–728. [CrossRef]

42. Medvedovski, E.; Jiang, J.; Robertson, M. Iron boride-based thermal diffusion coatings for tribo-corrosion oil production applications. *Ceram. Int.* **2016**, *42*, 3190–3211. [CrossRef]
43. Bangaru, N.V.; Krutenat, R.C. Diffusion coatings of steels: Formation mechanism and microstructure of aluminized heat-resistant stainless steels. *J. Vac. Sci. Technol. B* **1984**, *2*, 806–815. [CrossRef]
44. Fitzer, E.; Maurer, H.J. Diffusion and precipitation phenomena in aluminized and chromium-aluminized iron- and nickel-base alloys in materials and coatings to resist high temperature corrosion. In *Materials and Coatings to Resist High Temperature Corrosion*; Holmes, D.R., Rahmel, A., Eds.; Applied Science Publishers Ltd.: London, UK, 1978; pp. 253–269.
45. Wang, D. Corrosion behavior of chromized and/or aluminized 214Cr-1Mo steel in medium—BTU coal gasifier environments. *Surf. Coat. Technol.* **1988**, *36*, 49–60. [CrossRef]
46. Bai, C.; Ger, M.; Wu, M. Corrosion behaviors and contact resistances of the low-carbon steel bipolar plate with a chromized coating containing carbides and nitrides. *Int. J. Hydrogen Energy* **2009**, *34*, 6778–6789. [CrossRef]
47. Meier, G.H.; Cheng, C.; Perlkins, R.A.; Bakker, W. Diffusion chromizing of ferrous alloys. *Surf. Coat. Technol.* **1989**, *39–40*, 53–64. [CrossRef]
48. Cheetham, A.K.; Day, P. *Solid State Chemistry: Techniques*; Oxford Science Publications: Oxford, UK, 1991.
49. Donald, H.; Jenkins, B. Thermodynamics of the relationship between lattice energy and lattice enthalpy. *J. Chem. Educ.* **2005**, *82*, 950–952.
50. Ladd, M. *Crystal Structures: Lattices and Solids in Stereoview*; Horwood Series in Chemical Science; Elsevier: Chichester, UK, 1999.
51. Grabke, H.J.; Schutze, M. *Oxidation of Intermetallics*; WILEY-VCH Verlag GmbH: Berlin, Germany, 1998.
52. Callister, W.; Rethwisch, D. *Fundamentals of Material Science and Engineering: An Integrated Approach*, 5th ed.; John Wiley & Sons, Inc.: Marblehead, MA, USA, 2011; pp. 170–199.
53. Shrier, L.; Burstein, G.; Jarman, R. *Corrosion: Metal/Environment Reactions*, 3rd ed.; Butterworth-Heinemann, Elsevier: Oxford, UK, 1994.
54. Ehrburger, P.; Baranne, P.; Lahaye, J. Inhibition of the oxidation of carbon-carbon composite by boron oxide. *Carbon* **1986**, *24*, 495–499. [CrossRef]
55. Fehlner, F.P. *Low Temperature Oxidation, the Role of Vitreous Oxides*; John Wiley & Sons, Inc.: Hoboken, NJ, USA, 1986.
56. Seitz, F. On the porosity observed in the Kirkendal effect. *Acta Metall.* **1953**, *1*, 355–369. [CrossRef]
57. Melendez, M.; McDonald, A. Development of WC-based metal matrix composite coatings using low-pressure cold gas dynamic spraying. *Surf. Coat. Technol.* **2013**, *214*, 101–109. [CrossRef]
58. Pint, B.A.; Zhang, Y.; Tortorelli, P.F.; Haynes, J.A.; Wright, I.G. Evaluation of iron-aluminide CVD coatings for high temperature corrosion protection. *Mater. High Temp.* **2001**, *18*, 185–192. [CrossRef]
59. Agüero, A.; Muelas, R.; Pastor, A.; Osgerby, S. Long exposure steam oxidation testing and mechanical properties of slurry aluminide coatings for steam turbine components. *Surf. Coat. Technol.* **2005**, *200*, 1219–1224. [CrossRef]
60. Pérez, F.J.; Hierro, M.P.; Pedraza, F.; Carpintero, M.C.; Gómez, C.; Tarín, R. Effect of fluidized bed CVD aluminide coatings on the cyclic oxidation of austenitic AISI 304 stainless steel. *Surf. Coat. Technol.* **2001**, *145*, 1–7. [CrossRef]
61. Busby, J.; Hash, M.; Was, G. The relationship between hardness and yield stress in irradiated austenitic and ferritic steels. *J. Nucl. Mater.* **2005**, *336*, 267–278. [CrossRef]
62. Bahadur, A.; Mohanty, O. Structural studies of hot dip aluminized coatings on mild steel. *Mater. Trans.* **1991**, *32*, 1053–1061. [CrossRef]

Article

Effect of Initial Surface Roughness on Cavitation Erosion Resistance of Arc-Sprayed Fe-Based Amorphous/Nanocrystalline Coatings

Jinran Lin [1,2,3,*], **Zehua Wang** [2], **Jiangbo Cheng** [2], **Min Kang** [1], **Xiuqing Fu** [1] **and Sheng Hong** [2,4]

[1] College of Engineering, Nanjing Agricultural University, Nanjing 210031, China; kangmin@njau.edu.cn (M.K.); fuxiuqing@njau.edu.cn (X.F.)
[2] College of Mechanics and Materials, Hohai University, Nanjing 211100, China; zhwang@hhu.edu.cn (Z.W.); jiangbochenghhu@hotmail.com (J.C.); hongsheng1988@126.com (S.H.)
[3] Jiangsu Jinxiang Transmission Equipment Co., Ltd., Huaian 223001, China
[4] Material Corrosion and Protection Key Laboratory of Sichuan Province, Zigong 643000, China
* Correspondence: ljr-8@163.com; Tel.: +86-25-5860-6580

Academic Editor: Shiladitya Paul
Received: 23 September 2017; Accepted: 1 November 2017; Published: 14 November 2017

Abstract: The arc spraying process was used to prepare Fe-based amorphous/nanocrystalline coating. The cavitation erosion behaviors of FeNiCrBSiNbW coatings with different surface roughness levels were investigated in distilled water. The results showed that FeNiCrBSiNbW coating adhered well to the substrate, and was compact with porosity of less than 2%. With increasing initial surface roughness, the coatings showed an increase in mass loss of cavitation erosion damage. The amount of pre-existing defects on the initial surface of the coatings was found to be a significant factor for the difference in the cavitation erosion behavior. The cavitation erosion damage for the coatings was a brittle erosion mode. The evolution of the cavitation erosion mechanism of the coatings with the increase of the initial surface roughness was micro-cracks, pits, detachment of fragments, craters, cracks, pullout of the un-melted particle, and massive exfoliations.

Keywords: cavitation; roughness; coatings; amorphous/nanocrystalline; Fe-based

---

1. Introduction

Cavitation erosion, often reported as a common phenomenon in the overflowing components of hydraulic machinery, is related to two main aspects of hydrodynamic and material [1,2]. On the one hand, hydrodynamic cavitation takes place when vapor bubbles in a liquid are exposed to a sudden fluctuation of localized pressure, which can cause repeated nucleation, growth, violent collapse of bubbles, and send micro-jet or shock wave through the liquid to the solid surfaces [3]. Thus, surface roughness is capable of controlling cavitation erosion by affecting nuclei concentration, turbulence level and surface pressure distribution [4,5]. On the other hand, it is important to prepare a high performance coating to reduce or avoid cavitation erosion damage by selecting a suitable surface treatment technique, since cavitation erosion usually happens on a material surface of the flow-handling components such as hydraulic turbines, offshore/mining machineries, valves, and ship propellers [6–8].

Fe-based amorphous/nanocrystalline coatings prepared by thermal spraying have been widely adopted by hydraulic machinery, power plants and coastal installations because of their desirable combination of relatively low material cost, high hardness and toughness, and outstanding corrosion and wear resistance [9–12]. In recent years, considerable efforts have been devoted to investigate the cavitation erosion behavior of Fe-based amorphous/nanocrystalline coatings [13–21]. According to [13,18], the high-velocity oxygen-fuel (HVOF) sprayed Fe-based amorphous/nanocrystalline coatings with dense structure and high microhardness were preferable to improve the resistance of cavitation erosion in deionized water. In addition, both porosity and microhardness affected the cavitation erosion resistance of HVOF sprayed coatings [18–20]. The influence of the HVOF spray parameters has also been investigated in relation to the microstructure and hardness of the coatings and, in turn, the cavitation erosion resistance [21]. Compared with HVOF spraying, arc spraying has the potential to develop high quality Fe-based amorphous/nanocrystalline coatings with lower costs because of its advantages of simple device, flexible operation and high efficiency [22]. In our earlier studies, FeCrBSiNb(Ni/W) amorphous/nanocrystalline coatings were synthesized successfully using the arc spraying process and their cavitation erosion behavior was reported [16,17]. We demonstrated that the cavitation erosion resistance of arc-sprayed Fe-based amorphous/nanocrystalline coating deteriorated with increasing annealing temperature [11]. Previous investigations mainly described in detail the relationship among microstructure, mechanical properties and cavitation erosion resistance of the Fe-based amorphous/nanocrystalline coatings. However, the role that initial surface roughness plays in cavitation erosion process of arc-sprayed Fe-based amorphous/nanocrystalline coatings is still not clear. In particular, thermal sprayed coating possessed of non-uniform microstructure and its top layers close to surface showed higher porosity than the underneath layers, which was different from bulk metal.

The aim of this study was to discover the effect of initial surface roughness on cavitation erosion behavior of FeNiCrBSiNbW amorphous/nanocrystalline coating prepared by arc spraying process. Moreover, the mechanisms controlling cavitation erosion of the coatings with different initial surface roughness were discussed.

2. Experimental Procedure

A self-designed Fe-based cored wire (Fe-Ni-Cr-B-Si-Nb-W) with a diameter of 2 mm was used as the feedstock. The chemical composition of the cored wire was the same in Ref [17]. Stainless steel 1Cr18Ni9Ti was selected as the substrate. Prior to coating, the substrate samples were pre-cleaned in acetone, dried in hot air, and then grit blasted with 16 mesh alumina to provide a fresh and rough surface for better adhesion by removing all rust and oxide skin. The substrate samples were cooled with compressed air jets during and after spraying. The arc spraying of the cored wire was carried out at a spraying voltage of 36 V, a spraying current of 120 A, a wire feed rate of 2.7 m·min^{-1}, a compressed air pressure of 700 kPa, a stand-off distance of 200 mm, and a gun traverse speed of 100 mm·s^{-1}. The thickness of tested coating was about 230 μm. Then, the coating samples were wire cut for microstructural characterization and cavitation erosion testing.

The microstructures of the coating were observed using a scanning electron microscope (SEM, Hitachi S-3400 N, Tokyo, Japan) equipped with an energy dispersive spectroscopy (EDS, EX250). Selected area electron diffraction (SAED) pattern and finer-scale microstructural characterization of the coating were performed using a transmission electron microscope (TEM, JEOL JEM-2010, Tokyo, Japan). Porosity was measured on SEM images with the help of image analysis. An average value was taken from 20 view-fields from different locations on polished cross section of the coating at a magnification of 500.

The cavitation erosion experiments of the coatings were carried out using a vibratory cavitation apparatus according to the ASTM G32-10 standard [23]. The operating parameters of the cavitation test were as follows: Frequency of vibration 19 ± 1 kHz, peak-to-peak amplitude 60 ± 5 μm, power of ultrasonic generator 250 W. The details of the cavitation erosion test apparatus and its specimen's dimension have been reported in Ref. [16]. In this work, three surface roughness levels of the coating samples were prepared by manual grinding using 80, 600 and 1000 grit silicon carbide papers, which cause a large difference in the initial surface roughness and magnify its effect on the cavitation erosion behavior. Moreover, the initial surface roughness values of the coating samples ground using 600 and 1000 grit silicon carbide papers were close to the surface roughness of steam turbine blades after surface finish [24]. Prior to the test, the coating samples with an average initial surface roughness values (R_a) of 0.89, 0.32 and 0.2 μm were cleaned with acetone in an ultrasonic bath, dried in hot air, and weighed by an analytical balance with an accuracy of 0.1 mg. Then, the screw specimen with the Fe-based amorphous/nanocrystalline coating on it was attached to the free end of the horn and immersed about 3 mm in distilled water. The beaker was surrounded by flowing cool water to keep the distilled water inside it at $25 \pm 5\,^\circ$C. After each test period, the specimen was removed from the tip, cleaned with acetone, dried in hot air and weighted. The eroded surface morphologies of the coatings were characterized by SEM. Each test was repeated thrice to ensure the reproducibility and validity of the experiment results.

3. Results and Discussion

3.1. Characterization of the Coating

Figure 1 shows the cross-sectional microstructure of the FeNiCrBSiNbW amorphous/nanocrystalline coating deposited by arc spraying. From the overall view of the coating in Figure 1a, it can be seen that the coating has a dense and typical lamellar structure with an average thickness of 230 μm. The whole coating exhibits an apparent good adherence to the substrate with presence of a uniform and compact interface located between the substrate and the coating. It is notable in Figure 1b that some pores, un-melted particles and microcracks are observed in the coating. The average porosity value of the coating is less than 2% by using image analysis. Besides the pores, appearing as black regions, the coating consists of bright white region, grey region and dark grey region. The grey region is primarily coating alloy with the chemical composition of $Fe_{71}Ni_5Cr_{15}B_3Si_3Nb_2W_1$ (at.%). The bright white region and the dark grey region are W-rich phase and iron oxide phase respectively, which is in accordance with the results observed in previous work [25].

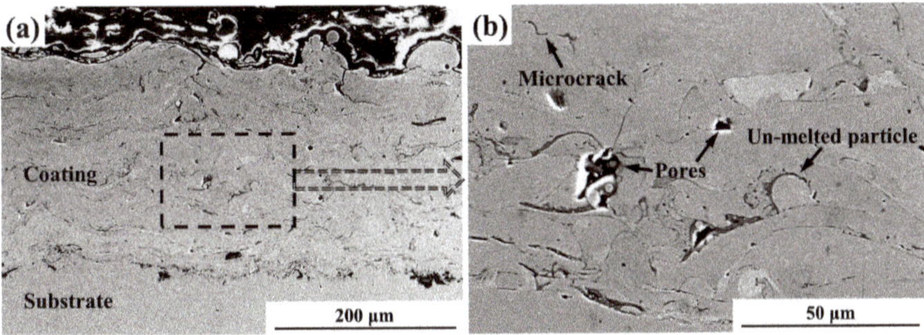

Figure 1. Scanning electron microscope (SEM) images of a transverse section of the as-sprayed coating: (**a**) an overall view morphology, and (**b**) a magnification of the rectangular frame in Figure 1a.

To get insight into the detailed microstructural information of the FeNiCrBSiNbW amorphous/nanocrystalline coating, TEM was utilized. Figure 2 shows a typical bright field TEM image of the coating, indicating the coexistence of amorphous phase and nanocrystalline grains. The nanocrystalline grains with dimensions ranging from 70 nm to 130 nm are uniformly distributed in amorphous matrix. This is confirmed by the diffraction spots, and the diffused halo rings in the selected area electron diffraction (SAED) pattern, as shown in the inset of Figure 2. The presence of amorphous phase is due to both the multicomponent alloy system with large glass formation ability and the high cooling rates of the in-flight particles. The formation of nanocrystalline grains is attributed to crystallization of the original amorphous region due to the preferred oxidation on the in-flight particles surface and the thermal fluctuation leading to localized heating during successive spraying [26,27].

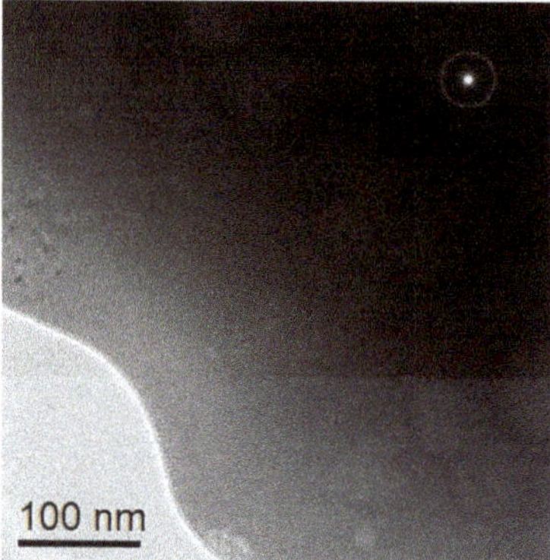

Figure 2. Transmission electron microscope (TEM) images of typical microstructure of the as-sprayed coating: Coexistence of amorphous phase and nanocrystalline grains.

3.2. Cavitation Erosion Behavior of the Coatings

Figure 3 shows the relationship between cumulative mass loss and cavitation erosion time for the FeNiCrBSiNbW amorphous/nanocrystalline coatings with three surface roughness levels. The results show that there is a considerable difference in mass loss rate of the as-sprayed coatings under three different surface conditions (i.e., 80, 600, and 1000 grit grinding), although the mass losses of all three coating specimens increase with increasing the test time. The incubation period is not observed for all three coating specimens, which reflects that the surface conditions in this study put the coatings one step ahead into the cavitation erosion damage process. The coating after 80 grit grinding exhibits the greater mass loss (21.8 mg) while the coating after 1000 grit grinding has the lower mass loss (10.3 mg) after 120 min test. In addition, the mass loss data of the coating after 600 grit grinding are very close to that of the coating after 1000 grit grinding during the first 30 min, but gradually higher than those of the coating after 1000 grit grinding when the test time exceeds 30 min. This may be because that a higher surface roughness level is helpful to increase the density of bubble nucleation near the surface of the coating specimen [28,29], which causes more impacts of the collapsing bubble on the surface and then much more mass loss of the coating.

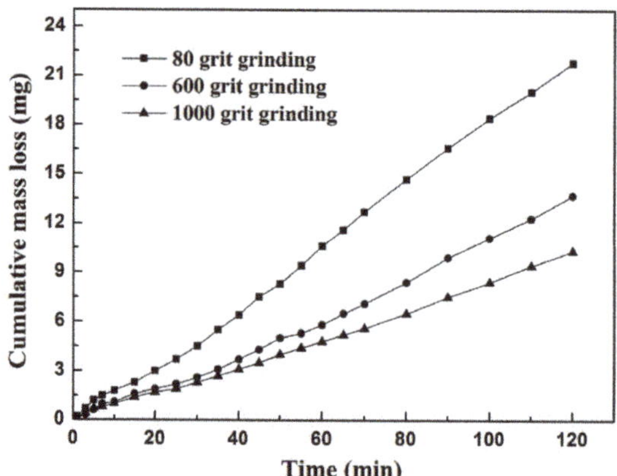

Figure 3. The relationship between cumulative mass loss and cavitation erosion time for the as-sprayed coatings with three surface roughness levels.

Figure 4 shows the initial surface morphological features of the FeNiCrBSiNbW amorphous/nanocrystalline coatings under three different surface conditions before the cavitation erosion test. As shown in Figure 4a, many scratches, pits and asperities are observed on the surface of the coating after 80 grit grinding. In Figure 4b, it can be noticed that the surface of the coating after 600 grit grinding is relatively smooth with some small scratches and a small number of irregular pits. The surface of the coating after 1000 grit grinding (Figure 4c) contains tiny scratches and limited pits compared to Figure 4a,b. This is consistent with the surface roughness result, where the Ra of the coating after 1000 grit grinding (0.2 µm) is about 22.9% and 62.5% that of the coatings after 80 grit grinding (0.89 µm) and 600 grit grinding (0.32 µm), respectively. Pre-existing irregular defects such as scratches, pits and asperities those resulted from surface preparation could act as raisers for stresses induced due to the direct impact and the lateral outflow of the micro-jets during successive cavitation process [30], which may suggest that the initial surface roughness plays a great role on the cavitation erosion damage of the coatings. This is verified by the SEM images of eroded surfaces as shown in Figure 5. Other than the initial surface roughness of the coating, the microstructure and phase composition were also important characteristics in determining the cavitation erosion resistance of the coatings. According to our previous study [11], the formation of oxides on the coating surface and between intersplats and the reduction of amorphous phase content would contribute to the decrease of the cavitation erosion resistance. The negative effect of the coating porosity on the cavitation erosion resistance has been indicated by other researchers [13,31].

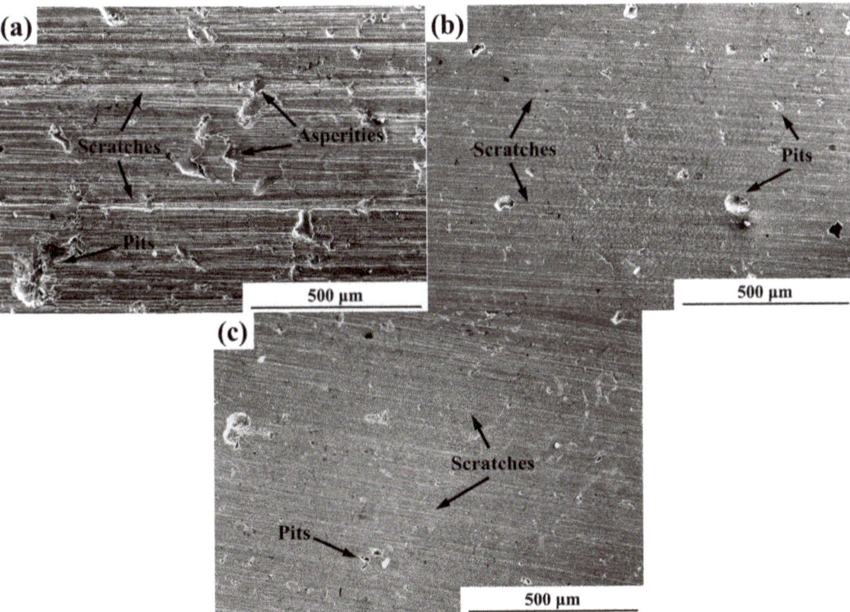

Figure 4. SEM images of the surface morphology of as-sprayed coatings under three surface conditions before the cavitation erosion test: (**a**) 80 grit grinding; (**b**) 600 grit grinding; and (**c**) 1000 grit grinding.

Figure 5 shows typical morphological features from the eroded surfaces of the FeNiCrBSiNbW amorphous/nanocrystalline coatings under three different surface conditions after cavitation erosion for 120 min in distilled water. As shown in Figure 5a, the surface of the coating after 80 grit grinding is roughened with a 0.5 mm crater in the center and numerous massive exfoliations of material in a widespread dispersion on the eroded surface, which give contribution to the overall higher mass loss of the coating. This may be associated with the combination of the initial pits that connected by cracks. Besides, a few initial scratches due to grinding are also observed on the eroded surface. Figure 5b shows a magnified micrograph of the crater with diameter of about 0.5 mm in Figure 5a, where pullout of the un-melted particle, cracks between intersplats, and cleavage fracture with river pattern are detected. These results reveal that the cavitation erosion damage for the coating is a brittle erosion mode. In Figure 5c, it can be noticed that the eroded surface of the coating after 600 grit grinding is relatively smooth with some initial scratches due to grinding. The crater is much shallower compared to Figure 5a, although the area of the crater is relatively larger. Further observation shows that micro-cracks between intersplat and un-melted particle, detachment of fragments, and small amount of pits are recorded on the eroded surface, as shown in Figure 5d. This indicates that the ground surface with lower surface roughness level effectively inhibits the propagation of cracks, hinders the pullout of the un-melted particle, and then delays the cavitation erosion progress of the coating. Figure 5e shows that the eroded surface of the coating after 1000 grit grinding is relatively uniform, including large cracks and number of craters with different shapes. There are also some small pits, micro-cracks, cleavage fracture with river pattern, layer detachment, and a large number of initial tiny scratches and pits due to grinding on the eroded surface (Figure 5f), indicating that it needs more time to roughen the surface and create initiate preferential cavitation erosion initiation sites under the present surface condition. The characteristic of layer detachment on the eroded surface of the coating after 1000 grit grinding reveals that the coating was destroyed mainly in the form of delamination, which was proven in our previous study [17]. From the above results, it may suggest that micro-cracks, pits, detachment of fragments, craters, cracks, pullout of the un-melted particle, and massive exfoliations contribute to the evolution of the cavitation erosion mechanism of the FeNiCrBSiNbW amorphous/nanocrystalline coatings with the increase of the initial surface roughness.

It has been proved by experimental results in the present work that cavitation erosion damage of the coatings became more serious as the initial surface roughness increased. However, it could be expected that the effect of initial surface roughness on the cavitation erosion behavior of the coating is not only dependent on the microstructure of material, but also dependent on the response of surface quality to the impact of micro-jets and shock waves. On the one hand, both the direct impact and the lateral outflow of the micro-jets on the coating surface are influenced by the initial surface roughness. In this manner, the surface of the coating after 80 grit grinding contains more pre-existing defects, as shown in Figure 4a, where nucleation, growth and collapsing of bubbles are more likely to occur [32]. In addition, the rougher surface with more pre-existing defects will entrap the lateral outflow of the water impacts and, in turn, the more cavitation erosion damage of the coating. On the other hand, the cavitation erosion process of the coatings with three initial surface roughness levels may be different. If the impact force of micro-jets and shock waves exceeds the yield stress of the coating, the impacts can plastically deform both the surface and the sub-surface and cause the occurrence of cracks originated from pre-existing defects. It is expected that by increasing the initial surface roughness, there should be an increase in the number of pits per unit of time on the surface of the coating during the cavitation erosion process. In the case of the coating after 1000 grit grinding, the limited pits on the surface merge relatively slowly and form an erosion crater. However, a large amount of pits on the surface of the coating after 80 grit grinding tend to spread over the cavitation erosion region and merge rapidly forming number of craters. With the rebound and implosion of subsequent cavitation bubbles, the un-melted particle peels off, leading to the massive exfoliation of material, as can be seen in Figure 5a,b. It is noteworthy that the amount of pre-existing defects on the initial surface of the FeNiCrBSiNbW amorphous/nanocrystalline coatings has a significant effect on their cavitation erosion behavior.

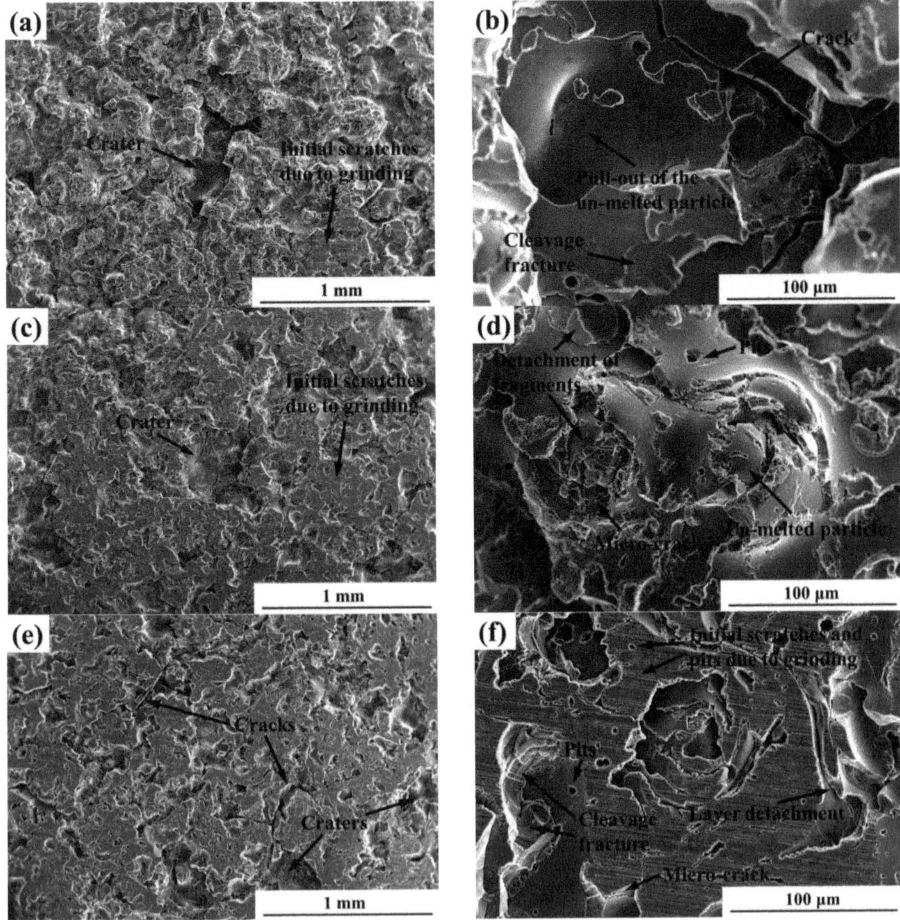

Figure 5. SEM images of the surface morphology of as-sprayed coatings under three surface conditions eroded for 120 min: (**a**,**b**) 80 grit grinding; (**c**,**d**) 600 grit grinding; and (**e**,**f**) 1000 grit grinding.

4. Conclusions

A FeNiCrBSiNbW coating with amorphous phase and nanocrystalline grains was fabricated using the arc spraying process. The coating had a dense structure with porosity of less than 2% and was adhering well to the substrate. The amount of pre-existing defects on the initial surface of the coatings had a significant effect on the cavitation erosion behavior. With increment of initial surface roughness, the cumulative mass losses of the coatings increased. The coating after 1000 grit grinding exhibited higher cavitation erosion resistance than that of the coating after 80 grit grinding in distilled water. The mechanisms involved in the cavitation erosion process of the coatings with the increase of the initial surface roughness were micro-cracks, pits, detachment of fragments, craters, cracks, pullout of the un-melted particle, and massive exfoliations.

Acknowledgments: The research was supported by the Fundamental Research Funds for the Central Universities (Grant Nos. KYZ201660 and RCQD16-04), the China Postdoctoral Science Foundation (Grant No. 2017M621665), the National Natural Science Foundation of China (Grant No. 51609067), the Natural Science Foundation of Jiangsu Province of China (Grant No. BK20150806), and the Opening Project of Material Corrosion and Protection Key Laboratory of Sichuan Province (Grant No. 2016CL08). The authors also gratefully acknowledge the financial support from the Policy-Induced Project of Jiangsu Province for the Industry–University–Research Cooperation (Grant No. BY2015071-02).

Author Contributions: Jinran Lin and Zehua Wang conceived and designed the research. Jinran Lin wrote the paper. Jinran Lin and Sheng Hong carried out the experiments. Jiangbo Cheng, Min Kang and Xiuqing Fu entered the discussion.

Conflicts of Interest: The authors declare no conflict of interest.

References

1. Karimi, A.; Martin, J.-L. Cavitation erosion of materials. *Int. Met. Rev.* **1986**, *31*, 1–26. [CrossRef]
2. Wu, C.L.; Zhang, S.; Zhang, C.H.; Zhang, H.; Dong, S.Y. Phase evolution and cavitation erosion-corrosion behavior of FeCoCrAlNiTi$_x$ high entropy alloy coatings on 304 stainless steel by laser surface alloying. *J. Alloys Compd.* **2017**, *698*, 761–770. [CrossRef]
3. Li, D.; Kang, Y.; Wang, X.C.; Ding, X.L.; Fang, Z.L. Effects of nozzle inner surface roughness on the cavitation erosion characteristics of high speed submerged jets. *Exp. Therm. Fluid Sci.* **2016**, *74*, 444–452. [CrossRef]
4. Chiu, K.Y.; Cheng, F.T.; Man, H.C. Evolution of surface roughness of some metallic materials in cavitation erosion. *Ultrasonics* **2005**, *43*, 713–716. [CrossRef] [PubMed]
5. Hutli, E.; Nedeljkovic, M.S.; Bonyar, A.; Radovic, N.A.; Llic, V.; Debeljkovic, A. The ability of using the cavitation phenomenon as a tool to modify the surface characteristics in micro- and in nano-level. *Tribol. Int.* **2016**, *101*, 88–97. [CrossRef]
6. Wang, Y.; Stella, J.; Darut, G.; Poirier, T.; Liao, H.L.; Planche, M.P. APS prepared NiCrBSi-YSZ composite coatings for protection against cavitation erosion. *J. Alloys Compd.* **2017**, *699*, 1095–1103. [CrossRef]
7. Woo, Y.B.; Lee, S.J.; Jeong, J.Y.; Kim, S.J. Evaluation on cavitation characteristics of CoNiCrAlY/ZrO$_2$-Y$_2$O$_3$ coating layer by atmospheric pressure plasma coating process. *Mater. Res. Bull.* **2014**, *58*, 78–82. [CrossRef]
8. Hong, S.; Wu, Y.P.; Zhang, J.F.; Zheng, Y.G.; Zheng, Y.; Lin, J.R. Synergistic effect of ultrasonic cavitation erosion and corrosion of WC-CoCr and FeCrSiBMn coatings prepared by HVOF spraying. *Ultrason. Sonochem.* **2016**, *31*, 563–569. [CrossRef] [PubMed]
9. Cheng, J.B.; Liang, X.B.; Xu, B.S.; Wu, Y.X. Characterization of mechanical properties of FeCrBSiMnNbY metallic glass coatings. *J. Mater. Sci.* **2009**, *44*, 3356–3363. [CrossRef]
10. Wu, Y.P.; Lin, P.H.; Wang, Z.H.; Li, G.Y. Microstructure and microhardness characterization of a Fe-based coating deposited by high-velocity oxy-fuel thermal spraying. *J. Alloys Compd.* **2009**, *481*, 719–724. [CrossRef]
11. Lin, J.R.; Wang, Z.H.; Lin, P.H.; Cheng, J.B.; Zhang, X.; Hong, S. Effects of post annealing on the microstructure, mechanical properties and cavitation erosion behavior of arc-sprayed FeNiCrBSiNbW coatings. *Mater. Des.* **2015**, *65*, 1035–1040. [CrossRef]
12. Wang, Y.; Li, K.Y.; Scenini, F.; Jiao, J.; Qu, S.J.; Luo, Q.; Shen, J. The effect of residual stress on the electrochemical corrosion behavior of Fe-based amorphous coatings in chloride-containing solutions. *Surf. Coat. Technol.* **2016**, *302*, 27–38. [CrossRef]
13. Wu, Y.P.; Lin, P.H.; Chu, C.L.; Wang, Z.H.; Cao, M.; Hu, J.H. Cavitation erosion characteristics of a Fe-Cr-Si-B-Mn coating fabricated by high velocity oxy-fuel (HVOF) thermal spray. *Mater. Lett.* **2007**, *61*, 1867–1872.
14. Hahn, M.; Fischer, A. Characterization of thermally sprayed micro- and nanocrystalline cylinder wall coatings by means of a cavitation test. *Proc. Inst. Mech. Eng. Part J J. Eng. Tribol.* **2009**, *223*, 27–37. [CrossRef]
15. Hahn, M.; Fischer, A. Characterization of thermal spray coatings for cylinder running surfaces of diesel engines. *J. Therm. Spray Technol.* **2010**, *19*, 866–872. [CrossRef]
16. Wang, Z.H.; Zhang, X.; Cheng, J.B.; Lin, J.R.; Zhou, Z.H. Cavitation erosion resistance of Fe-based amorphous/nanocrystal coatings prepared by high-velocity arc spraying. *J. Therm. Spray Technol.* **2014**, *23*, 742–749. [CrossRef]

17. Lin, J.R.; Wang, Z.H.; Lin, P.H.; Cheng, J.B.; Zhang, X.; Hong, S. Microstructure and cavitation erosion behavior of FeNiCrBSiNbW coating prepared by twin wires arc spraying process. *Surf. Coat. Technol.* **2014**, *240*, 432–436. [CrossRef]

18. Hou, G.L.; Zhao, X.Q.; Zhou, H.D.; Lu, J.J.; An, Y.L.; Chen, J.M.; Yang, J. Cavitation erosion of several oxy-fuel sprayed coatings tested in deionized water and artificial seawater. *Wear* **2014**, *311*, 81–92. [CrossRef]

19. Kim, Y.J.; Jang, J.W.; Lee, D.W.; Yi, S. Porosity effects of a Fe-based amorphous/nanocrystals coating prepared by a commercial high velocity oxy-fuel process on cavitation erosion behaviors. *Met. Mater. Int.* **2015**, *21*, 673–677. [CrossRef]

20. Zheng, Z.B.; Zheng, Y.G.; Sun, W.H.; Wang, J.Q. Effect of heat treatment on the structure, cavitation erosion and erosion-corrosion behavior of Fe-based amorphous coatings. *Tribol. Int.* **2015**, *90*, 393–403. [CrossRef]

21. Qiao, L.; Wu, Y.P.; Hong, S.; Zhang, J.F.; Shi, W.; Zheng, Y.G. Relationships between spray parameters, microstructures and ultrasonic cavitation erosion behavior of HVOF sprayed Fe-based amorphous/nanocrystalline coatings. *Ultrason. Sonochem.* **2017**, *39*, 39–46. [CrossRef] [PubMed]

22. Cheng, J.B.; Wang, B.L.; Liu, Q.; Liang, X.B. In-situ synthesis of novel Al-Fe-Si metallic glass coating by arc spraying. *J. Alloys Compd.* **2017**, *716*, 88–95. [CrossRef]

23. *ASTM G32-10-Standard Test Method for Cavitation Erosion Using Vibratory Apparatus*; ASTM International: West Conshohocken, PA, USA, 2010.

24. Jonas, O.; Steltz, W.; Dooley, B. *Steam Turbine Efficiency and Corrosion: Effects of Surface Finish, Deposits*; Report 1003997; Electrical Power Research Institute (EPRI): Palo Alto, CA, USA, 2001.

25. Lin, J.R.; Wang, Z.H.; Lin, P.H.; Cheng, J.B.; Zhang, J.J.; Zhang, X. Microstructure and corrosion resistance of Fe-based coatings prepared by twin wires arc spraying process. *J. Therm. Spray Technol.* **2014**, *23*, 333–339. [CrossRef]

26. Sharma, P.; Majumdar, J.D. Surface characterization and mechanical properties evaluation of Boride-dispersed Nickel-based coatings deposited on copper through thermal spray routes. *J. Therm. Spray. Technol.* **2012**, *21*, 800–809. [CrossRef]

27. Cheng, J.B.; Zhao, S.; Liu, D.; Feng, Y.; Liang, X.B. Microstructure and fracture toughness of the FePSiB-based amorphous/nanocrystalline coatings. *Mater. Sci. Eng. A* **2017**, *696*, 341–347. [CrossRef]

28. Heni, W.; Vonna, L.; Fioux, P.; Vidal, L.; Haidara, H. Ultrasonic cavitation test applied to thin metallic films for assessing their adhesion with mercaptosilanes and surface roughness. *J. Mater. Sci.* **2014**, *49*, 6750–6761. [CrossRef]

29. Jiang, N.N.; Liu, S.H.; Chen, D.R. Effect of roughness and wettability of silicon wafer in cavitation erosion. *Chin. Sci. Bull.* **2008**, *53*, 2879–2885. [CrossRef]

30. Friction, Lubrication, and Wear Technology. In *ASM Handbook (Volume 18)*; ASM International: Materials Park, OH, USA, 1992; pp. 408–435.

31. Santa, J.F.; Espitia, L.A.; Blanco, J.A.; Romo, S.A.; Toro, A. Slurry and cavitation erosion resistance of thermal spray coatings. *Wear* **2009**, *267*, 160–167. [CrossRef]

32. Scardina, P.; Edwards, M. Prediction and measurement of bubble formation in water treatment. *J. Environ. Eng.* **2001**, *11*, 968–973. [CrossRef]

Article

Phase Evolution and Microstructure Analysis of CoCrFeNiMo High-Entropy Alloy for Electro-Spark-Deposited Coatings for Geothermal Environment

Sigrun N. Karlsdottir [1], Laura E. Geambazu [2], Ioana Csaki [2,*], Andri I. Thorhallsson [1], Radu Stefanoiu [2], Fridrik Magnus [3] and Cosmin Cotrut [2]

[1] Department of Industrial Engineering, Mechanical Engineering and Computer Science, University of Iceland, Hjardarhagi 2-6, 107 Reykjavík, Iceland; snk@hi.is (S.N.K.); andri.isak.thorhallsson@gmail.com (A.I.T.)
[2] University Politehnica Bucharest, Splaiul Independentei 313, Bucharest 060042, Romania; laura.geambazu@gmail.com (L.E.G.); radu.stefanoiu@upb.ro (R.S.); cosmin.cotrut@upb.ro (C.C.)
[3] Science Institute, University of Iceland, Dunhaga 3, 107 Reykjavik, Iceland; fridrikm@hi.is
* Correspondence: ioana.apostolescu@upb.ro

Received: 20 May 2019; Accepted: 18 June 2019; Published: 21 June 2019

Abstract: In this work, a CoCrFeNiMo high-entropy alloy (HEA) material was prepared by the vacuum arc melting (VAM) method and used for electro-spark deposition (ESD). The purpose of this study was to investigate the phase evolution and microstructure of the CoCrFeNiMo HEA as as-cast and electro-spark-deposited (ESD) coating to assess its suitability for corrosvie environments encountered in geothermal energy production. The composition, morphology, and structure of the bulk material and the coating were analyzed using scanning electron microscopy (SEM) coupled with energy-dispersive spectroscopy (EDS), and X-ray diffraction (XRD). The hardness of the bulk material was measured to access the mechanical properties when preselecting the composition to be pursued for the ESD coating technique. For the same purpose, electrochemical corrosion tests were performed in a 3.5 wt.% NaCl solution on the bulk material. The results showed the VAM CoCrFeNiMo HEA material had high hardness (593 HV) and low corrosion rates (0.0072 mm/year), which is promising for the high wear and corrosion resistance needed in the harsh geothermal environment. The results from the phase evolution, chemical composition, and microstructural analysis showed an adherent and dense coating with the ESD technique, but with some variance in the distribution of elements in the coating. The crystal structure of the as-cast electrode CoCrFeNiMo material was identified as face centered cubic with XRD, but additional BCC and potentially σ phase was formed for the CoCrFeNiMo coating.

Keywords: high-entropy alloy; coating; electro-spark deposition; microstructure; corrosion; geothermal environment; XRD

---

1. Introduction

The concept of complex compositionally alloys, also referred to as high-entropy alloys, has been recently proposed waiving the idea of solute and solvent and adopting the concept of a mixture of multi-principal elements in an equimolar or nearly equimolar ratio. Conceptually, this is a radical departure from the conventional notions that opens up a vast alloy design space yet to be fully explored [1]. Due to their high mixing entropy, these alloys tend to form a simple solution-like phase [2–5] and show a variety of desired properties such as high hardness and improved oxidation and corrosion resistance [6–11].

With an extended lifetime of geothermal power plants around the world, there is an increased need to have cost-effective solutions available for the maintenance of power plant components to ensure sufficient efficiency of the plant. The main maintenance problems are due to the wear, erosion, and corrosion of plant components due to the corrosive nature and high temperatures of the geothermal steam [12]. Geothermal steam generally contains dissolved gasses, the most common being carbon dioxide (CO_2) and hydrogen sulfide (H_2S), which are highly corrosive. Other corrosive species are, for example, chloride (Cl^-) and sulfur ions (SO_4^{2-}) [12,13]. One of the most crucial power plant components of geothermal power plants is the turbine, which uses the geothermal steam to produce electricity. Geothermal turbine components can experience various corrosion problems, such as the erosion corrosion of turbine blade materials and wear of rotor materials [14].

The cost of maintenance for a turbine can be lowered if the repair could be performed by a simple and viable technique with a decreased frequency [15]. Coatings of new corrosion-resistant complex materials could be performed by electro-spark deposition (ESD) to enhance the corrosion resistance of the turbine used in the geothermal environment. ESD technology is simple and effective for the deposition of metallic materials on relatively small repair areas. Usually, the ESD coating process is performed on installations equipped with a manual electrode holder, and the electrode itself has the shape of a rod. The advantages of this method include the high adhesion of the resulting coatings, the possibility of local processing of large-sized parts, its relative simplicity, a low energy consumption, high environmental compatibility, and the possibility of process automation. ESD has been successfully used to produce protective coatings for nickel alloys against oxidation [16–19].

In this paper, the potential of using the new compositionally complex alloy (high-entropy alloy) CoCrFeNiMo as a protective coating on a steel substrate is studied. Previously, this alloy was tested in a geothermal environment, well known as a highly aggressive environment, and the results were extremely encouraging [20]. The bulk material in this study was produced as an electrode for the ESD device, and the coating was fabricated with the new compositionally complex alloy. The use of such alloys in highly aggressive environments could be a solution to increase the efficiency of various processes by lowering the maintenance operation frequency needed by each process due to corrosion and erosion problems. The CoCrFeNiMo alloy was designed to have good adhesion to the steel substrate, high hardness, and improved corrosion properties. In this study, we present the results obtained for corrosion in saline water and hardness testing of the bulk alloy, and the phase evolution and microstructure analysis of the bulk material and the coating fabricated by electro-spark deposition with an electrode manufactured with the CoCrFeNiMo alloy. The main purpose of this study was to investigate the phases present in the CoCrFeNiMo coating and to establish if such a coating could be suitable for protecting the steel parts interacting with harsh environments such as corrosive geothermal steam. The novelty of this paper is the complex phase-change study from the bulk material to the coating using electro-spark deposition with an in-house-built high-entropy alloy electrode.

2. Materials and Methods

2.1. Materials and Processing

Alloy ingots were prepared by vacuum arc melting the mixture of high purity metals Co, Cr, Fe, Ni, and Mo under high-purity argon gas on water cooled Cu hearth (Material Research Furnaces, Allenstown, NH, USA). The alloys were re-melted and flipped about five times to ensure the homogeneity of the ingot. The ingots were prepared in a cylindrical shape with a diameter of 23.20 mm and height of 5.28 mm, and the weight was 17.87 g.

Table 1 presents the composition of the HEA processed in the liquid state.

Table 1. Nominal composition for CoCrFeNiMo.

Elements (at%)	Co	Cr	Fe	Ni	Mo
CoCrFeNiMo	20	20	20	20	20

The as-cast sample was machined to obtain the electrode that was then used for the electro-spark deposition on a silicon steel substrate, using a Spark Depo 300 machine (DJK Europe GmbH, Prague, Czeck Republik). The coating was performed using an argon rate of 3 L/min. Electro-spark deposition (ESD) is a pulsed-arc microwelding process using short-duration, high-current electrical pulses to deposit an electrode material on a metallic substrate. A fused, metallurgically bonded coating could be applied with a low total heat input, and the bulk substrate material remains near ambient temperatures. The short duration of the electrical pulse allows an extremely rapid solidification of the deposited material and results in an exceptionally fine-grained, homogeneous coating.

The parameters used in the electro-spark deposition process are summarized in Table 2.

Table 2. Parameters used for CoCrFeNiMo electrospark deposition.

Sample	Capacitance (μF)	Voltage (V)	Frequency (Hz)	Atmosphere
CoCrFeNiMo	20	100	3.5	Argon

2.2. Microstructural Analyses

A field emission scanning electron microscope (FE-SEM) Zeiss Supra 25 (Zeiss, Cambridge, UK) was used for the microstructural analysis, and X-ray energy dispersive spectroscopy (XEDS) equipment with a Si (Li) X-ray detector and back-scattered electron (BSE) detector. INCA Energy 300 software (Oxford Instruments, Oxford, UK) was used for the chemical composition analysis. The surface of the as-cast and electro-spark-deposited (ESD) specimens were analyzed. The cross-section of the ESD specimen was mounted in thermosetting phenol formaldehyde resin (Bakelite), and grinded and polished with SiC abrasive paper (down to 1200 grit) for SEM and EDS analysis.

2.3. X-Ray Diffraction (XRD) Analysis

XRD was performed to investigate the phase change for each fabrication method. An X'pert Pro MRD system (Panalytical, Almelo, the Netherlands) was used to obtain the XRD pattern for vacuum arc re-melting (VAM) as-cast samples, with a Gobel mirror mounted on the incident side and a parallel plate collimator on the diffracted side. Measurements were performed in both symmetric θ–2θ geometry as well as in grazing incidence geometry for increased surface sensitivity. XRD analysis was also performed on the coating to examine the phases present after the coating procedure.

2.4. Electrochemical Corrosion Testing

The electrochemical corrosion tests were done at 25 °C in NaCl 3.5% solution. This test was performed to evaluate the corrosion resistance of the CoCrFeNiMo alloy in a chloride containing solution, since geothermal fluid commonly contains Cl^- ions, which can facilitate the corrosion process by penetrating through the oxide films formed. The corrosion resistance was determined with the linear polarization technique by measuring the potential in the open circuit during 6 h and drawing the potentiodynamic curves from −1 V (vs. OCP—open circuit potential) to +V (vs. SCE—saturated calomel electrode), with a scanning rate of 1 mV/s. Tests for corrosion resistance assessment were performed with a potentionstat/galvanostat PARSTAT 4000 (Princeton Applied Research, Tulsa, OK, USA) with a low current module VersaSTAT LC and the potentiodynamic curves acquired with the

VersaStudio v2.50.3 software. An electrochemical cell with a saturated calomel electrode was used in the testing with a platinum electrode for registering and the investigated sample as the working electrode.

2.5. Hardness Measurement

The hardness of the CoCrFeNiMo high-entropy alloy was tested along the diameter of the samples in three profiles in 20 different points, and the mean value was calculated. The hardness was measured with a Shimadzu Vickers hardness device (Shimadzu, Columbia, USA), with a 0.1-kgf load.

3. Results

3.1. Microstructures of As-Cast CoCrFeNiMo

The microstructure of the as cast CoCrFeNiMo revealed a homogeneous structure containing one single phase and several small segregated compounds, as can be seen in Figures 1 and 2. Figure 1a shows an SEM image of the microstructure of the as-cast CoCrFeNiMo alloy where the small segregated compounds are visible. The table in Figure 1 gives the elements in weight % (wt %) of the small segregated compounds detected in the EDS analysis. The area analyzed is identified with a white box in the SEM image. The compounds are rich in chromium (Cr), iron (Fe), and oxygen (O), but also contain the constitutive elements of the HEA alloy. The compounds are around 2–5 μm in width and many of them have a hexagonal shape.

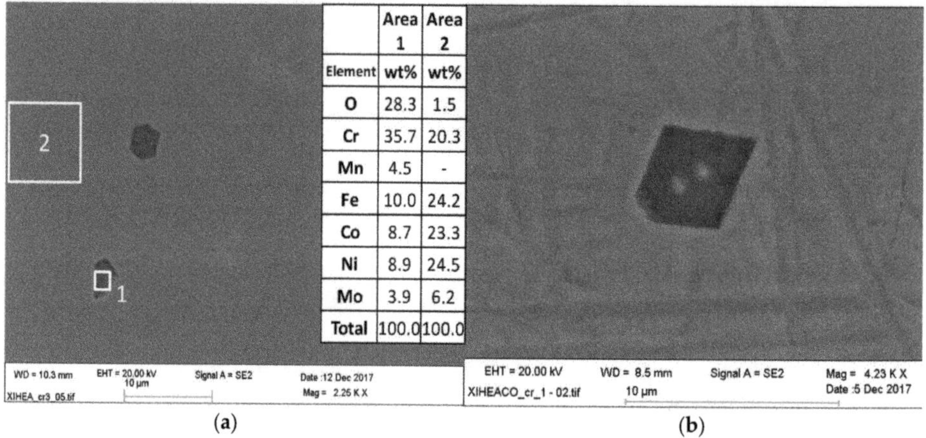

Element	Area 1 wt%	Area 2 wt%
O	28.3	1.5
Cr	35.7	20.3
Mn	4.5	-
Fe	10.0	24.2
Co	8.7	23.3
Ni	8.9	24.5
Mo	3.9	6.2
Total	100.0	100.0

(a)   (b)

Figure 1. (**a**) Scanning electron microscopy (SEM) image showing the microstructure of the as-cast CoCrFeNiMo material revealing segregated compounds within the bulk and a table with energy-dispersive spectroscopy (EDS) analyses of areas highlighted with white boxes in image (**a**); and (**b**) SEM image of the compound at higher magnification.

Figure 2 shows a SEM image and the corresponding EDS maps for the as-cast CoCrFeNiMo bulk material. The mapping of the elements with EDS reveals the homogeneous distribution of the high-entropy alloy components.

Figure 2. Scanning electron microscopy (SEM) image and EDS maps of the as-cast CoCrFeNiMo bulk material.

3.2. XRD Results for As-Cast CoCrFeNiMo

The XRD pattern of the as-cast CoCrFeNiMo material shows a face centered cubic (FCC) phase to be present (Figure 3).

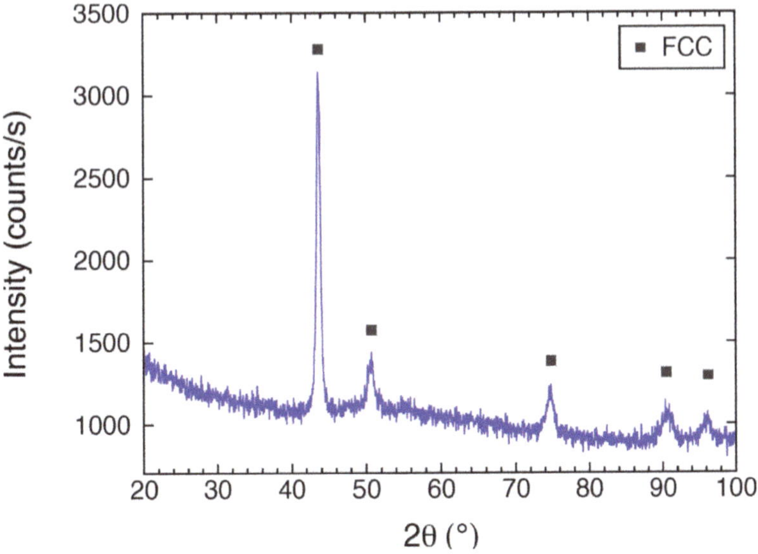

Figure 3. The X-ray diffraction (XRD) pattern for the as-cast CoCrFeNiMo material.

The valence electron concentration calculated for CoCrFeNiMo was 7.8 and indicates the presence of the FCC phase [6]. This is in good agreement with the XRD analysis shown in Figure 3. The formula used for calculating the valence electron concentration was as follows [6]:

$$VEC = \sum_{i=1}^{n} c_i VEC_i \tag{1}$$

where VEC_i is the valence electron concentration of each component and c_i is the concentration of each component of the CoCrFeNiMo high-entropy alloy.

3.3. Corrosion Behavior of As-Cast CoCrFeNiMo

A standard electrochemical technique, the potentiodynamic polarization test, was used to study the corrosion behavior of the CoCrFeNiMo as-cast material. The variation of the potential in open circuit E_{OC} and the potentiodynamic polarization curve for the as-cast CoCrFeNiMo sample are shown in Figure 4.

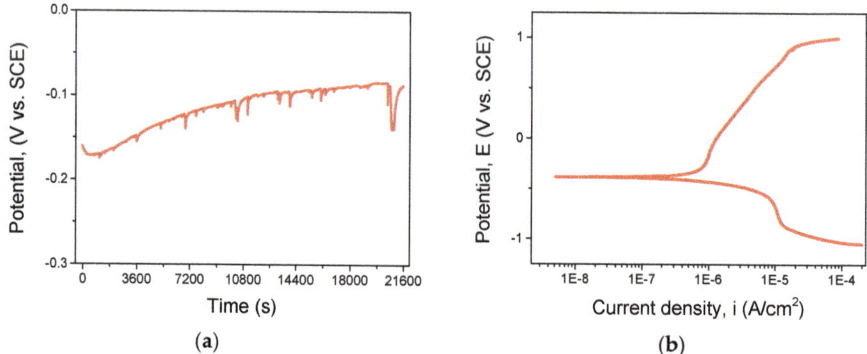

Figure 4. (**a**) Open circuit potential variation and (**b**) potentiodynamic polarization curve for the bulk CoCrFeNiMo alloy tested electrochemically in 3.5% NaCl solution.

Table 3 presents the electrochemical parameters for the CoCrFeNiMo alloy electrochemically tested in 3.5% NaCl solution at room temperature. The following parameters have been determined to characterize the corrosion resistance of the investigated samples: open circuit potential after 6h (E_{OC}), corrosion potential (E_{corr}), corrosion current density (i_{corr}), cathodic slope (β_c), and anodic slope (β_a). With the aid of the parameters determined with the Tafel extrapolation technique, the polarization resistance (R_p) was calculated.

Table 3. Electrochemical parameters of CoCrFeNiMo alloy in the chloride-containing solution (3.5% NaCl) at room temperature.

Sample	E_{oc} (mV)	E_{corr} (mV)	i_{corr} (μA/cm^2)	β_c (mV)	β_a (mV)	R_p (k$\Omega \times$ cm^2)	CR (mm/year)
VAR_HEA	−88	−360	0.017	207.13	92.96	1748.61	0.0072

The polarization resistance of the investigated alloy was also calculated based on the Stern-Geary equation (Equation (2)) [21,22].

$$R_p = 2.3 \frac{\beta_a + |\beta_c|}{\beta_a |\beta_c|} i_{corr} \tag{2}$$

where β_a is theanodic slope, β_c thecathodic slope, and i_{corr} the corrosion current density (μA/cm^2).

The corrosion rate was calculated according to ASTM G102-89 (2004) [23] with the following equation:

$$CR = K_i \frac{i_{corr}}{\rho} EW \tag{3}$$

where CR is the corrosion rate (mm/year), K_i is 3.27×10^{-3} is the material density (g/cm^3), i_{corr} the current density of material (μA/cm^2), and EW the equivalent weight (g).

It can be seen from Figure 4 that the E_{OC} shift to more electropositive values demonstrates that the alloy can develop a passive film on the surface after its deterioration. Some small deterioration of this passive film developed in the immersion in the chloride-containing solution (3.5% NaCl) can be identified by the potential drop during the immersion. The polarization curve shows an interval of passivation of the alloy until the value of −88 mV, where the breakdown potential appears. The corrosion current density value (i_{corr}) was 0.017 μA/cm^2, and the corrosion potential was around −360 mV. The corrosion rate in the NaCl 3.5% solution was calculated from to be 0.0072 mm/year, which is very low. The corrosion behavior of different HEA alloys was investigated recently by Shi et al. [24]. Their findings showed that the corrosion potential of FeCoNiCr was −460 mV after testing in the same solution (3.5% NaCl), while that of Co1.5CrFeNi1.5Ti0.5Mo0.1 was −380 mV. It was also stated that Mo-free Co1.5CrFeNi1.5Ti0.5 HEA alloy was susceptible to pitting, but the Mo added HEA alloys had a much higher passivation region.

The values obtained in the present study can also be compared with the results from testing stainless steel 630 in the same electrochemical conditions; E_{corr} for 630 stainless steel in NaCl 3.5% solution was measured to be −655.6 mV and i_{corr} 12.53 μA/cm^2 [25]. This indicates that CoCrFeNiMo HEA could be a useful material for components working in saline water such as geothermal brine or even harsher environments. In a previous study, the corrosion rate for the bulk CoCrFeNiMo high-entropy alloy was measured after in-situ testing in geothermal steam at the Reykjanes power plant [20]. The value obtained was 0.00033 mm/year, which is a very low corrosion rate after exposure to geothermal steam at 200 °C and 17 bar containing H_2S and CO_2 gases.

3.4. Hardness Measurements of As-Cast CoCrFeNiMo

A good indicator of improvement in mechanical properties for high-entropy alloys is hardness. The hardness value for CoCrFeNiMo was determined after a series of 20 measurements along the diameter of the sample, and the mean value was calculated. The mean hardness value of the sample was 593 HV, higher than that of 304 stainless steel (201 HV) that was measured in comparison. The high hardness value obtained for the CoCrFeNiMo high-entropy alloy and the good corrosion results in 3.5% NaCl and geothermal steam encouraged us to use the bulk alloy as an electrode for a coating using the electro-spark deposition process.

3.5. Microstructural and Chemical Composition Analysis of ESD-Prepared CoCrFeNiMo Coatings

Figure 5a,b shows a low magnification SEM image of the electro-spark-deposited CoCrFeNiMo alloy coating surface. The coating is dense and relatively homogenous. Figure 4b shows the corresponding EDS analysis of the area identified with a white box in Figure 5b. The chemical composition analysis verifies the presence of the constituent elements of the coatings, i.e., Co, Cr, Fe, Ni, and Mo, but also a minor amount of O, Al, and Si. The contrast in the SEM image indicates some variance in the distribution of elements in the coating.

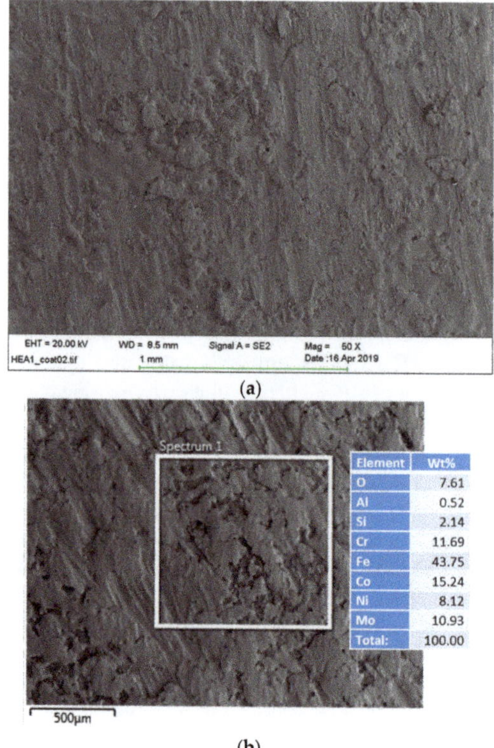

Figure 5. (a,b) SEM images of the surface of the CoCrFeNiMo coating; (b) includes a table with results from EDS analysis of the area identified with a white box in the SEM image.

Figure 6 shows SEM images of the surface of the deposited CoCrFeNiMo coating at higher magnification. At higher magnification the morphology of the coating is more apparent; large smooth islands are distributed on the surface surrounded by lower areas with more particulates.

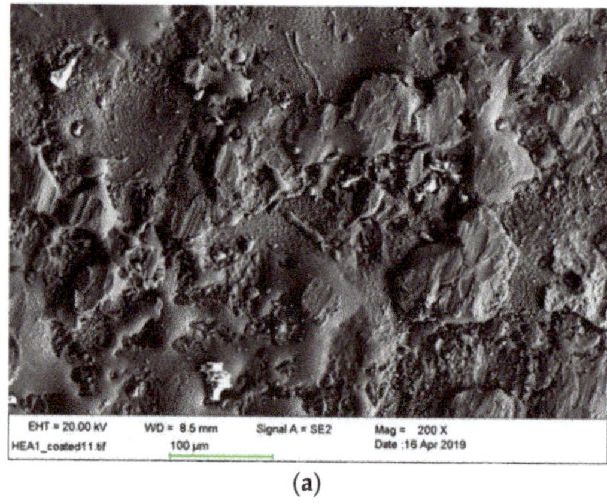

Figure 6. *Cont.*

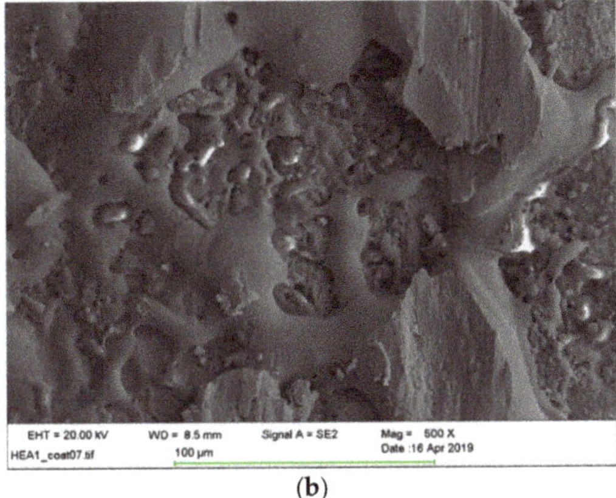

(b)

Figure 6. Surface view of the deposited CoCrFeNiMo coating; SEM images at (**a**) lower magnification and (**b**) higher magnification.

To analyze the distribution of elements in the coating in connection with the morphology, EDS elemental maps were generated of the area shown in Figure 6. The SEM image of the area and the corresponding maps for the elements are shown in Figure 7.

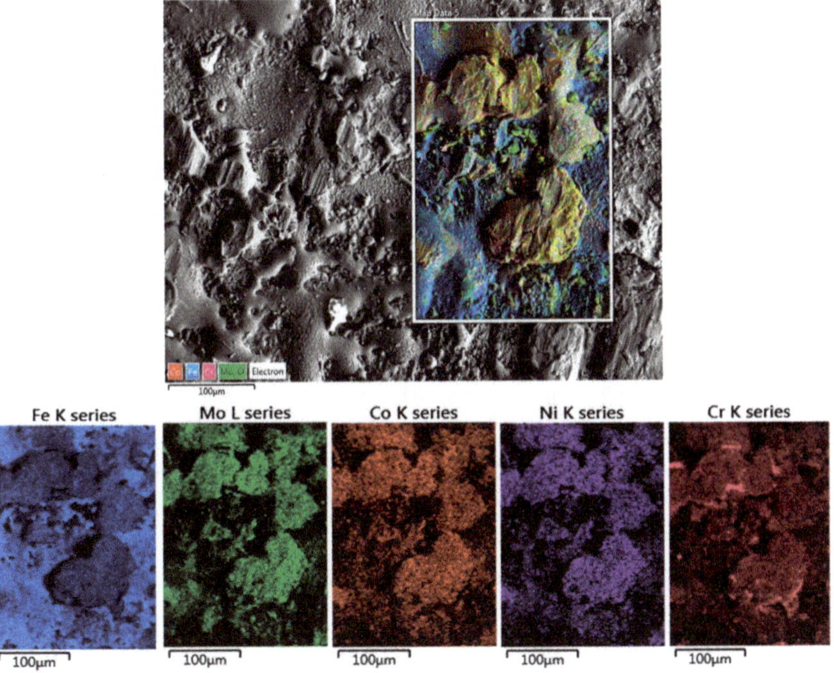

Figure 7. EDS elemental maps generated of the area shown in Figure 6.

It is evident from the generated maps that the elements of the HEA electrode material (Fe, Mo, Co, Ni, Cr) are present, but there is some variation in the distribution of the elements, as can be seen in the maps.

The microstructure and chemical composition of the cross-section of the deposited CoCrFeNiMo coating was also studied through SEM and EDS analysis. Figure 8 shows the BSE image of the cross-section of the coating and the EDS analysis of different parts of it. There was also compositional variation in the cross-section of the CoCrFeNiMo coatings as can be seen in Figure 8; different areas in the coating are marked 1 to 4, from the innermost part to the surface. The amount of Cr in the coating increases with the distance from the substrate; the light gray phase in the coating is rich in Fe, while the darker phase is more abundant in Cr. The oxygen content increases during the ESD process. In the first area, area 1, the iron content increases because area 1 is the most influenced by the substrate, the bonding between the layer, and the substrate being realized in that particular area. As the layer thickness increases, the substrate influence decreases as well. The increase in oxygen amount is due to the coating process. The ESD process sometimes promotes oxide formation due to the air atmosphere. The coating is realized layer by layer and sometimes part of the chromium segregates and forms an oxide layer during the process, which in the end can have a protective role for the coating itself [24].

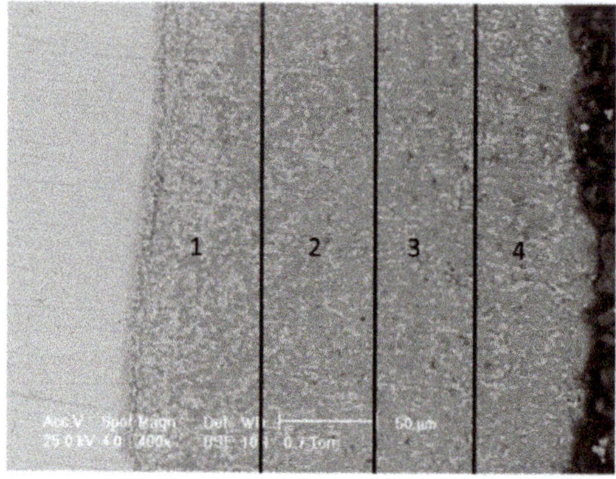

Elements	Co	Cr	Fe	Ni	Mo	Si	Ca	Ti	O
Substrate	-	16.5	66.4	10.4	3.9	0.8	0.9	1.0	-
Area 1	3.2	17.7	44.7	10.9	4.3	-	-	-	19.9
Area 2	1.6	18.6	37.0	9.8	4.9	-	-	-	28.1
Area 3	1.6	18.8	34.2	10.6	5.0	-	-	-	29.7
Area 4	2.4	18.5	29.4	11.7	6.8	-	-	-	31.2

Figure 8. Back scattered electron (BSE) image of the cross-section of the CoCrFeNiMo coating and results from EDS analysis of the areas labelled in the BSE image.

3.6. XRD Analysis of CoCrFeNiMo Coating

Figure 9 shows the XRD pattern of the CoCrFeNiMo ESD coating and, for comparison, the pattern for the bulk material.

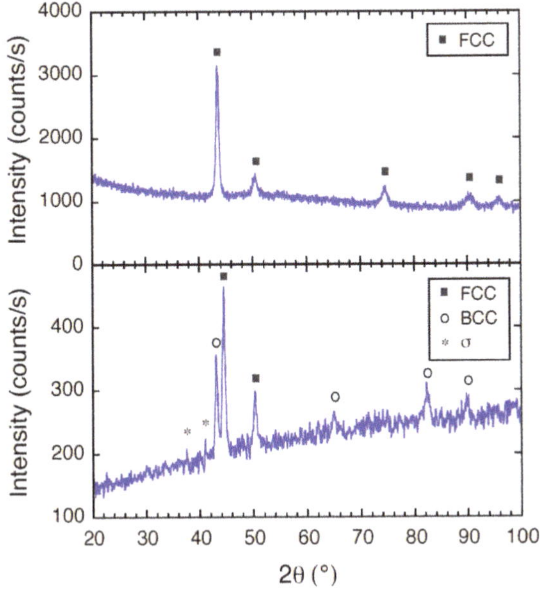

Figure 9. XRD pattern for CoCrFeNiMo ESD coating (bottom graph) and as-cast material (top graph) for comparison.

The XRD of the coating reveals that two main phases are present after the electro-spark deposition (ESD). The dominant phase in the bulk material is FCC, as predicted by the VEC calculation presented in Section 3.2. The formation of the FCC phase is promoted when most of the binary constituents crystallize in the FCC structure, as CoNi, CoFe, and FeNi. The complete solubility of CoNi and sufficient solubility of FeNi could stabilize the FCC phase in the bulk material [26–30]. The BCC phase after the ESD process could be due to the presence of Mo in this high-entropy alloy. Additionally, the small peaks around 40° in the XRD pattern could potentially be due to the formation of a sigma phase that can form in CoCrFeNiMo HEA materials when Mo is added to the mix [1,31]. These aspects are further discussed in Section 4.

4. Discussion

A unique feature of high-entropy alloys is the ability of different crystal structure elements to combine and form a single phase. The CoCrFeNiMo high-entropy alloy consists of elements of similar sizes (Co, Cr, Fe, Ni,) and the same valences and electronegativity. Mo is different in size but close to the other elements in terms of valence and electronegativity. Co is hexagonal close packed at room temperature but transforms to FCC when the temperature increases (after 450 °C it becomes FCC). The phase transformation in HEA could be governed to a higher extent by the binary constituents' pairs that evolve first, rather than by the individual elements themselves. The bulk material CoCrFeNiMo contains pairs of CoNi, NiCr, NiFe, and NiMo with an FCC structure [32]. The CoFe pair presents a BCC structure, and CoCr, CoMo, FeCr, and FeMo tend to stabilize the σ phase. In the case of coatings, the interface is influenced by the substrate. It has been found that the sigma phase can form from the BCC phase without precipitation of the FeCr system [33]. The electro-spark deposition process ensures good bonding because the first layers are mixed with the substrate (as shown in Figure 8 zone 1). This could explain the appearance of a mixture of phases and variance in the composition in the coated layer. The addition of molybdenum tends to stabilize the formation of a BCC structure and appearance of a σ phase. The formation of the σ phase has also been reported in CoCrFeNiMox alloys with increasing Mo content [1,31]. Increasing Mo content has also been shown to transform the cast structure into a eutectic structure [34]. Despite that, the detected phase of the bulk alloy is FCC, while the XRD pattern (Figure 9b) reveals a mixture of phases for the coating. In the coating process, a high temperature is generated, which can modify the phases in the structure. The results from the XRD and the microstructural and EDS analyses presented in Figure 8, areas 1, 2, 3, and 4 confirm this. The increase in Mo content transforms the eutectic microstructure into a hypo/hyper eutectic structure. This preference is probably due to a higher driving force for the formation of a stable σ phase of the FeCrMo type. Cr is an important constituent in HEAs and, as already discussed in the context of equiatomic alloys, Cr stabilizes the BCC structure and can promote the formation of the σ phase particularly in the presence of Fe, Co, and Ni. The bulk alloy processing method, due to the rapid cooling process, stabilized the FCC phase in the CoCrFeNiMo high-entropy alloy. The binary pairs CoFe and CrMo in the high entropy alloy tend to stabilize the BCC phase. The BCC phase is present in the coating due to the mechanism of rapid cooling during the electro-spark deposition process. The coating process is based on metallurgical bond formation due to an electrical discharge, which increases the temperature for the electrode and substrate, and then the layer is cooled almost instantaneously. The rapid cooling of the high entropy alloy on the substrate favors the BCC phase formation. Cr, Fe, Mo have a BCC structure at room temperature and usually promote the BCC phase formation. The σ phase is sometimes observed in Cr-containing steels and has a typical composition of equiatomic FeCr with a tetragonal structure. The σ phase has also been observed with equiatomic CoCr or FeMo in binary CoCr and FeMo alloys. A large number of HEAs containing Fe and/or Co together with higher amounts of Cr and/or Mo have shown the formation of the σ phase at various stages in their processing. In HEAs, the σ phase is also a multicomponent solid solution. The formation of the σ phase is an indication that different types of solid solutions in HEAs could form depending on the interaction and atomic size difference between elements and not just the configurational entropy alone. The σ phase is in fact a topologically close-packed phase in which components with larger atomic size occupy one specific set of lattice sites while smaller atoms occupy another set so as to get a higher number of bonds to lower its overall free energy, although the interactions (or enthalpy of mixing) between components are low [27]. The electro-spark deposition using the heated electrode and electric current to ensure the spark deposition modified the phases in the alloy and promoted the formation of new phases for the coating. The substrate influence and the coating process induced the mixture of phases in the coating, forming FCC, BCC, and potentially the σ phase in the coated layer.

The corrosion rate of the CoCrFeNiMo HEA bulk material calculated from the electrochemical parameters obtained in the corrosion test in the salt solution at room temperature was very low (0.0072 mm/year) and thus encouraging for the further study of this alloy for the chloride-containing geothermal fluid commonly encountered in geothermal energy utilization. The corrosion resistance of the CoCrFeNiMo bulk material can be attributed to the passivation of the alloy by the formation of thin chromium-oxide-rich surface film, and also to the presence of Mo, which can decrease the susceptibility to corrosion, particularly pitting corrosion, by acting as a barrier at the surface against electrochemical attack. Mo-free containing HEA alloys have been found to pit in 1 M NaCl, while the Mo-containing alloys were not susceptible to pitting [25]. Whether the formation of additional phases, such as the BCC phase, in the fabricated CoCrFeNiMo HEA coating in this study has a large effect on the corrosion resistance of the coatings compared to the bulk material is being studied in our ongoing project dedicated to developing HEA coatings for geothermal environments. It has been reported that both BCC and FCC phases can be formed in HEA coatings; Ye at al. [35] prepared $Al_xCoCrCuFeNi$ HEA coatings with laser cladding on AISI 1045 steel, which were proven to have both FCC and BCC structures. A relatively homogenous distribution of elements was observed, and the coatings possessed better corrosion resistance than the 314L stainless steel tested in the same conditions (0.05 mol/L HCl) [36]. This encourages us to continue the research using 1 M NaCl solutions at room temperature, and in simulated and in-situ in geothermal environments at elevated temperatures.

The hardness was measured to assess the mechanical properties of the CoCrFeNiMo HEA in this study, and the results were promising. The value obtained, greater than that of stainless steel, could positively influence the mechanical properties of the coating and wear resistance. In our further study, this will be investigated, as well as the potential evaluation of the fracture toughness at micro-scale with the experimental approach reported in the literature [35].

5. Conclusions

- Vacuum-arc-melted CoCrFeNiMo HEA was fabricated and used for producing an electro-spark-deposited coating.
- Testing of the as-cast material revealed high hardness (593 HV) and low corrosion rates (0.0072 mm/year), which is promising for the high wear and corrosion resistance needed for the harsh geothermal environment.
- Adherent and dense coating was obtained, but some variance in the distribution of elements was observed in the coating, with Cr increasing with the distance from the substrate.
- The crystal structure of the as-cast electrode CoCrFeNiMo material was identified as FCC with XRD, but additional phases were formed in the CoCrFeNiMo coating, such as BCC and potentially the σ phase.

Author Contributions: S.N.K., conceptualization, analysis of data, methodology, writing of manuscript; I.C., conceptualization, analysis of data, methodology, writing of manuscript; L.E.G., contributed with resources, fabrication, and methodology; A.I.T., performed SEM and EDS analysis and analyzed the data, contributed to XRD analysis, reviewing of paper; R.S., fabrication of bulk material; C.C., performed the corrosion tests and analysis; F.M., performed analysis of data and XRD tests.

Funding: This work is part of the H2020 EU project Geo-Coat: Development of novel and cost-effective corrosion resistant coatings for high temperature geothermal applications. Call H2020-LCE-2017-RES-RIA-TwoStage (Project No. 764086).

Acknowledgments: The authors would like to acknowledge the resources and collaborative efforts provided by the consortium of the Geo-Coat project No. 764086. The authors want to thank Florin Miculescu for his help in analyzing the microstructure of the cross-sectioned samples of the coated layer with SEM and EDS analyses.

References

1. Liu, W.H.; Lu, Z.P.; He, J.Y.; Luan, J.H.; Wang, Z.J.; Liu, B.; Liu, Y.; Chen, M.W.; Liu, C.T. Ductile CoCrFeNiMox high entropy alloys strengthened by hard intermetallic phases. *Acta Mater.* **2016**, *116*, 332–342. [CrossRef]
2. Zou, Y.; Ma, H.; Spolenak, R. Ultra strong ductile and stable high-entropy alloys at small scales. *Nat. Commun.* **2015**, *6*, 7748. [CrossRef] [PubMed]
3. Zhang, Y.; Stocks, G.M.; Jin, K.; Lu, C.; Bei, H.; Sales, B.C.; Wang, L.; Beland, L.K.; Stoller, R.E.; Samolyuk, G.D.; et al. Influence of chemical disorder on energy dissipation and defect evolution in concentrated solid solution alloys. *Nat. Commun.* **2015**, *6*, 8736. [CrossRef] [PubMed]
4. Tsai, M.H. Three strategies for the design of advanced high-entropy alloys. *Entropy* **2016**, *18*, 252. [CrossRef]
5. Zhang, Y.; Zhou, Y.J.; Lin, J.P.; Chen, G.L.; Liaw, P.K. Solid-solution phase formation rules for multi-component alloys. *Adv. Eng. Mater.* **2008**, *10*, 534–538. [CrossRef]
6. Guo, S.; Liu, C.T. Phase stability in high entropy alloys: formation of solid solution phase or amorphous phase. *Prog. Nat. Sci. Mater. Int.* **2011**, *21*, 433–446. [CrossRef]
7. Yang, X.; Zhang, Y. Prediction of high-entropy stabilized solid-solution in multi-component alloys. *Mater. Chem. Phys.* **2012**, *132*, 233–238. [CrossRef]
8. Guo, S.; Hu, Q.; Ng, C.; Liu, C.T. More than entropy in high-entropy alloys: Forming solid solutions or amorphous phase. *Intermet.* **2013**, *41*, 96–103. [CrossRef]
9. Ren, M.X.; Li, B.S.; Fu, H.Z. Formation condition of solid solution type high entropy alloy. *Trans. Nonferrous Metall. Soc. China* **2013**, *23*, 991–995. [CrossRef]
10. Liu, W.H.; Wu, Y.; He, J.Y.; Zhang, Y.; Liu, C.T.; Lu, Z.P. The phase competition and stability of high-entropy alloys. *JOM* **2014**, *66*, 1973–1983. [CrossRef]

11. Wang, Z.; Guo, S.; Liu, C.T. Phase selection in high-entropy alloys: From nonequilibrium to equilibrium. *JOM* **2014**, *66*, 1966–1972. [CrossRef]

12. Karlsdottir, S.N. Corrosion, scaling and material selection in geothermal energy production. In *Comprehensive Renewable Energy*, 1st ed.; Sayigh, A., Ed.; Elsevier: Amsterdam, The Netherlands, 2012; pp. 239–256.

13. Karlsdottir, S.N.; Hjaltason, S.M.; Ragnarsdottir, K.R. Corrosion behaviour of materials in hydrogen sulphide abatement system at Hellisheiði geothermal power plant. *Geothermics* **2017**, *70*, 222–229. [CrossRef]

14. Matsuda, H. Maintenance for reliable geothermal turbine. *GRC Trans.* **2006**, *30*, 755–760.

15. Brown, A.; Müller, S.; Dobrotková, Z. *Renewable Energy: Markets and Prospects by Technology*; IEA: Paris, France, 2011; 66p.

16. Kudryashov, A.E.; Potanin, A.Y.; Lebedev, D.N.; Sukhorukova, I.V.; Shtansky, D.V.; Levashov, E.A. Structure and properties of Cr-Al-Si-B coatings produced by pulsed electrospark deposition on a nickel alloy. *Surf. Coat. Technol.* **2016**, *285*, 278–288. [CrossRef]

17. Xie, Y.J.; Wang, M.C. Isothermal oxidation behaviour of electrospark deposited MCrAlX-type coatings on a Ni-based superalloy. *J. Alloys Compd.* **2009**, *480*, 454–461. [CrossRef]

18. Xie, Y.J.; Wang, M.C. Epitaxial MCrAlY coating on a Ni-base superalloy produced by electrospark deposition. *Surf. Coat. Technol.* **2006**, *201*, 3564–3570. [CrossRef]

19. Podchernyaeva, I.A.; Panasyuk, A.D.; Teplenko, M.A.; Podol'skii, V.I. Protective coatings on heat-resistant nickel alloys (Review). *Powder Metall. Met. Ceram.* **2000**, *39*, 434–444. [CrossRef]

20. Csáki, I.; Stefanoiu, R.; Karlsdottir, S.N.; Geambazu, L.E. Corrosion behavior in geothermal steam of CoCrFeNiMo high entropy alloy. In Proceedings of the CORROSION 2018, Phoenix, AZ, USA, 15–19 April 2018.

21. Stern, M. A method for determining corrosion rates from linear polarization data. *Corrosion* **1958**, *14*, 60–64. [CrossRef]

22. Sheir, L.L.; Jarman, R.A.; Burstein, G.T. *Corrosion: Corrosion Control*; Butterworth-Heinemann: Boston, MA, USA, 1994; pp. 19–38.

23. *ASTM G4-01(2014): Standard Guide for Conducting Corrosion Tests in Field Applications*; ASTM International: West Conshohocken, PA, USA, 2008.

24. Shi, Y.; Yang, B.; Liaw, P.K. Corrosion-resistant high entropy alloys: A review. *Metals* **2017**, *7*, 43. [CrossRef]

25. Sohi, M.H.; Ebrahimi, M.; Raouf, A.H.; Mahboubi, F. Comparative study of the corrosion behavior of plasma nitrocarburised AISI 4140 steel before and after post-oxidation. *Mater. Des.* **2010**, *31*, 4432–4437. [CrossRef]

26. Praveen, S.; Murty, B.S.; Kottada, R.S. Alloying behaviour in multi-component AlCoCrCuFe and NiCoCrCuFe high entropy alloys. *Mater. Sci. Eng.* **2012**, *534*, 8389. [CrossRef]

27. Liu, W.H.; Wu, Y.; He, J.Y.; Nieh, T.G.; Lu, Z.P. Grain growth and the Hall-Petch relationship in a high-entropy FeCrNiCoMn alloy. *Scr. Mater.* **2013**, *68*, 526–529. [CrossRef]

28. Otto, F.; Dlouhý, A.; Somsen, C.; Bei, H.; Eggeler, G.; George, E.P. The influences of temperature and microstructure on the tensile properties of a CoCrFeMnNi high-entropy alloy. *Acta Mater.* **2013**, *61*, 5743–5755. [CrossRef]

29. Zhu, C.; Lu, Z.P.; Nieh, T.G. Incipient plasticity and dislocation nucleation of FeCoCrNiMn high-entropy alloy. *Acta Mater.* **2013**, *61*, 2993–3001. [CrossRef]

30. Tsai, K.Y.; Tsai, M.H.; Yeh, J.W. Sluggish diffusion in Co-Cr-Fe-Mn-Ni high entropy alloys. *Acta Mater.* **2013**, *61*, 4887–4897. [CrossRef]

31. Shun, T.T.; Chang, L.Y.; Shiu, M.H. Microstructure and mechanical properties of multiprincipal component CoCrFeNiMox alloys. *Mater. Charact.* **2012**, *70*, 63–67. [CrossRef]

32. Hansen, M. *Constitution of Binary Alloys, Metallurgy and Metalurgical Enginerring Series*; McGraw-Hill: New York, NY, USA, 1958; pp. 966–982.

33. Tsai, M.-H.; Tsai, K.-Y.; Tsai, C.-W.; Lee, C.; Juan, C.-C.; Yeh, J.-W. Criterion for Sigma phase formation in Cr- and V-containing high-entropy alloys. *Mater. Res. Lett.* **2013**, *1*, 207–212. [CrossRef]

34. Zhu, J.M.; Fu, H.M.; Zhang, H.F.; Wang, A.M.; Li, H.; Hu, Z.Q. Microstructures and compressive properties of multicomponent AlCoCrFeNiMox alloys. *Mater. Sci. Eng.* **2010**, *527*, 6975–6979. [CrossRef]

35. Ye, X.; Ma, M.; Cao, Y.; Liu, W.; Gu., Y. The property research on high-entropy alloy AlxFeCoNiCuCr coating by laser cladding. *Phys. Procedia* **2011**, *12*, 303–312. [CrossRef]
36. Ast, J.; Ghidelli, M.; Durst, K.; Göken, M.; Sebastiani, M.; Korsunsky, A.M. A review of experimental approaches to fracture toughness evaluation at the micro-scale. *Mater. Des.* **2019**, *173*, 107762. [CrossRef]

Article

Effect of Microstructural Modifications on the Corrosion Resistance of CoCrFeMo$_{0.85}$Ni Compositionally Complex Alloy Coatings

Francesco Fanicchia [1], Ioana Csaki [2],*, Laura E. Geambazu [2], Henry Begg [1] and Shiladitya Paul [1]

[1] TWI Ltd., Granta Park, Cambridge CB21 6AL, UK; francesco.fanicchia@twi.co.uk (F.F.);
    henry.begg@twi.co.uk (H.B.); shiladitya.paul@twi.co.uk (S.P.)
[2] Faculty of Engineering, University Politehnica Bucharest, Splaiul Independentei 313,
    060042 Bucharest, Romania; laura.geambazu@gmail.com
* Correspondence: ioana.apostolescu@upb.ro; +40-749-504-540

Received: 30 August 2019; Accepted: 23 October 2019; Published: 24 October 2019

Abstract: A compositionally complex alloy (CCA) was developed in powder form and applied as a coating onto a carbon steels substrate by using thermal spray. The purpose of this study was to investigate the effect of microstructural modification induced by using two different powder production methods, mechanical alloying and gas atomisation, onto the corrosion resistance of the coatings for a CoCrFeMo$_{0.85}$Ni composition. The evolution of microstructure from powders to coatings was analysed using scanning electron microscopy coupled with energy-dispersive spectroscopy and X-ray diffraction. In order to evaluate the corrosion performance of the coatings, electrochemical corrosion tests were performed in a 3.5 wt % NaCl solution at pH = 4. The study demonstrates that the powder production method has a significant influence on the phase composition and, in turn, corrosion behaviour of the resulting coating, with the gas atomising route imparting better corrosion resistance properties. Nevertheless, the appearance of the face-centered cubic (FCC) phase characteristic of the CoCrFeMo$_{0.85}$Ni alloy within the coating produced from the mechanically alloyed powder, opens the possibility for this powder manufacturing technique to effectively produce compositionally complex alloys.

Keywords: compositionally complex alloy; high entropy alloy; coating; thermal spray; microstructure; corrosion

---

1. Introduction

Compositionally complex alloys (CCAs, also referred to as high entropy alloys in literature) are multicomponent equi- or near-equiatomic alloys which form mostly solid solutions including random solid solutions and partially ordered ones. Depending on the property required, e.g. hardness, corrosion resistance, etc., a particular phase in the multiphase alloy is targeted for a specific application [1–3]. Successful manufacture of bulk CCAs by using traditional casting techniques can prove challenging due to large differences in both density and melting point of individual constituent elements. Therefore, solid state processing via mechanical alloying of powder and subsequent consolidation may present a more viable manufacturing route. The requirement for near-equiatomic mixtures can also lead to potentially high cost, if more exotic or costly elements are required. Deposition of CCAs as a coating onto an inexpensive substrate may provide a low-cost means of exploiting advantageous materials properties.

The corrosion resistance in NaCl solution has been studied by Qiu et al. [4], on a laser clad AlCrFeCuCo alloy. The alloy's microstructure showed a combination of face-centered cubic (FCC) and body-centered cubic (BCC) phases and exhibited excellent corrosion resistance, which, however, was observed to vary with process deposition parameters. More recently, the microstructure and corrosion behavior of a plasma sprayed (CoCrFeNi)$_{95}$Nb$_5$ alloy was assessed by Wang et al. [5]

and consisted of Laves phase within an FCC solid solution. Elemental segregation was observed, with higher melting point elements (Cu, Nb) enriching the interdendritic regions. Better corrosion performance than previously studied coatings was measured in 3.5 wt % NaCl solution, with the Cu and Nb-rich regions representing the areas of preferential corrosion. In another work by Gao et al. [6], CCA of the CoCrFeNiAl$_{0.3}$ composition was produced by using radio frequency magnetron sputtering and tested in 3.5 wt % NaCl solution. The coatings consisted of a polycrystalline FCC structure with homogeneous element distribution and showed increased hardness and improved corrosion performance as compared to wrought 304 stainless steel. Some CCAs have demonstrated excellent performance in both H$_2$SO$_4$ and NaCl solutions. Similar to conventional alloys, it is interesting to note that Cr, Ni, Co, Ti in CCAs enhance corrosion resistance in acid solutions, Mo tends to inhibit pitting corrosion, whereas Al and Mn display a negative effect [7]. Wang et al. [8] applied thermal spray technology to fabricate coatings of the Ni$_x$Co$_{0.6}$Fe$_{0.2}$Cr$_y$Si$_z$AlTi$_{0.2}$ composition. Results indicated that the hardness of the CCAs prepared by using the thermal spraying in combination with annealing at 1100 °C for 10 h was significantly increased compared to that of the cast alloy. Moreover, the alloy exhibited excellent corrosion resistance, resulting from the presence of the Cr$_3$Si phase and several other (unidentified) phases. More recently, the effect of grain refinement and elemental partitioning onto the strength and corrosion resistance of a friction stir processed Cu-containing CCA has been evaluated by Nene et al. [9]. Their work shows that grain refinements are an important factor to enhance the strength of the alloy, while the partition of elements such as Cu within the main FCC matrix would worsen the corrosion resistance. It is clear from these studies that the corrosion behavior of CCA coatings is intrinsically linked to their microstructure, which, in turns, depends on the physical-chemical properties of the reagents (e.g., powders) used to produce the coated system.

The aim of this paper is therefore to investigate the relationship between powder and coating microstructure and its effect on the corrosion performance of the latter. As a case study, powders of the CoCrFeMo$_{0.85}$Ni composition have been produced by using both solid state processing (mechanical alloying) and gas atomization and deposited by using high-velocity oxygen fuel (HVOF) spray on carbon steel substrates. The phase evolution during the various stages of CCA coating production has been evaluated by using scanning electron microscopy (SEM), energy-dispersive X-ray spectroscopy (EDX) and X-ray diffraction (XRD), while the corrosion performance was evaluated in 3.5 wt % NaCl solution through the linear polarization resistance technique.

2. Materials and Methods

2.1. Materials and Processing

Mechanically alloyed powders were produced from powders of pure elements Co, Cr, Fe, Ni and Mo (Laboratorium®, Bucharest, Romania), processed with a planetary ball mill (Fritsch–Pulverisette 6®, Idar-Oberstein, Germany) for an effective time of 210 min. Elemental powders were placed in stainless steel vial with stainless steel balls in a 10:1 ball to powder weight ratio for this particular composition. The wet milling process was selected, in 2% n-heptane, in order to increase the alloying ratio and decrease the tendency of the powders to adhere onto the balls or vials. From the overall batch of powder produced, a −56 + 20 µm size distribution was extracted by using mechanical sieving. Gas atomized powders were produced under argon atmosphere (HERMIGA 75/5 VI EAC, Phoenix Scientific Industries Ltd., Brighton, UK), with an estimated cooling rate of 10^5–10^6 °C/s. A powder of final size distribution −48 + 15 µm was finally obtained by means of mechanical sieving. The composition of the materials employed in the work is reported in Table 1 for both powders (reported as nominal weight of elements used for mechanical alloying and gas atomisation) and substrate (as provided by the manufacturer).

Table 1. Nominal composition (wt %) of powders (both mechanically alloyed and gas atomised) and substrates used in the experiments.

Element [wt %]	Co	Cr	Fe	Ni	Mo	C	Mn	Si	P	S	N	Cu	Ti	Al	V
CoCrFeMo$_{0.85}$Ni	19.19	16.94	18.19	19.12	26.56	–	–	–	–	–	–	–	–	–	–
Carbon Steel	–	0.3	bal	0.3	0.08	0.2	0.9	0.35	0.025	0.015	0.012	0.3	0.02	0.03	0.02

The depositions were performed by using a Praxair Surface Technologies (Indianapolis, USA) Tafa Model 5220 HP/HVOF® gun by using kerosene as fuel gas and nitrogen as powder carrier. The set of deposition process parameters, employed for both mechanically alloyed (MA) and gas atomised (GA) powders, is reported in Table 2.

Table 2. High-velocity oxygen fuel process parameters, employed to deposit both mechanically alloyed (MA) and gas atomised (GA) powders.

Process Parameter	Units	Value
Oxygen flow	[slpm]	834
Kerosene flow	[slpm]	0.33
Nitrogen flow	[slpm]	12.27
Standoff distance	[mm]	360
Number of passes	[–]	20

Substrates measuring 25 mm × 25 mm × 6 mm were grit blasted with 60 mesh brown alumina and degreased with acetone prior to deposition.

2.2. Microstructural Analyses

A Zeiss® Supra 25, equipped with an Energy Dispersive X-Ray Spectroscopy detector (EDX, Oxford Instruments®, Oxford, UK) was employed as field emission scanning electron microscope (FE-SEM) for microstructural and chemical analysis. Coating surfaces and cross-sections were analysed before and after corrosion testing. Cross-sections were prepared by mounting specimens in thermosetting phenol formaldehyde resin (Bakelite) and polishing with both SiC abrasive paper and diamond paste to a final roughness of approximately 0.025 μm Ra. The phase composition of both powders and coatings was studied by using X-ray diffraction (XRD) and a Bruker® D8 Advance diffractometer 40 kV, 30 mA, Cu-Kα radiation with λ = 1.5406 Å.

2.3. Electrochemical Corrosion Testing

Coating degradation was assessed by means of linear polarisation resistance test, by using an ACM Instruments® Gill 16 electrochemical system. (Lancaster, UK) In the electrochemical test cell, a saturated calomel electrode (SCE) and platinated titanium electrode were used as reference and auxiliary (counter) electrode respectively. Results were analysed using the ACM® Core Running software (v5.0) A 3.5 wt % NaCl aqueous solution at room temperature (25 °C) and constant pH = 4 was employed. The pH was controlled by adding HCl 7.5 M solution during the experiments. Measurements were taken at −/+ 10 mV ΔE versus the corrosion potential, at a 10 mV/min scanning rate for 168 h with measurement taken every 15 min. This time was selected as blistering was observed to occur within the coating from mechanically alloyed powder, for which it has been decided to end both tests at 168 h (i.e., 7 days).

3. Results

3.1. CCA Composition Design

The composition of the CCA alloy employed in this work was designed according to theoretical calculations of solid solution formation parameter (Ω), entropy of mixing (ΔS_{mix}), enthalpy of mixing (ΔH_{mix}), atomic size difference (δ) and valence electron concentration (VEC) [10]:

$$\Omega = \frac{T_m \Delta S_{\text{mix}}}{\Delta H_{\text{mix}}} \tag{1}$$

$$\Delta S_{\text{mix}} = -R \sum_{i=1}^{n} c_i ln c_i \tag{2}$$

$$\Delta H_{\text{mix}} = \sum_{i=1,\, i \neq j}^{n} 4 \Delta H_{ij}^{\text{mix}} c_i c_j \tag{3}$$

$$\delta = \sqrt{\sum_{i=1}^{n} c_i \left(1 - \frac{r_i}{\bar{r}}\right)} \tag{4}$$

$$VEC = \sum_{i=1}^{n} c_i VEC_i \tag{5}$$

where n is the number of components, i and j refer to a specific component, c is the atomic fraction, T_m is the melting point of the alloy (defined by the mixing rule), R is the ideal gas constant, ΔH_{ij}^{mix} is the enthalpy of mixing of elements i and j, r is atomic radius and $\bar{r} = \Sigma c_i r_i$ is the average radius of the alloy.

These thermodynamic parameters can be used in an attempt to predict the phases and crystal structure of a CCA. According to the criteria of phase stability formation calculated by Zhang et al. [11], if $\Omega \geq 1.1$ and $\delta \leq 6.6$ then the solid solution forms. Takeuchi et al. [12] studied the effects of changes in the mixing entropy, ΔS_{mix}, on the crystal structure formation. Their findings suggest that the high mixing entropy provided by CCAs could overcome the enthalpy of formation of strong intermetallic compounds, thereby suppressing their generation in favor of a random solid solution. Additionally, the VEC seems to play a decisive role in determining whether FCC or BCC crystal structure solid solution forms in CCAs. Specifically, a high VEC (≥ 8) favors the formation of FCC-type solid solutions, while a lower VEC (<6.87) favors the formation of BCC-type solid solutions, according to Guo et al. [13].

In this work, the CCA composition was selected based on its likelihood to provide significant performance in terms of both tribological and corrosion resistance. For tribological performance (e.g., resistance to erosion), a mixture of FCC and BCC structures would be expected to provide adequate performance as the FCC phase is expected to present a ductile nature while the opposite holds for a BCC crystal structure [14]. Resistance to corrosion is instead expected to be enhanced by elements such as Ni, Mo, and Cr, due to the formation of passive films on the alloys surface [15]. Therefore, based on the above guidelines and following previous promising work on the CCA of the CoCrFeMo$_x$Ni class as material for corrosion protection in corrosive geothermal steam [16], this alloy class was selected for further analysis. After an iterative design study on several possible Mo$_x$ concentrations, the CoCrFeMo$_{0.85}$Ni composition was identified as CCA for this study. The thermodynamic parameters in Equations (1) to (5) above, calculated for the CoCrFeMo$_{0.85}$Ni alloy, are reported in Table 3. Calculations take into account only the component type and not the fabrication method, so the values are identical for both GA and MA powder.

Table 3. Calculated thermodynamic parameters for the compositionally complex alloy (CCA) alloy of CoCrFeMo$_{0.85}$Ni composition selected in this study.

Parameter	Units	Value
VEC	–	7.86
δ	–	4.268
ΔS_{mix}	J/K mol	13.36
ΔH_{mix}	kJ/mol	−4.574
Ω	–	5.992

3.2. Microstructure of Powders and Coatings

SEM micrographs of the MA and GA powders are reported in Figure 1. As expected, two completely different morphologies are revealed in the two powders: an irregular shape, with particles of angular and blocky geometry (MA) and homogeneous spherical particles (GA).

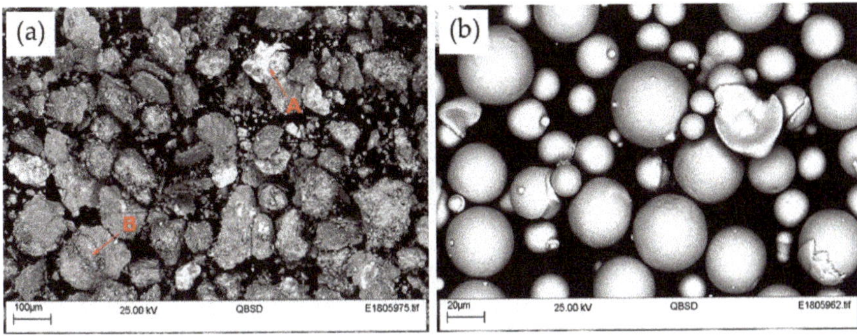

Figure 1. SEM micrograph of (**a**) mechanically alloyed (MA) powder and (**b**) gas atomized (GA) powder. The presence of particles of different composition is clearly visible in the MA powder.

This morphological variation can be attributed to the differences in the mechanical and thermodynamic processes involved in the two manufacturing methods. Mechanical alloying (MA) uses mechanical impact which generates high localized pressure, to break and bond together particles of the elements to be alloyed. The gas atomisation method (GA) instead, relies on the in-flight solidification of particles directly atmoised from a melt of the alloy composition. The spherical geometry then naturally arises due to its lower Gibbs free energy over that of other possible geometries. Localised EDX analysis on the mechanically alloyed powder demonstrates that the bright regions (marked as A in Figure 1a) are composed of Mo with only traces of the other CCA elements, Co, Cr, Fe and Ni. A variable composition is instead probed in other grey regions in the powder, such as region B, where no clear compositional pattern could be found. This compositional variation, which would suggest incomplete alloying, was somewhat expected at the short total alloying time employed in this study, 210 min, selected in an attempt to provide a final powder size distribution processable by means of HVOF. It is therefore likely that a composition close to the CCA could be achieved at further mechanical alloying processing time, however with a corresponding reduction in particles' nominal size distribution. Bulk EDX patterns for both powders show similar elemental distribution, although a reduced concentration of Mo is measured in the gas atomised powder, likely due to the differences in vapour pressure between the elements during the melting and solidification of this latter process. EDX analysis performed on the GA powder did not reveal any localised area of a composition different from the theoretical CCA. XRD spectra for the two powders are presented in Figure 2.

The scans revealed a clear distinction between the phases identified in the two powders. The MA powder showed peaks characteristic of the single elements composing the CCA (i.e., Ni, Cr, Fe FCC and Mo BCC), thus suggesting incomplete alloying as previously suggested by EDX analysis. The elemental peaks did not appear to be present in the GA powder, where instead new peaks corresponding to the FCC and BCC phases of the $CoCrFeMo_{0.85}Ni$ composition appeared, thus suggesting the attainment of solid solution. The fact that the measured 2θ location for these peaks corresponds to the same angles was also identified by Shun et al. [17] on the same alloy composition. Moreover, as in [17], peaks corresponding to the intermetallic σ and π phases were also measured. It is likely that the FCC phase was stabilized and rich in elements showing a stable FCC structure (Co, Fe and Ni), while the BCC and the small amount of σ phase could be stabilized by Mo.

The microstructure of the resulting MA and GA coatings, obtained from HVOF deposition of the powders in Figure 1, is reported in Figure 3. Average thicknesses for the two coatings have been measured as ~130 and ~250 μm for MA and GA respectively.

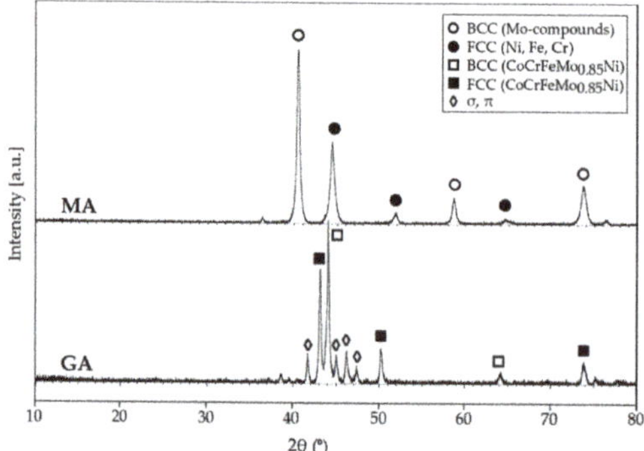

Figure 2. XRD scans for the mechanically alloyed (MA, top) and gas atomised (GA, bottom) powders. The MA powders show peaks related to the single elements of the CCA, while new face-centered cubic (FCC) and body-centered cubic (BCC) peaks are observed in the GA powder, suggesting that alloying has occurred.

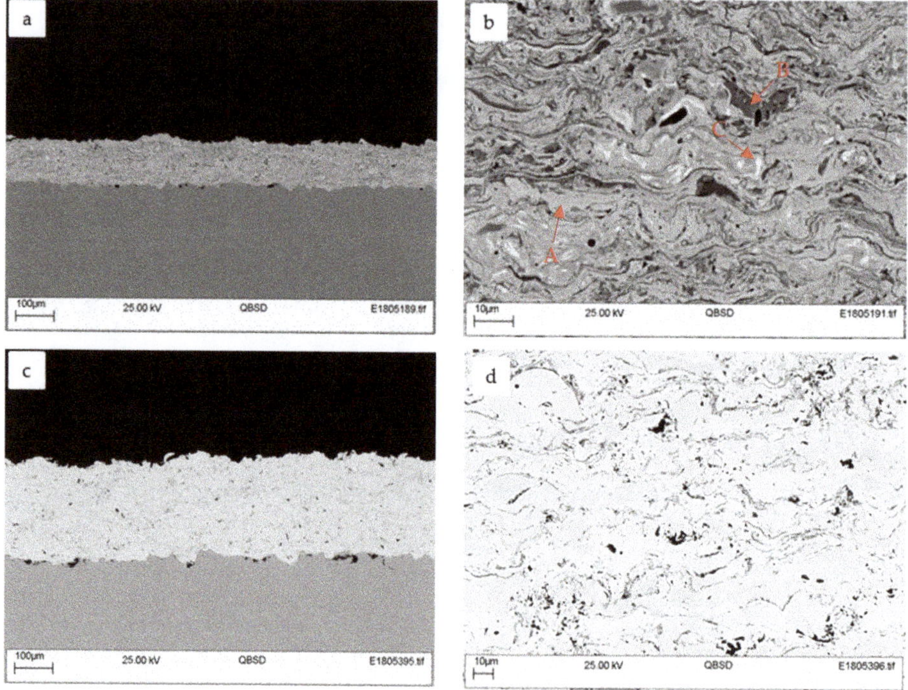

Figure 3. SEM micrographs of coatings cross-section for (**a,b**) MA and (**c,d**) GA coatings. For the MA powder, the inhomogeneity of the initial powder was also observed in the resulting coating, while a more uniform phase distribution was observed for the GA system.

The cross-sections show some interesting features. There seems to be a clear difference in terms of phase distribution in the two coatings. While the GA system showed a homogeneous distribution of what appeared to be a single phase, splats of different composition were observed in the MA system. An EDX analysis of the regions of differing contrast within the MA coating (Figure 3b) showed that the light grey matrix (labelled as A in Figure 3b) exhibited peaks for all of the CCA components and is thus likely to represent the FCC (or BCC) structure characteristic of the $CoCrFeMo_{0.85}Ni$ composition. The darker grey area (labelled as B in Figure 3b), exhibited similar peaks, but indicated a lower concentration of the element Mo compared with region A. Mo was also the main element composing the lightest region (labelled C in Figure 3b), which would suggest the presence of powder particles of elemental Mo within the initial powder. Due to the extremely high solidification rates experienced during HVOF deposition, it is unlikely that these Mo-rich regions could be formed by elemental segregation during solidification.

In order to further investigate the nature of the phases observed, XRD analysis was performed on the two coatings, as reported in Figure 4.

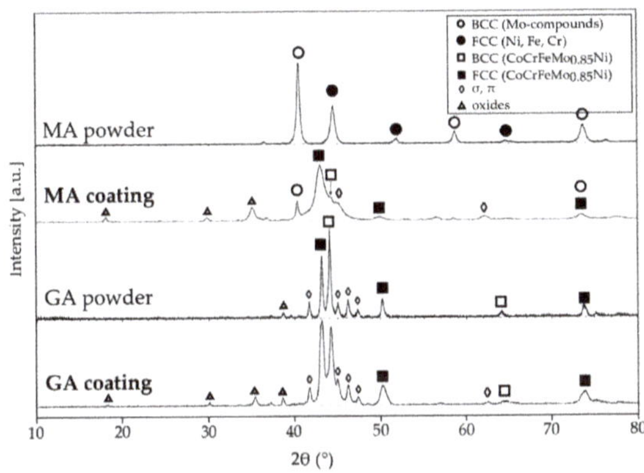

Figure 4. XRD scans for mechanically alloyed (MA) and gas atomized (GA) coatings and powders, showing how FCC and BCC structures are present within the MA coating besides these being absent in the original MA powder.

The phases present in the MA coating are a mixture of FCC and BCC phases characteristic of the CCA composition, together with some residual BCC linked to the Mo component and newly developed σ/μ phases. This suggests that alloying occurred in-flight during the thermal spray operation. The coating procedure, with exposure of the powder at very high temperature followed by rapid cooling, promoted the appearance of a small amount of BCC phase stabilized by the FeCr pair. The same phenomenon was observed in the CoCrFeMoNi high entropy alloy processed by using vacuum arc remelting and the BCC phase was present in the structure due to the rapid cooling process [18]. The coating also shows the appearance of oxide phases, naturally generated within the oxidising atmosphere of the thermal spray process. Conversely, similar phases were observed in both GA coating and powder, suggesting that limited modifications occurred in-flight, during the thermal spray operation. The only difference between the GA coating and powder was the appearance of oxide peaks, as expected from the thermal spray operation.

3.3. Electrochemical Corrosion Testing

Photographs of the surfaces of the coated specimens are presented in Figure 5 for both MA and GA before and after exposure to the corrosion environment.

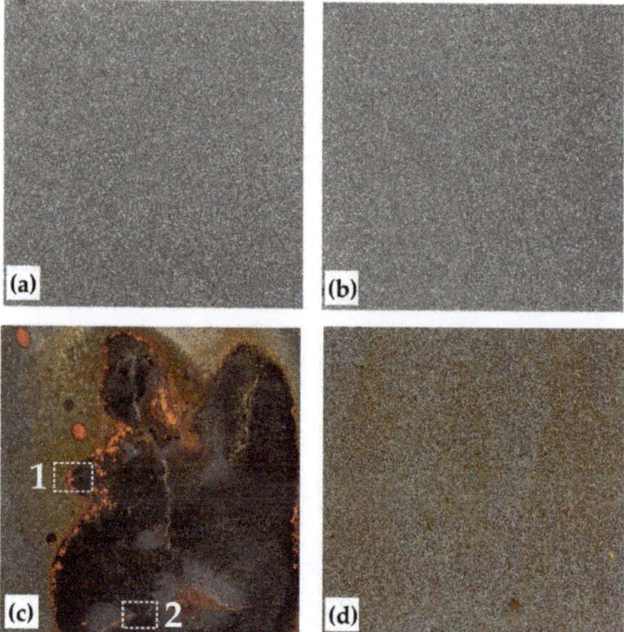

Figure 5. Photographs of the exposed surface of MA (**a,c**) and GA (**b,d**) coatings before (**a,b**) and after exposure (**c,d**). A notable structural damage linked to development of corrosion products is observed on the surface of the MA coating, while a more uniform type of attack is noted on the GA specimen.

While the appearance of the as-deposited coatings was comparable between the two coatings (Figure 5a,b), the condition post exposure showed significant differences. In the MA coating, corrosion products and cracks were observed, while no cracking was visible in the GA system. However, the presence of localised discoloured spots was visible in the GA system. In the case of MA system, it is proposed that permeation of the corrosive solution through the coating has occurred. The permeation of the solution would have led to corrosion of the underlying carbon steel substrate, thereby generating corrosion products. The likely corrosion products for carbon steel would be mixtures of compounds of the steel constituents, which have a Pilling–Bedworth (P–B) ratio >1.5. Pilling–Bedworth ratio is the ratio of the volume of the metal oxide formed from a volume of the corresponding metal (from which the oxide is created). The high P–B ratio indicates that the growth of the corrosion product under the coating would lead to tensile stresses on the top surface of the coating which might crack once the stresses reach a critical level. A more in-depth analysis (SEM-EDX) of regions 1 and 2 in Figure 5c is presented in Figure 6a,b respectively.

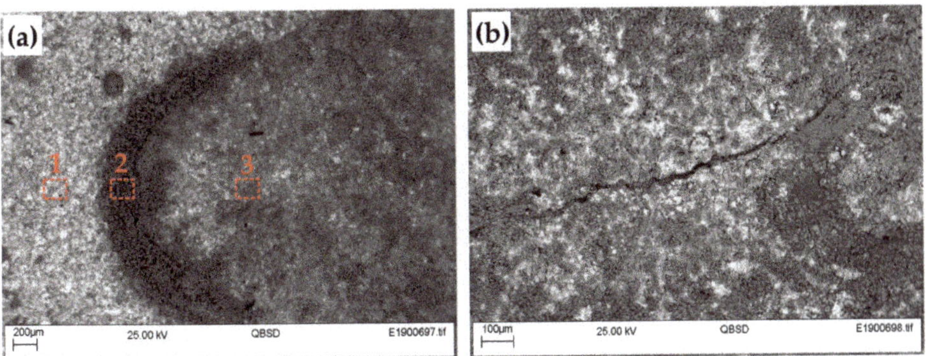

Figure 6. SEM micrographs of MA coating after exposure (Figure 5c), corresponding to the highlighted regions (**a**) 1, showing the presence of different corroded regions and (**b**) 2, showing cracks on the surface.

Figure 6a shows regions of different appearance, corresponding to various corrosion products. An EDX analysis of the three highlighted regions shows the lighter region on the left-hand side of the figure (region 1) containing primarily the constituents of the CCA (i.e., Co, Cr, Fe, Ni and Mo), with Fe having the highest concentration, together with elements O, Cl and Na. The presence of Cl and Na were likely due to the adhering salt solution on the rough coating surface. The darker region 2 is largely dominated by the element Fe, with the additional presence of Cl and O and some traces of the elements of the CCA. Finally, region 3 exhibited peaks comparable to region 2, but with a higher concentration in CCA elements, likely due to a reduced thickness of the oxide layer in the former region.

Two types of cracks can be identified in the corroded MA specimen, as is clear from the cross-section micrographs in Figure 7. Figure 7a shows a vertical crack within the coating at a location of coating detachment. This type of crack is likely generated by mechanical stresses created by the growth of corrosion product at the substrate/coating interface. The substrate/coating delamination is linked to the low toughness of the corrosion product generated at the substrate/coating interface, as depicted by the SEM micrograph in Figure 7b, which shows the location of crack initiation. The effect of stress-induced coating cracking from growth of corrosion products has already been observed [19].

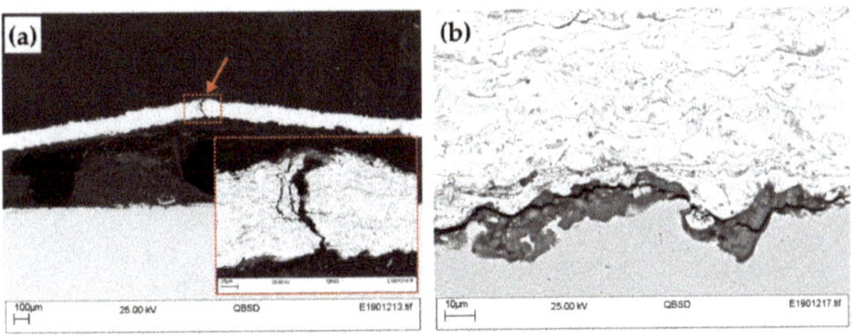

Figure 7. Cross-sectional SEM micrograph of the MA specimen in Figure 5c, showing delaminated coating due to interfacial growth of corrosion product at the substrate/coating interface (**a**) and cracking within in the corrosion product, in an area away from the delaminated region (**b**).

In the case of the GA coating (Figure 5d), although localised darker areas were observed, the discolored appearance suggests that corrosion products have started to form, even if not covering the entire surface. An in-depth SEM-EDX analysis of the coating surface is presented in Figure 8.

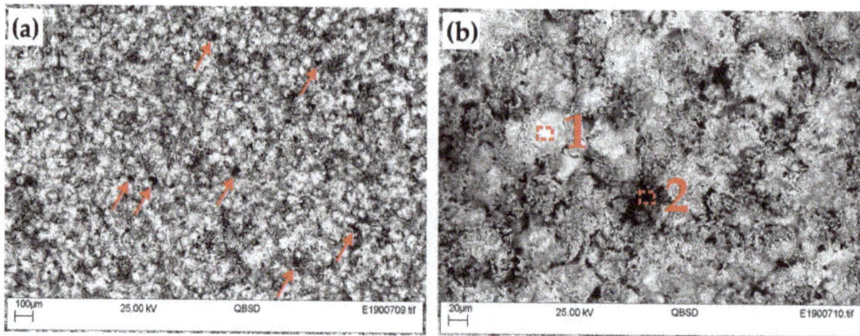

Figure 8. SEM micrograph of the exposed surface of the GA specimen in Figure 5d at (**a**) low and (**b**) high magnification, showing the presence of light areas, likely representing the splats of the as-deposited coating and darker areas, probably areas of preferential corrosion on the coating surface.

The low-magnification micrograph, Figure 8a, shows the presence of darker areas on the coating surface (highlighted by red arrow in the figure), representing a combination of initial surface porosity and localised areas of corrosion (as also seen in Figure 5d). EDX analysis performed at high-magnification (Figure 8b), showed that the lighter particles (region 1), of flat splat-like morphology, contain the CCA elements in the theoretical stoichiometry, while Fe has the highest concentration amongst the other CCA elements in the dark areas (region 2). This would suggest that corrosion was occurring preferentially in those regions. An in-depth analysis of this effect is provided by an SEM-EDX analysis of the specimen cross-section (Figure 9).

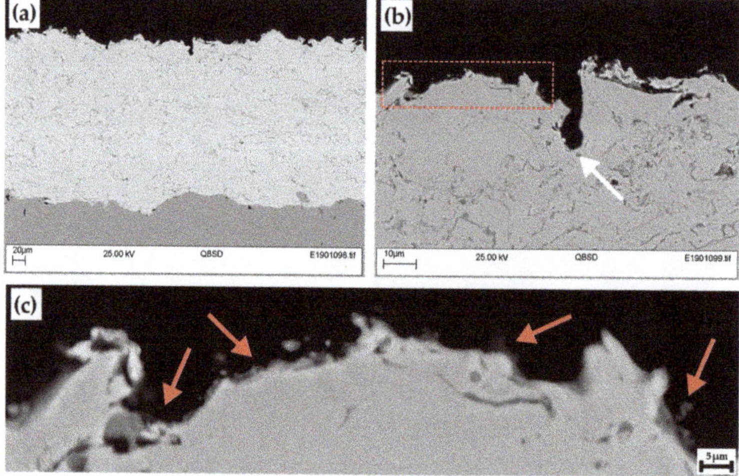

Figure 9. Cross-sectional SEM micrograph of the GA specimen in Figure 5d, demonstrating (**a**) no coating permeation at low magnification, and (**b,c**) the presence of a thin layer of corrosion product at the surface of the coating.

The low-magnification micrograph (Figure 9a) showed no evidence of through-coating permeation, while higher magnification micrographs (Figure 9b,c) showed evidence of corrosion products. These corrosion products could not be uniquely identified by using XRD or EDX analysis due to their small dimensions. It is worth noting the feature identified by a white arrow in Figure 9b, where the shape observed is unlikely to be representative of an as-deposited coating, due to the tall vertical wall observed at the right side. This, together with analysis of current density data showing several peaks, suggests that pitting occurred in the specimen at such locations. Areas on the bottom of features in rough specimens generate localized areas of higher Cl⁻ concentration, therefore locally increasing the dissolution kinetics of the metal. It is suggested that in this sample, the passive film naturally generated on the surface breaks down at localized areas of higher roughness. If the local breakdown region is then re-passivated, and if the solution is agitated so as to take away loosely bound corrosion products from the surface, it is possible that a continuous depassivation/re-passivation cycle is repeated. This may explain the absence of corrosion product on the bottom of the identified feature.

Finally, an understanding of the corrosion mechanism can be appreciated also by analysing the corrosion potential over time for the two specimens as well as the substrate alone (Figure 10). The two coatings show different behaviours. The corrosion potential of the MA coating showed a decreasing trend throughout the test, which approached asymptotically the corrosion potential of the substrate. This would confirm that the corrosive solution had reached the interface between coating and substrate, causing the latter to preferentially corrode. The potential gap between MA coating and substrate is given by the fact that it is likely that not only is the substrate corroding but a mixture between coating and substrate. The gas atomized coating (GA), conversely, showed a corrosion potential of increasing trend and of much higher magnitude compared with the GA coating and substrate. This increasing trend could be justified either by the formation and growth of a passive film on the coating surface or closing of open porosity (naturally present within the as-deposited coating) or a combination of the two. This possible phenomenon needs to be further verified by further testing.

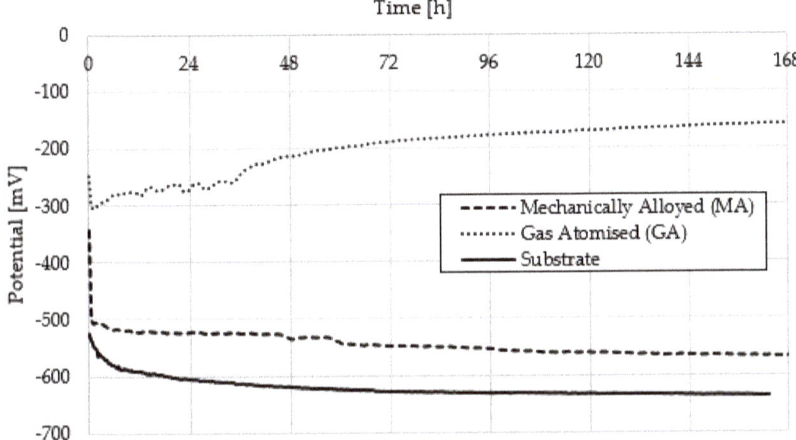

Figure 10. Graph of corrosion potential over time for MA, GA and substrate only specimens demonstrating corrosion at the substrate/coating interface for specimen MA, and possibly passivation and closure of open porosity for specimen GA.

Further insights can be provided by analyzing the linear polarization resistance (LPR) curves of the systems tested (Figure 11).

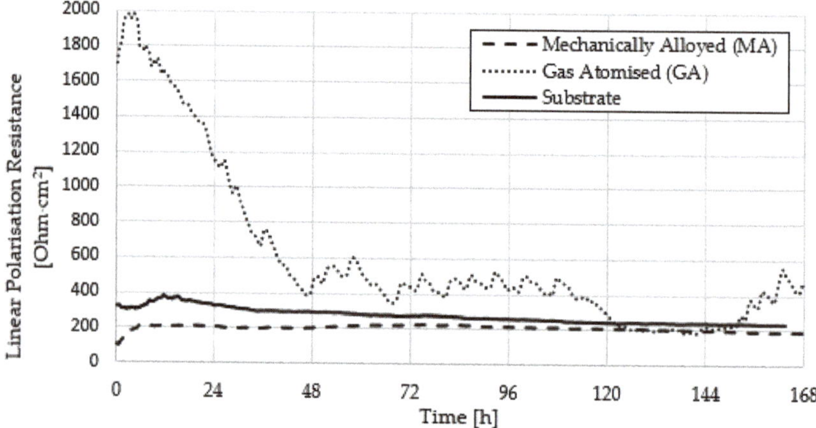

Figure 11. Graph of linear polarisation resistance over time for MA, GA and substrate only specimens.

The values measured for the MA and substrate systems exhibit a rather constant behavior throughout the exposure, suggesting that uniform corrosion is occurring. The low absolute LPR values measured for these two systems show that little resistance to the corrosive environment at the conditions employed. A different behavior is observed in the GA system, with an initially higher LPR followed by a decreasing trend with exposure time. This trend suggests the initial generation of a passive film onto the coating surface, followed by breakdown as exposure progresses. The local spikes observed in the curve of the GA system suggests local passive film formation/breakdown, which could be corresponding to the events observed at the areas of higher localised roughness (see Figure 9). The nature and formation mechanism of passive film onto Cr^- and Mo-containing alloys has received considerable interest during the years. For instance, Duarte et al. [15] have evaluated the passivation scheme for amorphous and nanocrystalline $Fe_{50}Cr_{15}Mo_{14}C_{15}B_6$ stainless-type glass-forming alloy. According to the study, although the crystallinity and nature of the corrosive environment have a great effect on the nature of the passive film, in general the Fe element is observed to corrode first, followed by a surface enrichment in Cr due to the passive film generation. The growth and/or breakdown of the passive film is then a complex phenomenon, driven by the specific alloy composition due to the formation of Cr-depleted and Mo-rich regions within the material. Although this cannot be conclusively proven by this work, it is likely that the more homogeneous nature of the phases present within the GA-type coating could lead to the consistent formation of a Cr-rich passive layer on the exposed splats' surface, with local breakdowns only linked to the rough nature of the surface, generating localized areas of high ions concentration. Conversely, it is likely that the inhomogeneous nature of the MA coating, with splats of single elements composition, leads to the quick preferential corrosion of Fe-rich regions already within the initial stages of exposure, thus irreversibly damaging the coating structure and generating preferential paths for corrosive medium permeation. In this case, the rapid corrosion kinetics of the Fe-rich phases highly overcomes the possible formation of passive films on other areas of the coating. It is worth noting that open porosity was present in both as-deposited MA and GA coating systems at comparable levels (see Figure 3). The fact that permeation was observed only within specimen MA, and the fact that multiple phases (as opposed to the solid solution of specimen GA) are present within this coating, would suggest that galvanic effects between phases have a major effect in generating preferential corrosion paths.

Finally, it is worth noting that the LPR approach employed for corrosion evaluation, only acts as a preliminary approach to corrosion evaluation. Additional electrochemical tests are currently being undertaken on freestanding coatings to understand the fundamental passivation/depassivation mechanisms which determine the corrosion behaviour of the studied systems.

4. Discussion

The mutual solubility between solvent and solute components in a binary alloy system could be judged by using Hume-Rothery rules, namely, crystal structure, atomic size difference, valence and electronegativity. In fact, all these factors also influence the interaction between different elements and make the enthalpy of mixing either negative (attractive interaction leading to ordering and the formation of intermetallic compounds), positive (repulsive interaction leading to clustering and segregation) or near zero (leading to the formation of disordered solid solutions). The competition between enthalpy of mixing and entropy of mixing further affects the solubility between two components. When solubility is limited, terminal solid solutions (rich in one specific component) can be obtained. When a solid solution forms at all compositions, the system is called isomorphous. However, continuous solid solutions in binary alloy system are not common because the conditions for their formation are very strict to fulfill [20].

In compositionally complex alloys, the high entropy developed enhances the mixing of elements and therefore could stabilize the solid solution formation over intermetallic compounds. Conversely, when a large positive mixing enthalpy occurs only between some specific elements of the alloy, strong elemental segregation might occur [21]. The phases expected from CALPHAD® calculations on the CoCrFeMo$_{0.85}$Ni alloy at room temperature are an FCC phase and σ, or σ + μ phases [17,22] The formation of σ phase is an indication that different types of solid solutions in CCAs could form depending on the interaction and atomic size difference between elements and not just the configurational entropy alone. The σ phase is in fact a topologically close-packed phase in which components with larger atomic size occupy one specific set of lattice sites while smaller atoms occupy another set so as to get a higher amount of bonding to lower their overall free energy, although their interactions (or enthalpy of mixing) between components are small [18] Within the CCA of composition CoCrFeMo$_x$Ni, both σ and μ phases have been proven to be (Mo,Cr)-rich phases, with the μ phase likely generated from σ due to high stresses generated during solidification [17]. The presence of these hard intermetallic phases has been proven to strengthen the overall alloy with ductile FCC CoCrFeNi matrix [22]. If it could be demonstrated that this strengthening effect holds true, and if the corrosion resistance of the same alloy is proven satisfactory, the CCA studied in this project can be effectively employed as an erosion-corrosion barrier for geothermal applications [18].

An in-depth analysis of XRD data (Figure 4) showed that both coatings produced in this work, MA and GA, present a mixture of FCC and BCC phases, together with a reduced presence of a combination of σ and μ. The presence of the BCC phase is important for applications involving tribological resistance as this phase could provide improved mechanical properties, such as higher yield strength, than that of FCC CCAs. The presence of the BCC phase in the CoCrFeMo$_{0.85}$Ni alloy was not observed by Shun et al. on alloys produced by using arc melting. Strictly speaking, CALPHAD calculations performed by Liu at al. [22] on the same alloy class show that only FCC, σ and μ phases are the predicted stable phases at room temperature for the specific composition under analysis and a BCC stability region appears only for mole fractions of Mo higher than ~0.4 and high temperatures (>1500 °C). The appearance of BCC phase, at room temperature, in the coatings produced by using thermal spray in this work would thus suggest that a metastable BCC phase is generated at high temperature in Mo-rich regions (created by local-solute segregation during solidification) and then retained after the material has solidified, due to the extremely high solidification rate characteristic of the thermal spray technique [23]. For the MA coating, a small peak corresponding to the BCC phase of the CCA alloy appears overlapped with a shoulder containing contributions from the BCC CCA phase and σ/μ phases (Figure 4). Note that the peaks of this overlapped region do not coincide with the location of the FCC (Ni, Fe and Cr) peak, originally present in the MA powder. Moreover, it is interesting to note that some of the BCC phase characteristic of the Mo-rich compounds, originally present in the MA powder, is also retained in the coating system, suggesting that not all of the Mo particles present in the powder have alloyed during the process.

However, perhaps the most interesting feature observed in the MA coating system, compared with its parent powder, is the appearance of the FCC peak characteristic of the CCA composition.

The presence of this peak suggests that, although alloying was not observed in the original powder, this has been developed during the spray operation. This is a surprising finding as one would expect, from an observation of the SEM micrograph in Figure 1a that the loosely bounded powder flakes would detach from each other in-flight. The presence of the FCC phase in the coating instead suggests that a degree of bonding has been attained by the flakes in the MA powder, with the consequence that component mixing can occur in-flight (Figure 12).

As-manufactured powder

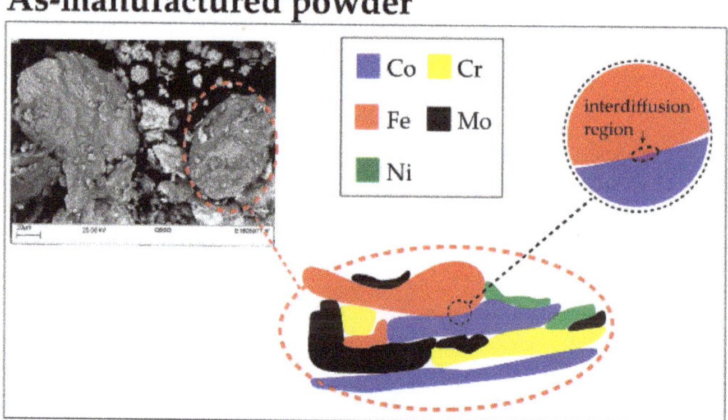

During thermal spray

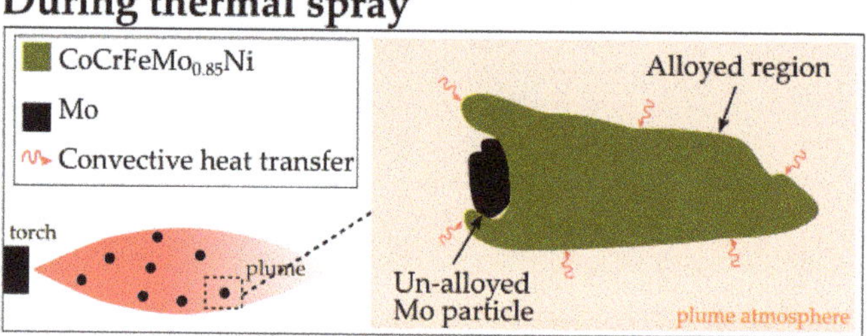

Figure 12. Schematic representation of proposed alloying process occurring for MA powders during the thermal spray process. The original powders are composed of portions of single elements, bound together by localized regions where interdiffusion has occurred in the mechanical alloying process. These intediffusion regions allow the portions of single elements to be kept in close proximity in-flight during the thermal spray process. The heat transferred convectively by the thermal spray plume is then sufficient to melt the single elements, thus generating an alloy of the CCA composition.

A more simplistic situation is provided by the GA system, where the same phases were observed both in the powder and coating. The gas atomization process was sufficient to generate alloying for the composition under analysis. The presence of the metastable BCC phase within the GA powder would suggest the influence of solidification kinetics in the process, due to the extremely high solidification rates attained during gas atomization. It should be mentioned that phase type prediction has been here performed in a qualitative manner from constituent-elements features and/or unlike-atomic pairs features, thus in accordance with the Hume-Rothery rules. These rules, while being comprehensive, do not allow for an accurate prediction because CCAs involve multi-elements and many unlike atomic pairs. The thermodynamic parameters calculated for the CCA composition of this work (see Table 3), are at the border between an alloy with pure FCC and a combination of FCC + BCC phases and therefore a limited quantity of BCC phase was expected. As observed from the results of this work, the phases present in the final alloy material did not always match what predicted, even by using thermodynamic calculations performed by powerful instruments such as CALPHAD and considerations on the kinetics of phases' formation must also be accounted for. Nevertheless, the appearance of the CCA solid solution FCC and BCC phases in the coating produced from mechanically alloyed powder is very promising as it opens the possibility to manufacture coatings of any specific CCA composition by means of mechanical alloying, as opposed to more traditional and complex technologies, such as gas atomization. However, the poor corrosion results obtained by the mechanically alloyed coating, compared with the gas atomized one, also suggest that more work is required in order to optimize the coating microstructure and attain a more homogeneous phase distribution within the final system.

5. Conclusions

The powder-to-coating phase evolution and its influence on the corrosion properties of a compositionally complex alloy (CCA) of the composition $CoCrFeMo_{0.85}Ni$ has been assessed by means of scanning electron microscopy-electron dispersive X-ray spectroscopy (SEM-EDX), X-ray diffraction (XRD) and linear polarisation resistance (LPR). Mechanical alloying (MA) and gas atomisation (GA) have been employed as powder production techniques and coatings were deposited, at the same deposition conditions for the two powders, by means of high-velocity oxygen fuel spray (HVOF).

The research has highlighted the following main findings:

- For the same CCA composition, the powder manufacturing process has a significant influence on the microstructure of the final coating, even when manufactured using the same deposition conditions. While a homogeneous phase distribution was observed in the coating produced from GA powder, splats of different composition were observed in the coating produced via the MA route.
- The corrosion performance of the two coatings depended significantly on the coating microstructure. Despite similar levels of porosity, less corrosion resistance was observed in the MA coating compared with the GA coating. This was attributed to galvanic effects provided by the non-homogeneous phase distribution within the MA coating. Therefore, further optimisation of the mechanical alloying powder production parameters is suggested in order to improve the corrosion resistance properties of CCAs manufactured by this route.
- Both MA and GA coatings are characterised by FCC, BCC and a combination of $\sigma + \mu$ phases characteristic of the CCA composition. However, the coating generated from the MA powder also shows the presence of original BCC phase characteristic of Mo-rich compounds. The presence of CCA-characteristic phases within the MA coating was surprising as these phases were not detected within the original powder. This opens the possibility to employ the mechanical alloying route of powder production for the manufacture of CCA coatings as opposed to more costly techniques such as gas-atomisation.

Author Contributions: Data curation, I.C.; Investigation, F.F. and L.E.G.; Supervision, S.P.; Writing—original draft, F.F.; Writing—review & editing, I.C. and H.B.

Funding: This work is part of the H2020 EU project Geo-Coat: "Development of novel and cost-effective corrosion resistant coatings for high temperature geothermal applications" funded by the H2020 EU project no. 764086.

Acknowledgments: The authors would also like to acknowledge the resources and collaborative efforts provided by the consortium of the Geo-Coat project.

References

1. Zhang, Y.; Zuo, T.T.; Tang, Z.; Gao, M.C.; Dahmen, K.A.; Liaw, P.K.; Lu, Z.P. Microstructure and properties of high entropy alloys. *Prog. Mater. Sci.* **2014**, *61*, 1–93. [CrossRef]
2. Koundinya, N.T.B.N.; Babu, C.S.; Sivaprasad, K.; Susila, P.; Babu, N.; Baburao, J. Phase evolution and thermal analysis of nanocrystalline AlCrCuFeNiZn high entropy alloy produced by mechanical alloying. *J. Mater. Eng. Perform.* **2013**, *2*, 3077–3084. [CrossRef]

3. Tsai, K.Y.; Tsai, M.H.; Yeh, J.W. Sluggish diffusion in Co–Cr–Fe–Mn–Ni high-entropy alloys. *Acta Mater.* **2013**, *61*, 4887–4897. [CrossRef]

4. Qiu, X.W.; Zhang, Y.P.; He, L.; Liu, C. Microstructure and corrosion resistance of AlCrFeCuCo high entropy alloy. *J. Alloy. Compd.* **2013**, *549*, 195–199. [CrossRef]

5. Wang, W.; Qi, W.; Xie, L.; Yang, X.; Jiantao, L.; Zhang, Y. Microstructure and Corrosion behavior of (CoCrFeNi)95Nb5 high-entropy alloy coating fabricated by plasma spraying. *Materials* **2019**, *12*, 694. [CrossRef]

6. Gao, L.; Liao, W.; Zhang, H.; Surjadi, J.U.; Sun, D.; Lu, Y. Microstructure, mechanical and corrosion behaviors of CoCrFeNiAl0.3 high entropy alloy (HEA) films. *Coatings* **2017**, *7*, 156. [CrossRef]

7. Chen, Y.Y.; Duval, T.; Hung, U.D.; Yeh, J.W.; Shih, H.C. Microstructure and electrochemical properties of high entropy alloys-a comparison with type-304 stainless steel. *Corros. Sci.* **2005**, *47*, 2257–2279. [CrossRef]

8. Wang, L.M.; Chen, C.C.; Yeh, J.W.; Ke, S.T. The microstructure and strengthening mechanism of thermal spray coating NixCo0.6Fe0.2CrySizAlTi0.2 high entropy alloys. *Mater. Chem. Phys.* **2011**, *126*, 880–885. [CrossRef]

9. Nene, S.; Frank, M.; Liu, K.; Sinha, S.; Mishra, R.; McWilliams, B.; Cho, K. Corrosion-resistant high entropy alloy with high strength and ductility. *Scr. Mater.* **2019**, *166*, 168–172. [CrossRef]

10. Ye, Y.; Wang, Q.; Lu, J.; Liu, C.; Yang, Y. High-entropy alloy: Challenges and prospects. *Mater. Today* **2016**, *19*, 349–358. [CrossRef]

11. Zhang, Y.; Zhou, Y.J.; Lin, J.P.; Chen, G.L.; Liaw, P.K. Solid-Solution phase formation rules for multi-component alloys. *Adv. Eng. Mater.* **2008**, *10*, 534–538. [CrossRef]

12. Takeuchi, A.; Inoue, A. Classification of bulk metallic glasses by atomic size difference, heat of mixing and period of constituent elements and its application to characterization of the main alloying element. *Mater. Trans. JIM* **2005**, *46*, 2817–2829. [CrossRef]

13. Guo, S.; Ng, C.; Lu, J.; Liu, C.T. Effect of valence electron concentration on stability of fcc or bcc phase in high entropy alloys. *J. Appl. Phys.* **2011**, *109*, 1–6. [CrossRef]

14. Tang, Z.; Zhang, S.; Cai, R.; Zhou, Q.; Wang, H. Designing High Entropy Alloys with dual fcc and bcc solid-solution phases: Structures and mechanical properties. *Metall. Mater. Trans. A* **2019**, *50*, 1888–1901. [CrossRef]

15. Duarte, M.; Klemm, J.; Klemm, S.; Mayrhofer, K.; Stratmann, M.; Borodin, S.; Romero, A.; Madinehei, M.; Crespo, D.; Serrano, J.; et al. Elemental-Resolved Corrosion Analysis of Stainless-Type Glass-Forming Steels. *Science* **2013**, *341*, 372–376. [CrossRef]

16. Csaki, I.; Karlsdottir, S.N.; Stefanoiu, R.; Geambazu, L.E. *Corrosion Behaviour in Geothermal Steam of CoCrFeNiMo High Entropy Alloy*; NACE: Phoenix, AZ, USA, 2018.

17. Shun, T.T.; Chang, L.Y.; Shiu, M.H. Microstructure and mechanical properties of multiprincipal component CoCrFeNiMox alloys. *Mater. Charact.* **2012**, *70*, 63–67. [CrossRef]

18. Karlsdottir, S.N.; Geambazu, L.E.; Csaki, I.; Thorhallsson, A.I.; Stefanoiu, R.; Magnus, F.; Cotrut, C. Phase evolution and microstructure analysis of CoCrFeNiMo high-entropy alloy for electro-spark-deposited coatings for geothermal environment. *Coatings* **2019**, *9*, 406. [CrossRef]

19. Wang, Y.; Huang, Z.; Yan, Q.; Liu, C.; Liu, P.; Zhang, Y.; Guo, C.; Jiang, G.; Shen, D. Corrosion behaviors and effects of corrosion products of plasma electrolytic oxidation coated AZ31 magnesium alloy under the salt spray corrosion test. *Appl. Surf. Sci.* **2016**, *378*, 435–442. [CrossRef]

20. Murty, B.S.; Yeh, J.W.; Rangathan, S. *High Entropy Alloys*; Butterworth-Heinemann Elsevier: Oxford, UK, 2012; p. 14.

21. Liu, W.H.; Wu, Y.; He, J.Y.; Nieh, T.G.; Lu, Z.P. Grain growth and the Hall-Petch relationship in a high-entropy FeCrNiCoMn alloy. *Scr. Mater.* **2013**, *68*, 526–529. [CrossRef]

22. Liu, W.; Lu, Z.; He, J.; Luan, J.; Wang, Z.; Liu, B.; Liu, Y.; Chen, M.; Liu, C. Ductile CoCrFeNiMox high entropy alloys strengthened by hard intermetallic phases. *Acta Mater.* **2016**, *116*, 332–342. [CrossRef]

23. Fanicchia, F.; Maeder, X.; Ast, J.; Taylor, A.; Guo, Y.; Polyakov, M.; Michler, J.; Axinte, D. Residual stress and adhesion of thermal spray coatings: Microscopic view by solidification and crystallisation analysis in the epitaxial CoNiCrAlY single splat. *Mater. Des.* **2018**, *153*, 36–46. [CrossRef]

Article

Corrosion Resistance of Pipeline Steel with Damaged Enamel Coating and Cathodic Protection

Liang Fan [1], Signo T. Reis [2], Genda Chen [1,*] and Michael L. Koenigstein [2]

[1] Department of Civil, Architectural, and Environmental Engineering,
    Missouri University of Science and Technology, Rolla, MO 65409-0030, USA; lf7h2@mst.edu
[2] Roesch Inc., Belleville, IL 62226, USA; reis@mst.edu (S.T.R.); MKoenigstein@roeschinc.com (M.L.K.)
* Correspondence: gchen@mst.edu; Tel.: +1-573-341-4462

Received: 20 February 2018; Accepted: 9 May 2018; Published: 14 May 2018

Abstract: This paper presents the first report on the corrosion resistance of pipeline steel with damaged enamel coating and cathodic protection in 3.5 wt % NaCl solution. In particular, dual cells are set up to separate the solution in contact with the damaged and intact enamel coating areas, to produce a local corrosion resistance measurement for the first time. Enamel-coated steel samples, with two levels of cathodic protection, are tested to investigate their impedance by electrochemical impedance spectroscopy (EIS) and their cathodic current demand by a potentiostatic test. Due to its glass transition temperature, the enamel-coated pipeline can be operated on at temperatures up to 400 °C. The electrochemical tests show that cathodic protection (CP) can decelerate the degradation process of intact coating and delay the electrochemical reactions at the enamel-steel interface. However, CP has little effect on the performance of coating once damaged and can prevent the exposed steel from corrosion around the damaged site, as verified by visual inspections. Scanning electron microscopy (SEM) indicated no delamination at the damaged enamel–steel interface due to their chemical bond.

Keywords: pipe steel; enamel; cathodic protection; electrochemical impedance spectroscopy (EIS); scanning electron microscopy (SEM)

---

1. Introduction

Organic coatings, such as epoxy, are widely used in combination with supplementary cathodic protection (CP) to prevent steel pipelines from corrosion. When a coating has defects or is damaged during pipeline installations and operations, its steel substrate is directly exposed to the surrounding environment. In this case, the exposed steel can still be prevented from corrosion through CP as a secondary defense system [1,2]. However, the effect of CP makes the exposed metal surface strongly alkaline because of water reduction. This causes organic coating delamination through the hydrolysis of coating or coating-substrate interface [2,3].

Porcelain enamel, as an inorganic material, is chemically bonded to its substrate metal by fusing glass frits at a temperature of 750–850 °C. It can not only be finished with a smooth and aesthetic surface, but it also provides good chemical stability, high corrosion resistance, and excellent resistance to abrasion in an extreme erosion environment [4]. When applied to pipeline lining, enamel coating does not only extend the service life of steel pipes but also increases the pipelines operating temperature to 400 °C, with a safety factor of approximately 1.25 [5].

Our previous studies on steel samples with intact enamel coating [6,7] indicated that enamel coating could protect steel from corrosion in NaCl solution by providing an effective barrier to electrolyte penetration. In real-world operating conditions, solids may flow with fluids in a pipeline and generate abrasive forces; this can impact on the internal enamel coating, resulting in small-scale chipping and coating erosion [8]. The exposed steel would have been further protected by the CP if present. However, the corrosion resistance of steel pipes with damaged enamel coating, and the effect of CP on the interface condition between the enamel coating and its steel substrate have never been investigated.

Electrochemical tests are widely used to study the degradation process of coatings, however, electrochemical responses are concentrated on the local areas where coatings are damaged. This is because their impedance is much lower than that of the surrounding areas with intact coating. In this study, a dual-cell test setup was used to separate the 3.5 wt % NaCl solution in contact with the damaged and intact coating areas, during response measurements [9,10], using potentiostatic and electrochemical impedance spectroscopy (EIS) tests, respectively. Therefore, the potential effect of the damaged coating area on the corrosion process of the intact coating area, as alluded by epoxy coating, can be investigated. To help interpret the effect of CP on the condition of coating–substrate interfaces, coating microstructures were examined with scanning electron microscopy (SEM).

2. Materials and Methods

2.1. Sample Preparation

An API 5L X65 steel pipe (MRC Global, Houston, TX, USA), with an outer diameter of 323.85 mm and a wall thickness of 9.53 mm, was selected as the substrate metal in this study. The chemical composition of the steel provided by the vendor is presented in Table 1. The steel pipe was cut into 9 25 mm $\times$ 50 mm coupon samples. The cut samples were steel blasted for 1 min, to remove mill scales and rusts, and then cleansed with acetone.

Table 1. Chemical composition of steel pipe.

Element	C	Mn	P	S	Si	Cu	Ni	Cr	Mo	Al	V	Fe	Others
wt %	0.17	1.15	0.07	0.02	0.26	0.10	0.04	0.07	0.07	0.024	0.02	98	0.006

The steel coupons were coated with enamel slurry T-001 (Tomatec Product, Florence, KY, USA). The chemical compositions of T-001 glass frits were determined by X-ray Fluorescence (XRF, The Mineral Lab, Inc., Golden, CO, USA) as presented in Table 2. Prior to the coating of steel samples, the thermal properties of glass T-001, such as glass-transition temperature (T_g), softening temperature (T_s), and the coefficient of thermal expansion (CTE) were determined using the Orton automatic recording dilatometer (model 1500, Orton, Westerville, OH, USA).

Table 2. Chemical compositions of T-001 glass frits (wt %).

Elements	SiO_2	B_2O_3	Na_2O	CaO	MnO_2	Al_2O_3	TiO_2	K_2O	Fe_2O_3	MgO	BaO	Others
T-001	60.3	12.84	7.20	2.37	5.37	4.49	0.14	2.12	3.48	0.17	1.47	0.05

The enamel slurry was prepared by first milling glass frits, clay and certain electrolytes together, and then mixing them with water until the mixture was in a stable suspension state. The water, glass frits, and clay were then mixed in a proportion of 1.00:2.40:0.17 by weight. The enamel slurry was manually sprayed on the surface of each coupon sample. All samples were heated at 150 °C for 10 min, to drive away moisture; fired at 815 °C for 10 min; and finally cooled to room temperature. An optic microscope Hirox (Tokyo, Japan) was used to measure the coating surface roughness, finding an average value of 1 μm. The PosiTest, following ASTM D4541-09 [11], was used to measure the bond strength between the coating and the steel substrate, finding an average value of 17 MPa. Due to the roughness of the steel surface, the thickness of the enamel coating varied slightly at different locations with a standard deviation of 19 μm.

To study the effect that damage has on the corrosion resistance of enamel coating, one damage area, as shown in Figure 1, was created at the center of each enamel-coated sample using an impact test apparatus according to the ASTM Standard G14 [12]. The apparatus consists of a 0.91 kg steel rod with a hemispherical head and a vertical section of hollow aluminum tubing to guide the rod. The weight rod was dropped from a height of 84 cm to damage the coatings. A close-up view of Figure 1 shows the detail around the damaged area.

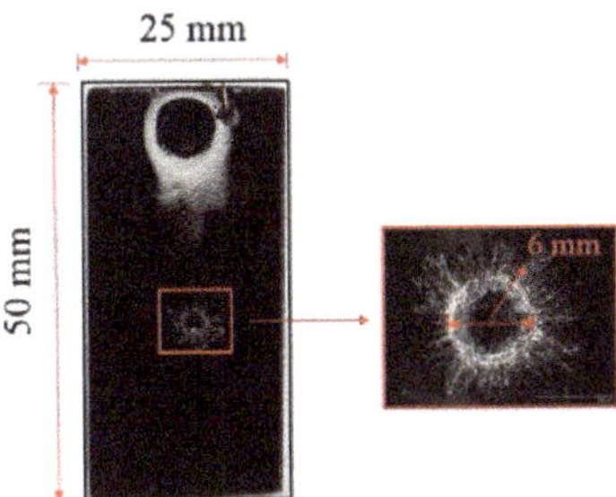

Figure 1. Impact-induced coating damage.

2.2. Characterization of Enamel Coatings

The coating microstructure was characterized with scanning electron microscopy (SEM, Hitachi S4700, Tokyo, Japan). As shown in Figure 1, each damaged enamel-coated sample was cold mounted in epoxy resin (EpoxyMount, Allied High Tech Products, Inc., Rancho Dominguez, CA, USA). A 10 mm-thick cross section was cut from the damaged coating area of the sample and abraded with carbide papers with grits of 80, 180, 320, 600, 800, and 1200. After abrading, all samples were cleansed with deionized water and dried at room temperature prior to SEM imaging.

2.3. Electrochemical Tests

As shown in Figure 2a, except for the surface of the enamel coating, each coupon sample was embedded into the epoxymount to test corrosion performance. The epoxymount was over 2 mm thick to ensure that the surface of the enamel coating was the response site during the electrochemical tests. As shown in Figure 2b, a PVC funnel (1 cm in diameter) was attached onto the coating surface, covering the damaged area. The sample was placed in a large plastic container with the funnel faced up. The funnel and container were filled with 3.5 wt % NaCl solution to ensure that the funnel was completely submerged. The solution was prepared by adding purified sodium chloride (Fisher Scientific, Inc., Waltham, MA, USA) into distilled water. CP was introduced for the entire coated area.

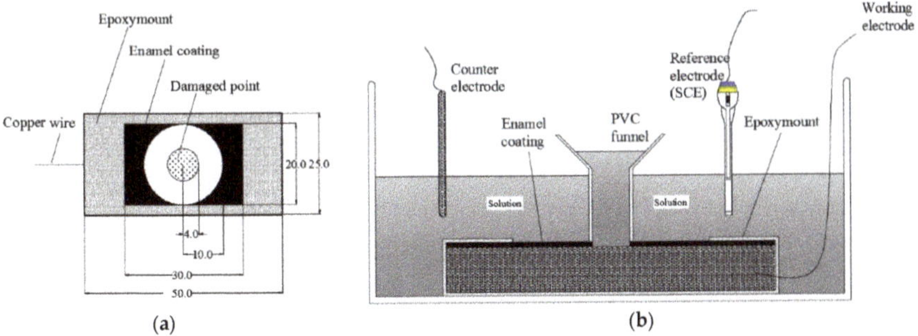

Figure 2. Schematic representation of the double electrochemical cell (unit: mm): (**a**) planar view of the sample with damaged coating and attached funnel; (**b**) side view of the electrochemical cell immersed in the bulk solution.

During electrochemical tests, the 3.5 wt % NaCl solution around the damaged coating area was separated, by the funnel, from the solution around the remaining intact coating area. If it were otherwise, the electrochemical responses would have been concentrated on the damaged area since its impedance would be much lower than that of the other area. Thus, the measured responses would be representative to neither the damaged coating area nor the other intact coating area. For the same reason, the damaged and intact coating areas were tested up to 10 and 70 days, respectively.

The electrochemical tests were conducted at room temperature every 5 days in a classic three-electrode system with a saturated calomel electrode (SCE) as the reference electrode, a graphite rod as the counter electrode, and a coupon sample as the working electrode. The three electrodes were connected to an Interface1000E Potentiostat (Gamry Instrument, Warminster, PA, USA) for measurement. The SCE and graphite rods were immersed in the large container for the intact enamel coating area, as shown in Figure 2, and in the funnel for the damaged enamel coating area (not shown in Figure 2 for clarity). One sample was subjected to zero cathodic potential (under the open circuit potential or OCP condition), another one to a cathodic potential of −0.85 vs. SCE/V, and the third one to a cathodic potential of −1.15 vs. SCE/V. Potentiostatic tests were first conducted to measure currents for 1000 s at −0.85 vs. SCE/V or −1.15 vs. SCE/V. EIS tests were then conducted under a sinusoidal potential wave (10 mV in amplitude and a frequency range of 10^5–10^{-2} Hz) around a cathodic potential of zero, −0.85 vs. SCE/V and −1.15 vs. SCE/V. EIS test data was simulated with classical electrical equivalent circuits (EEC) and analyzed with the software ZSimpWin (Version 3.21).

3. Results and Discussion

3.1. Thermal Properties

Figure 3 shows the thermal elongation of the enamel coating and pipe steel as a function of temperature. The steel has a measured CTE of 19.7 ppm/°C, while the enamel coating T-001 has a measured CTE of 13.0 ppm/°C. The CTE of the steel remained constant over a temperature range of 100–600 °C, while the CTE of the enamel coating was only constant over a range of 200–500 °C. The difference between the CTE of steel and enamel coating lead to an initial compressive stress on the coating during cooling; this can reduce cracking in enamel and is desirable in engineering applications. The glass transition temperature for enamel slurry T-001 is 506 °C, which allows the enamel-coated pipeline to operate at temperatures up to 400 °C, considering a safety factor of approximately 1.25.

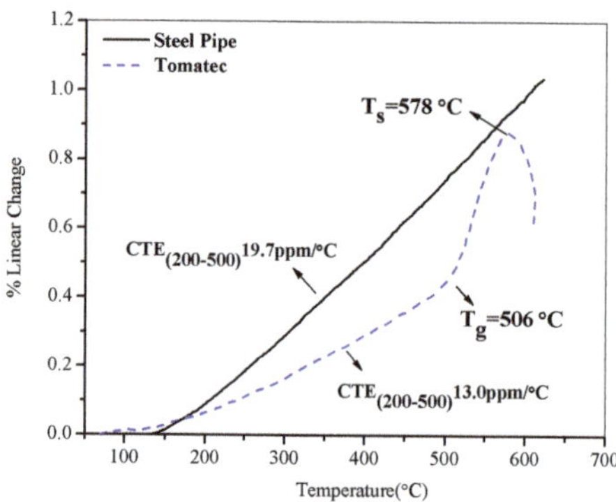

Figure 3. Thermal properties of enamel coating and pipe steel.

3.2. Coating Microstructure

Cross-sectional SEM images of enamel-coated steel samples, tested under the OCP and CP (−1.15 V/SCE) conditions, are presented in Figure 4. In general, the enamel coatings have amorphous structures with isolated air bubbles. Gaseous CO, CO_2, and H_2 are generated during the firing process of enameling. When cooled down, these gases were trapped as a thick layer of enamel solidifies; this generated the isolated air bubbles [13,14]. Figure 4a,b represents the stitched images of five SEMs taken along a radial direction of the damaged coating, as shown in the detailed damaged zone in Figure 1. Due to chipped coating falling off after impact tests, the coating thickness decreased gradually from 244 to 4 µm for samples tested under the OCP, and from 190.48 to 4 µm for samples tested under −1.15 V/SCE. However, the substrate surface is still covered with a thin layer of enamel coating at the center of damaged area, as indicated in Figure 1.

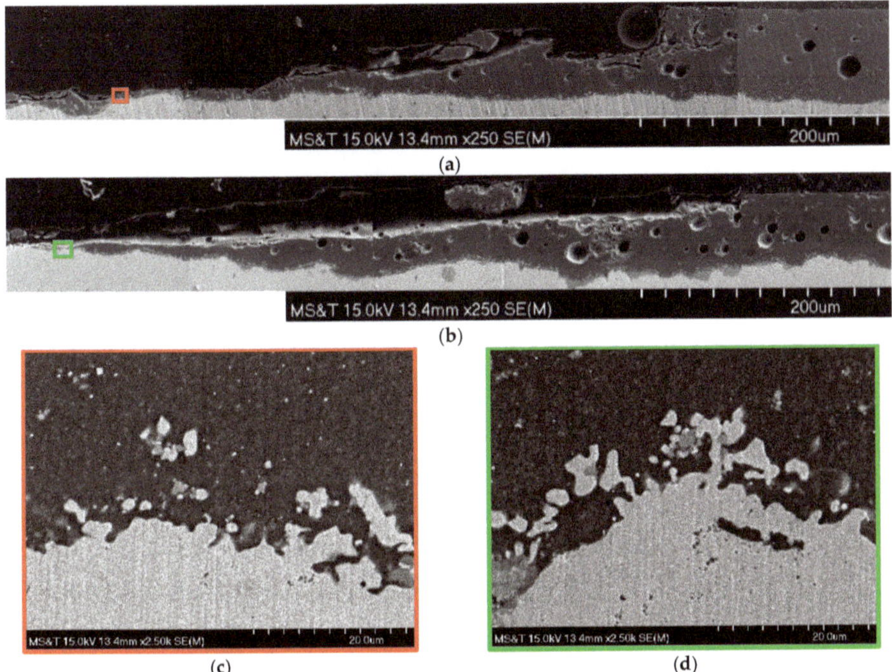

Figure 4. Cross-sectional SEM images of enamel-coated samples under the OCP (**a,c**) and −1.15 V/SCE (**b,d**) with a magnification of 250× (**a,b**) and 2500× (**c,d**).

Figure 4c,d shows magnified details of the enamel–steel interfaces from Figure 4a,b. They show the extensive formation of an island-like structure in the enamel coating, during the firing process. In essence, a durable steel enamel interface transition zone was formed [15]. The island-like structure is iron-alloys, formed as a result of the chemical reactions of metal oxides in the enamel and the carbon and iron in the steel. No delamination was found after the corrosion tests; thus, the CP did not affect the mechanical condition of the interface between the enamel coating and steel substrate.

3.3. EIS

Figure 5 shows the EIS Bode diagrams of 3 representative samples tested under a cathodic potential of −1.15 V/SCE and −0.85 V/SCE, and an OCP, respectively in intact enamel coating (Figure 5(a1,b1,c1)) and damaged enamel coating (Figure 5(a2,b2,c2)). Both the measured (Meas.) data in various symbols and their fitted (Ftd.) curves are presented in Figure 4.

On a log–log scale, the impedance of the sample tested under −1.15 V/SCE in the first 40 days decreased linearly with the frequency; this relation was independent of the day of testing, as indicated in Figure 5(a1). Starting from the 50th day, the impedance experienced a gradual decrease at a low frequency but remained over 10 GΩ cm^2 at a frequency of 0.02 Hz. The phase angles in the high and middle frequency ranges were close to 90° during the entire immersion time and increased with the frequency in the low frequency range.

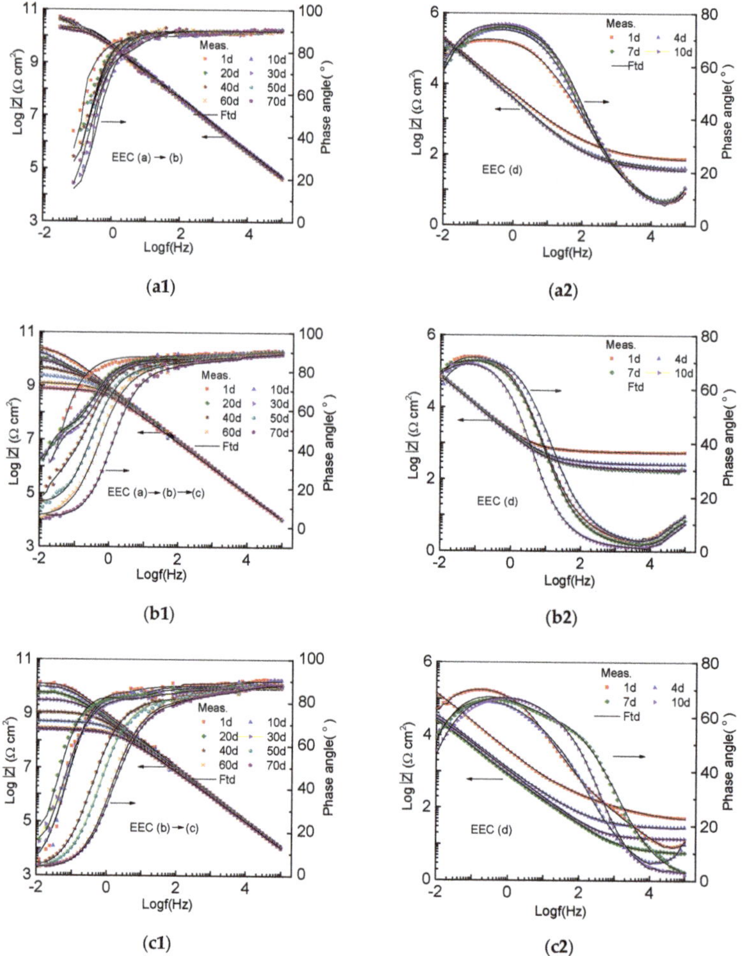

Figure 5. Bode diagrams of enamel-coated samples immersed in 3.5 wt % NaCl solution up to 70 days at (**1**) intact coating zone, and up to 10 days at (**2**) damaged coating zone under a cathodic potential of (**a**) −1.15 vs. SCE/V, (**b**) −0.85 vs. SCE/V, and (**c**) the OCP. d: day.

For the sample tested under a cathodic potential of −0.85 V/SCE, as shown in Figure 5(b1), the impedance on a log-log scale decreased linearly in the first 10 days and then, over time, showed a gradually-expanding horizontal platform in the low to middle frequency range. The impedance at a frequency of 0.02 Hz decreased from 24 GΩ cm^2 at the beginning to 0.76 GΩ cm^2 at the end of the test. The phase angle increased with the frequency from the low to middle frequency range and remained 90° until 70 days of immersion time in the high frequency range. The phase–frequency curves in the low frequency range shifted towards the middle frequency range over the immersion time.

The impedance and phase angle of the sample tested under the OCP, as shown in Figure 5(c1), showed a similar trend to the sample tested under a cathodic potential of −0.85 V/SCE, particularly towards the end of the corrosion test. However, the horizontal platform was further extended to the middle frequency range and the impedance at a frequency of 0.02 Hz was 0.26 GΩ cm² after 70 days of testing.

Figure 5(a2,b2,c2) shows the Bode diagrams of the samples tested in the damaged coating zone. Overall, the Bode diagrams of the samples tested under the CP and the OCP are similar, indicating comparable corrosion performances of all samples in the damaged zone. The impedance became stable after 4 days of immersion in the solution. Because of the damage made to the coating, the impedance at 0.02 Hz was approximately 0.1 MΩ cm², which is 10^6 times smaller than that of the samples tested in the intact coating zone. On a log-log scale, the impedance linearly decreased in the low frequency range and gradually approached an asymptotic value in the high frequency range. The maximum phase angle, lower than 80°, appeared in the low frequency range, indicating that corrosion had already taken place in the steel substrate.

Figure 6 shows four equivalent electrical circuit (EEC) models used to fit the EIS test data taken from different samples under various test conditions. In this study, a constant phase element (CPE) was used instead of a pure capacitor due to non-homogeneity in coating thickness and roughness [16,17], or the electrochemical reactivity of the steel substrate [18]. A CPE is defined by two parameters, Y and n, and its impedance is represented by:

$$Z_{\text{CPE}} = Y^{-1}(j\omega)^{-n} \tag{1}$$

where $j = \sqrt{-1}$ is the imaginary unit, Y is a CPE constant, ω is the angular frequency, and n ($0 \leq n \leq 1$) is an index that represents the deviation of the CPE from a corresponding pure capacitor [2].

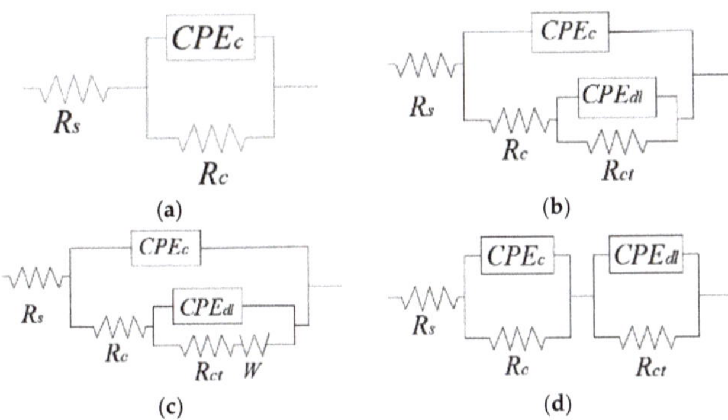

Figure 6. Equivalent electrical circuit (EEC) models for the samples tested (**a**) in the first 45 days under a cathodic potential of −1.15 V/SCE and in the first 10 days under a cathodic potential of −0.85 V/SCE on the intact coating zone; (**b**) from the 45th day to the end of test under a cathodic potential of −1.15 V/SCE and from the 15th day to the 45th day under a cathodic potential of −0.85 V/SCE on the intact coating zone; (**c**) from the 45th day to the end of test under a cathodic potential of −0.85 V/SCE on the intact zone; (**d**) on the damaged coating zone.

The EEC models used to fit into the EIS data from various tested samples are included in Figure 5. Model (a) [19,20] was used for the samples with intact coating tested under −1.15 V/SCE up to 40 days, taking into consideration the decrease in coating resistance and increase in coating capacitance as water begins to seep through the channels in enamel coating. Here, R_s represents the solution resistance, R_c and CPE_c represent the pore resistance and capacitance of the coating, respectively. After 40 days of immersion, when water and oxygen molecules arrived at the substrate surface and reacted with the steel substrate, the EIS data was fitted with Model (b) till the end of the corrosion tests [19–21]. Here, R_{ct} is the charge transfer resistance and CPE_{dl} is the double layer capacitance at the steel-electrolyte interface. However, only one capacitive loop was observed in the phase-frequency diagram. This is likely because the time constant associated with the dielectric properties of enamel was difficult to distinguish from that of the electrochemical reaction at the steel-electrolyte interface [20,22].

For the intact enamel coating zone under -0.85 V/SCE, Model (a) was used in the first 10 days of immersion, Model (b) was applied from the 15th day to the 45th day, and Model (c) was used till the last day of the test. A Warburg impedance W in Model (c) was included to take into account the diffusion behavior, which was induced by the accumulation of corrosion products at the active corrosion sites. For the intact coating zone under the OCP, Model (b) was used for tests up to 40 days and Model (c) for the remaining tests.

For all the damaged coating zones, two time constants can clearly be observed in the phase-frequency diagram, and thus Model (d) was used to fit the test data [23]. While Model (b) was applicable for the intact coating zone when the solution had penetrated through the channel in the coating and was in contact with the steel substrate, Model (d) was more appropriate for the damaged-coating zone, since the coating layer became thinner and the solution could penetrate into the coating more easily. The electrochemical reactivity occurred uniformly on the damaged coating surface.

Figure 7 shows the change of pore resistance R_c and capacitance CPE_c of the intact coatings. In general, pore resistance measures the ease of electrolyte penetration into the coating, which is related to the number and distribution of open pores and pinholes in the enamel coating. The coating capacitance indicates the extent of electrolyte diffusion into the coating, which is associated with the thickness and dielectric properties of the coating [24]. The R_c value of the samples tested under -1.15 V/SCE decreased from 57.6 to 4.92 GΩ cm^2, while the R_c value of the samples tested under -0.85 V/SCE and the OCP decreased more rapidly from 20.9 to 1.57 MΩ cm^2 over 70 days. The coating capacitance of all the samples increased with immersion time, since the electrolyte solution gradually penetrated into the coating, thus increasing the coating capacitance. All the samples tested under the CP have larger coating resistances than the samples under the OCP. Thus, the CP improved the coating performance [2]. The sample tested under -1.15 V/SCE had a larger coating resistance and a smaller coating capacitance than the respective values of the sample under -0.85 V/SCE. This result indicates that a higher cathodic potential used in tests does not adversely affect the coating properties; it can decelerate the degradation process of the coating.

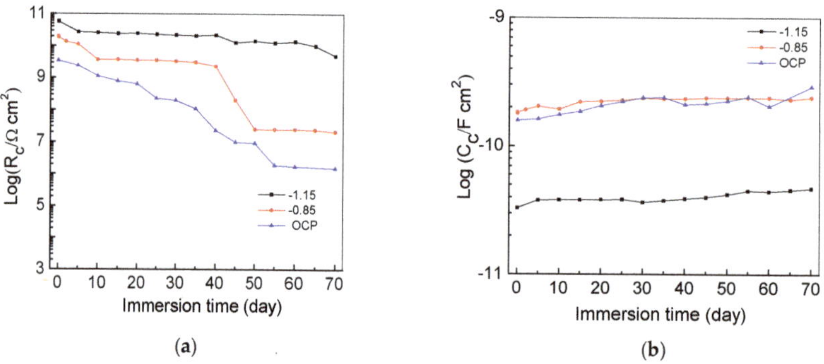

Figure 7. Properties of intact coating under various CP levels: (**a**) pore resistance R_c and (**b**) capacitance CPE_c.

The R_c values of the damaged coating decreased rapidly over the immersion time as shown in Figure 8, which was measured in days (d). Specifically, the R_c value of the samples under the CP dropped from approximately 400 to 150 Ω cm^2, while the R_c value of the samples under the OCP reduced more dramatically from 500 Ω cm^2 in one day, to 110 Ω cm^2 in 10 days; indicating the failure of coating in protecting the steel substrate. The CPE_c values of all the tested samples reached nearly the same value of 2 mF·cm^2 after 4 days of immersion. Therefore, after coating has been damaged, the CP has little effect on the coating performance.

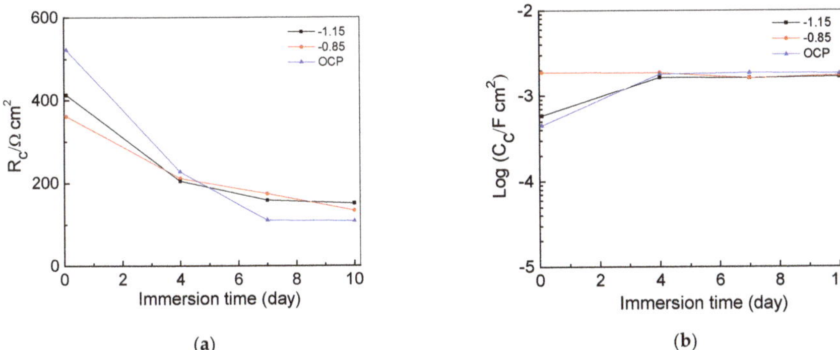

Figure 8. Damaged coating properties: (**a**) pore resistance R_c and (**b**) capacitance CPE_c.

Figure 9 displays the properties of the steel-electrolyte interface under intact coating: charge transfer resistance R_{ct} and double layer capacitance CPE_{dl}. Charge transfer resistance is the resistance against electrons transferring across the steel surface, which is inversely proportional to the corrosion rate [24]. For the samples tested under -1.15 V/SCE, -0.85 V/SCE and the OCP, the charge transfer resistances were reduced to 1.13, 0.7, and 0.14 GΩ cm^2, respectively, at the end of testing, after 70 days. This comparison indicated increasing electrochemical reactions on the steel-electrolyte interface over time, as the level of CP decreased. The double layer capacitance CPE_{dl} is also a measure of the ease of charge transfer across a steel-electrolyte interface. The CPE_{dl} of the samples tested under -1.15 V/SCE, -0.85 V/SCE and the OCP were increased to 6.523×10^{-11}, 1.613×10^{-10}, and 4.314×10^{-10} F cm^{-2}, respectively, at the end of testing, after 70 days. The sample tested under -1.15 V/SCE had the highest charge transfer resistance and the lowest double layer capacitance. Thus, the higher the cathodic potential, the more effectively the electrochemical reactions can be delayed at the steel-electrolyte interface [2].

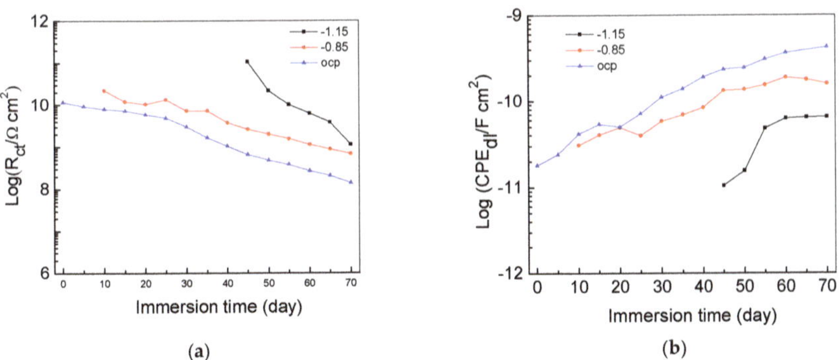

Figure 9. Properties of the steel-electrolyte interface under intact enamel coating: (**a**) charge transfer resistance R_{ct} and (**b**) double layer capacitance CPE_{dl}.

After the enamel coating was damaged, the charge transfer resistance of the samples tested under -1.15 V/SCE, -0.85 V/SCE and the OCP slightly decreased to 4.96×10^5, 3.78×10^5, and 6.67×10^4 Ω cm^2, respectively, after 10 days of immersion as shown in Figure 10. This is about 10^4 times smaller than that of the intact coating tested after 70 days of immersion. The double layer capacitances of the samples tested under -1.15 V/SCE, -0.85 V/SCE and the OCP also changed slightly, they were 1.37×10^{-4}, 6.08×10^{-4}, and 5.48×10^{-4} F cm^{-2} after 10 days of immersion, respectively. They were approximately 10^6 times larger than those of the samples with intact enamel coating tested after 70 days of immersion.

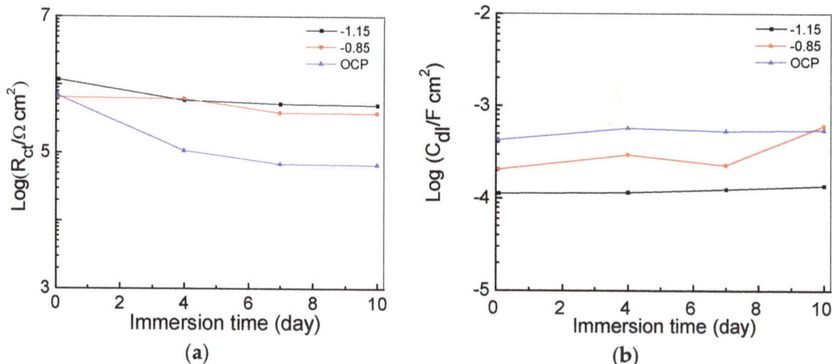

Figure 10. Properties of the steel-electrolyte interface under damaged enamel coating: (**a**) charge transfer resistance R_{ct} and (**b**) double layer capacitance CPE_{dl}.

3.4. Potentiostatic

Figure 11a shows the variation of currents taken from the intact enamel coating zone under −0.85 vs. SCE/V and −1.15 vs. SCE/V. Each dot represents one measurement of data per day till the end of testing, after 70 days. For both samples the current fluctuated around −0.2 nA from the beginning to 45 days of immersion. Then, the sample tested under −1.15 V/SCE decreased slowly to approximately −0.3 nA at the end of testing, while the sample tested under −0.85 V/SCE decreased dramatically to approximately −0.8 nA at the end. Similarly, Figure 11b presents the variations of currents on the samples with damaged enamel coating. The currents of both samples eventually reached approximately −5 µA after 10 days of immersion, which are about 10^4 times larger than those of the respective test samples with the intact enamel coating. This is because more electrochemically reactive spots were generated. In all test cases, the measured current is always negative, implying that the CP current can flow through the coating along electrolyte pathways to reach the metal substrate and protect the steel from corrosion [25].

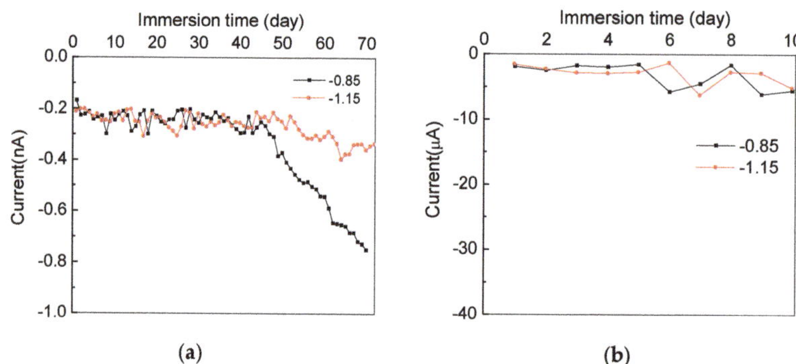

Figure 11. Variations of currents applied to various samples under −0.85 vs. SCE/V and −1.15 vs. SCE/V: (**a**) intact coating zone and (**b**) damaged coating zone.

3.5. Visual Observations after Corrosion Tests

At the conclusion of the corrosion tests, the damaged spots of all tested samples were visually examined. No corrosion products were observed on the damaged surface under a cathodic potential of −1.15 V/SCE as shown in Figure 12. Brown corrosion products can be clearly seen on the damaged point of the sample tested under the OCP.

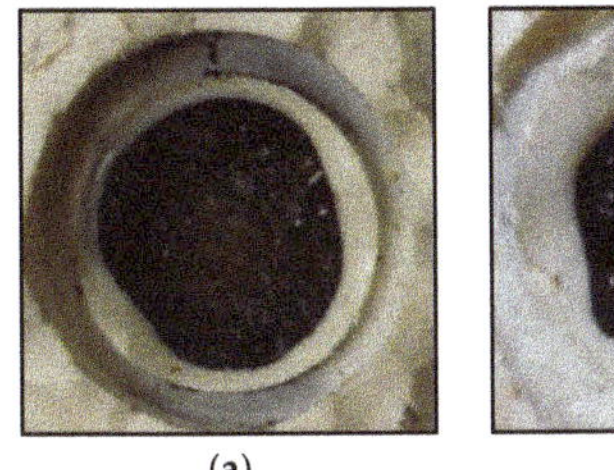 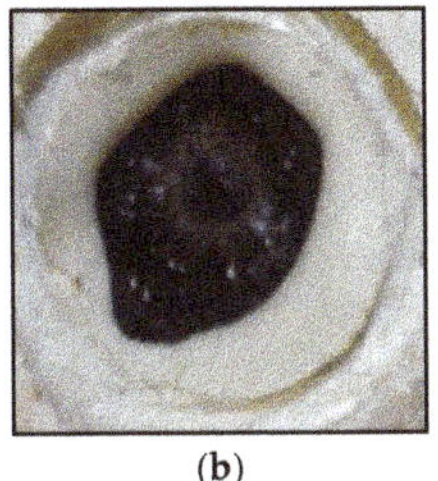 

| (a) | (b) | (c) |

Figure 12. Damaged surface conditions of the samples tested under (**a**) −1.15 vs. SCE/V, (**b**) −0.85 vs. SCE/V and (**c**) the OCP after corrosion tests.

4. Conclusions

Based on the experimental results and analysis from one representative sample in each test condition, the following conclusions can be drawn:

- The enamel coating is subjected to initial compression due to its lower CTE than steel, thus it is less susceptible to tensile cracks. In comparison with epoxy coating, the enamel coating has a higher glass transition temperature, and thus allows an increase of pipeline operation temperature up to 400 °C, with a safety factor of approximately 1.25.
- Enamel residual remained between anchor points of the steel substrate after the enamel coating had been chipped off, this was due to impact loading. During all the corrosion tests, no further delamination was found, and the CP did not change the coating properties and the mechanical condition at the coating-substrate interface.
- At the intact coating areas, the higher the potential (up to −1.15 V/SCE) applied in CP, the higher the coating resistance and charge transfer resistance. The CP does not cause debonding between the coating and its steel substrate, it decelerates the degradation process of the coating and delays the electrochemical reactions at the steel-electrolyte interface.
- The resistances of all the damaged coatings were less than 1 kΩ cm^2, indicating the loss of their barrier effect in protecting the steel substrate from corrosion. The introduction of CP does not improve the coating performance once damaged.
- The resistances against electrolyte penetration into the enamel coating and charge transfer through the steel-electrolyte interface in the intact and damaged enamel coating areas differed by at least 10^4 times after 70 days of testing. It is thus important to separate the electrochemical processes in the intact and damaged zones during corrosion tests.

Author Contributions: L.F. and G.C. conceived and designed the experiments; L.F. and S.T.R. performed the experiments; L.F. analyzed the data; M.L.K. contributed reagents and materials; L.F. prepared the manuscript.

Funding: The authors gratefully acknowledge the financial support provided by the U.S. Department of Transportation under Award No. DTPH5615HCAP10.

Conflicts of Interest: The authors declare no conflicts of interest.

References

1. Love, C.T.; Xian, G.; Karbhari, V.M. Cathodic disbondment resistance with reactive ethylene terpolymer blends. *Prog. Org. Coat.* **2007**, *60*, 287–296. [CrossRef]
2. Zhu, C.; Xie, R.; Xue, J.; Song, L. Studies of the impedance models and water transport behaviors of cathodically polarized coating. *Electrochim. Acta* **2011**, *56*, 5828–5835. [CrossRef]
3. Martinez, S.; Žulj, L.V.; Kapor, F. Disbonding of underwater-cured epoxy coating caused by cathodic protection current. *Corros. Sci.* **2009**, *51*, 2253–2258. [CrossRef]

4. Rossi, S.; Parziani, N.; Zanella, C. Abrasion resistance of vitreous enamel coatings in function of frit composition and particles presence. *Wear* **2015**, *332*, 702–709. [CrossRef]

5. Lazutkina, O.R.; Kostenko, M.G.; Komarova, S.A.; Kazak, A.K. Highly reliable energy-efficient glass coatings for pipes transporting energy carriers, liquids, and gases. *Glass Ceram.* **2007**, *64*, 93–95. [CrossRef]

6. Fan, L.; Tang, F.; Reis, S.T.; Chen, G.; Koenigstein, M.L. Corrosion resistance of transmission pipeline steel coated with five types of enamels. *Acta Metall. Sin. (Engl. Lett.)* **2017**, *30*, 390–398. [CrossRef]

7. Fan, L.; Tang, F.; Reis, S.T.; Chen, G.; Koenigstein, M.L. Corrosion resistances of steel pipes internally coated with enamel. *Corrosion* **2017**, *73*, 1335–1345. [CrossRef]

8. Lauer, R.S. Advancements in the abrasion resistance of internal plastic coatings. *Mater. Perform.* **2014**, *53*, 52–55.

9. Deflorian, F.; Rossi, S. An EIS study of ion diffusion through organic coatings. *Electrochim. Acta* **2006**, *51*, 1736–1744. [CrossRef]

10. Le Thu, Q.; Takenouti, H.; Touzain, S. EIS characterization of thick flawed organic coatings aged under cathodic protection in seawater. *Electrochim. Acta* **2006**, *51*, 2491–2502. [CrossRef]

11. *ASTM D4541–09(2009) Standard Test Method for Pull-Off Strength of Coatings Using Portable Adhesion Testers*; ASTM Standards: West Conshohocken, PA, USA, 2009.

12. *ASTM G14-04(2010)e1 Standard Test Method for Impact Resistance of Pipeline Coatings (Falling Weight Test)*; ASTM Standards: West Conshohocken, PA, USA, 2010.

13. Yang, X.; Jha, A.; Brydson, R.; Cochrane, R.C. An analysis of the microstructure and interfacial chemistry of steel–enamel interface. *Thin Solid Films* **2003**, *443*, 33–45. [CrossRef]

14. Samiee, L.; Sarpoolaky, H.; Mirhabibi, A. Microstructure and adherence of cobalt containing and cobalt free enamels to low carbon steel. *Mater. Sci. Eng. A* **2007**, *458*, 88–95. [CrossRef]

15. Liu, H.H.; Shueh, Y.; Yang, F.S.; Shen, P. Microstructure of the enamel-steel interface: Cross-sectional TEM and metallographic studies. *Mater. Sci. Eng. A* **1992**, *149*, 217–224. [CrossRef]

16. Presa, M.R.; Tucceri, R.I.; Florit, M.I.; Posadas, D. Constant phase element behavior in the poly(*o*-toluidine) impedance response. *J. Electroanal. Chem.* **2001**, *502*, 82–90. [CrossRef]

17. Yao, Z.P.; Jiang, Z.H.; Wang, F.P. Study on corrosion resistance and roughness of micro-plasma oxidation ceramic coatings on Ti alloy by EIS technique. *Electrochim. Acta* **2007**, *52*, 4539–4546. [CrossRef]

18. Orazem, M.E.; Tribollet, B. Equivalent circuit analogs. In *Electrochemical Impedance Spectroscopy*, 1st ed.; John Wiley & Sons: Hoboken, NJ, USA, 2008; p. 233. ISBN 978-0-470-04140-6.

19. Zhang, J.T.; Hu, J.M.; Zhang, J.Q.; Cao, C.N. Studies of water transport behavior and impedance models of epoxy-coated metals in NaCl solution by EIS. *Prog. Org. Coat.* **2004**, *51*, 145–151. [CrossRef]

20. Zhang, J.T.; Hu, J.M.; Zhang, J.Q.; Cao, C.N. Studies of impedance models and water transport behaviors of polypropylene coated metals in NaCl solution. *Prog. Org. Coat.* **2004**, *49*, 293–301. [CrossRef]

21. Hu, J.M.; Zhang, J.Q.; Cao, C.N. Determination of water uptake and diffusion of Cl^- ion in epoxy primer on aluminum alloys in NaCl solution by electrochemical impedance spectroscopy. *Prog. Org. Coat.* **2003**, *46*, 273–279. [CrossRef]

22. Walter, G.W. A review of impedance plot methods used for corrosion performance analysis of painted metals. *Corros. Sci.* **1986**, *26*, 681–703. [CrossRef]

23. Tang, F.J.; Chen, G.; Brow, R.K.; Volz, J.S.; Koenigstein, M.L. Corrosion resistance and mechanism of steel rebar coated with three types of enamel. *Corros. Sci.* **2012**, *59*, 157–168. [CrossRef]

24. Tang, F.J.; Cheng, X.M.; Chen, G.; Brow, R.K.; Volz, J.S.; Koenigstein, M.L. Electrochemical behavior of enamel-coated carbon steel in simulated concrete pore water solution with various chloride concentrations. *Electrochim. Acta* **2013**, *92*, 36–46. [CrossRef]

25. Ranade, S.; Forsyth, M.; Tan, M.Y.J. In situ measurement of pipeline coating integrity and corrosion resistance losses under simulated mechanical strains and cathodic protection. *Prog. Org. Coat.* **2016**, *101*, 111–121. [CrossRef]

Article

Experiences with Thermal Spray Zinc Duplex Coatings on Road Bridges

Ole Øystein Knudsen [1,*], **Cato Dørum** [2] **and Martin Gagné** [3]

1   SINTEF, P.O. Box 4760 Torgarden, NO-7465 Trondheim, Norway
2   Norwegian Public Roads Administration, 5008 Bergen, Norway; haakon.matre@vegvesen.no (H.M.); cato.dorum@vegvesen.no (C.D.)
3   Zelixir Inc., Toronto, ON M5A4R4, Canada; mgagne@zelixir.ca
*   Correspondence: ole.knudsen@sintef.no; Tel.: +47-9823-0420

Received: 12 May 2019; Accepted: 4 June 2019; Published: 8 June 2019

Abstract: Road bridges are typically designed with a 100-year lifetime, so protective coatings with very long durability are desired. Thermal spray zinc (TSZ) duplex coatings have proven to be very durable. The Norwegian Public Roads Administration (NPRA) has specified TSZ duplex coatings for protection of steel bridges since 1965. In this study, the performance of TSZ duplex coatings on 61 steel bridges has been analyzed. Based on corrosivity measurements on five bridges, a corrosivity category was estimated for each bridge in the study. Coating performance was evaluated from pictures taken by the NPRA during routine inspections of the bridges. The results show that very long lifetimes can be achieved with TSZ duplex coatings. There are examples of 50-year old bridges with duplex coatings in good condition. Even in very corrosive environments, more than 40-year old coatings are still in good condition. While there are a few bridges in this study where the coating failed after only about 20 years, the typical coating failures are due to application errors, low paint film thickness and saponification of the paint. Modern bridge designs and improved coating systems are assumed to increase the duplex coating lifetime on bridges even further.

Keywords: thermal spray zinc; duplex coatings; coating lifetime; coating maintenance

---

1. Introduction

Road bridges are typically designed with a lifetime of 100 years. However, lifetime extensions are normal and there are many bridges that are older than 100 years. Bridges are most likely to be replaced or decommissioned due to increased traffic capacity or the closing of the road than for exceeding the design lifetime. With such long lifetimes, the maintenance of the protective coating is a major contributor to the operational expenses. Hence, coating systems with very long durability are of great interest. Multi-layer paint coating systems are specified in many countries, but such coating systems have limited lifetime, especially in corrosive environments. According to the ISO 12944-1 coating selection standard, a very long lifetime is defined as more than 25 years [1]. In a 100 years construction life perspective (or more than 100 years), 25 years is relatively short and implies that the coating must be maintained several times.

In order to increase the coating lifetime and reduce the coating maintenance costs, The Norwegian Public Roads Administration (NPRA) introduced thermally sprayed zinc (TSZ) duplex coatings for coastal steel bridges in 1965 and for all bridges in 1977, replacing red lead coatings. Klinge published a study of the TSZ duplex coating on the Rombak Bridge in Norway, documenting almost 40-year coating lifetimes [2,3]. The bridge was coated in the field with a duplex coating consisting of a pure zinc TSZ coating of 100 μm and two layers of alkyd paint at 100 μm each. The bridge was opened to traffic in 1965 and for various reasons left partly uncoated until 1970, when the duplex coating was applied. The bridge is a 750 m long suspension bridge with a bolted truss work under the bridgeway.

The large numbers of overlapping joints in the truss work are typical corrosion traps, but the duplex coating even protected these from corrosion. In 2012, preventive coating maintenance was performed by washing and the application of a new topcoat due to the partial flaking of the old topcoat.

There is a long history of corrosion protection of steel bridges with TSZ coatings. The Ridge Avenue Bridge in Philadelphia was metalized with zinc already in 1938 [4]. In the UK, the Forth Road Bridge was installed with zinc metalized and painted steel in 1964, which was the largest metalized steel bridge at the time [5]. The coating performed very well in contrast to the parallel Forth Railway Bridge that was painted, probably with red lead. The experiences with the Forth Road Bridge were important when the NPRA started to specify TSZ duplex coatings in 1965. TSZ duplex coatings have later been extensively used in the Scandinavian countries, UK, France and, to some extent, in the USA and Canada. However, no extensive review of their performance has been published to our knowledge.

Duplex coatings consist of a metal coating, typically hot-dip galvanizing, electroplated zinc or TSZ, painted with a protective organic coating. TSZ with a sealer is usually not regarded as a duplex coating. Duplex coatings are now very common across many industry sectors. For example, in the automotive industry, duplex coatings are the norm for the manufacturing of about 50 million car bodies per year. TSZ duplex coatings for steel structures were already documented to perform well in the 19-year field test by the American Welding Society that was terminated in 1974 [6]. In his book from 1994, van Eijnsbergen published a comprehensive study of duplex coatings consisting of hot-dip galvanized steel and paint [7]. Based on his findings, he claimed that the durability of a duplex coating is longer than the sum of the durability of the metallic and the paint coatings:

$$D_{\text{duplex}} = K \cdot (D_{\text{zinc}} + D_{\text{paint}}) \tag{1}$$

where D is the durability or lifetime of the coats and K is a synergy factor, ranging from 1.5 in aggressive climates to 2.3 in less aggressive climates.

The objective of the present work has been to investigate the performance of zinc duplex coatings on 61 steel bridges in Norway as a function of corrosivity on site and the type of paint coating applied. Corrosivity was measured on four coastal bridges and one inland bridge in order to help estimate corrosivity on all the bridges. Typical coating failures and their causes are also discussed.

2. Materials and Methods

2.1. Corrosivity

Corrosivity was measured as material loss on steel panels according to ISO 9226 [8] on five road bridges, described in Table 1. The bridges were selected based on climatic conditions, height and availability for the deployment of samples. Gjemnessund and Sotra (43 and 14) are part of the coating condition study, while Nessundet, Tjeldsund and Hardanger are not because they have other coatings than investigated in this study. Annual temperature and precipitation averages and sailing clearance are given for each bridge in the table. Wind strength and direction will also affect corrosivity, but such data are not available and have not been measured in this study.

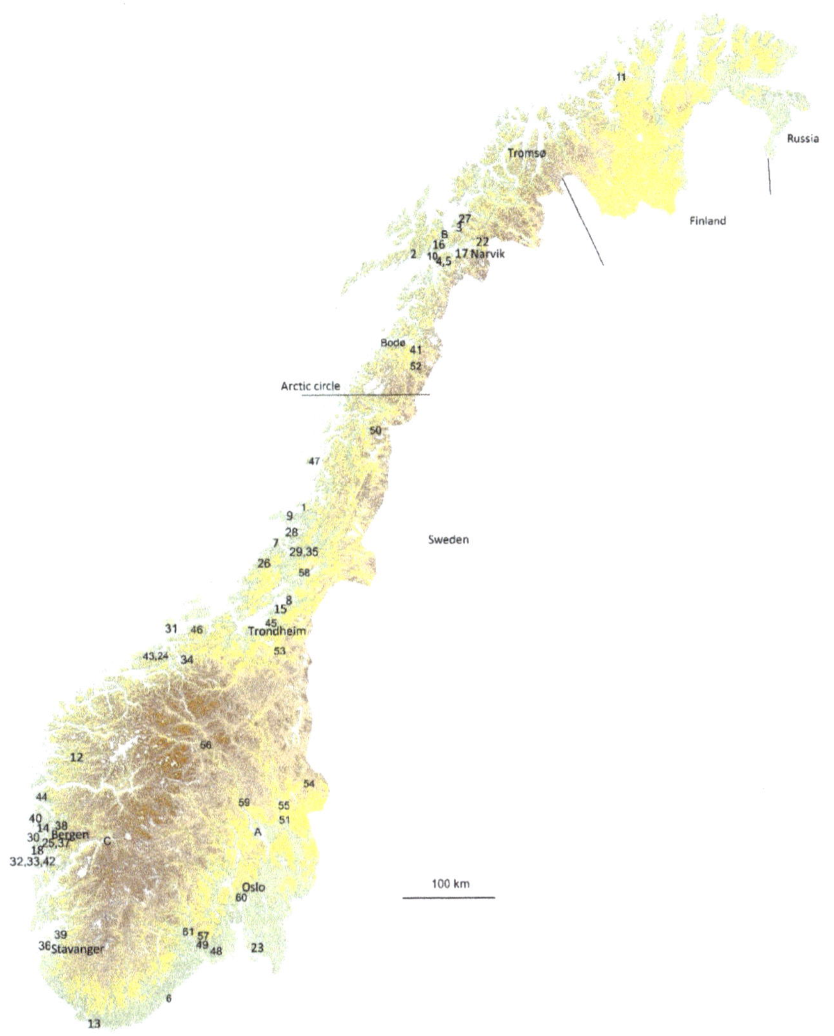

Figure 1. A map of Norway showing the location of the 61 bridges studied. Corrosivity was measured on bridges A, B, C, 14 and 43.

Panels of cold rolled mild steel, 150 mm × 100 mm × 2 mm in dimensions, were exposed for one year. Corrosivity was measured at the level of the bridgeway on all the bridges. For Gjemnessund, Sotra and Hardanger, corrosivity was also measured on one tower at various heights between 5 and 70 m above sea level. After retrieval, the samples were cleaned in a solution of 500 mL 37% HCl and 3.5 g hexamethylene-tetraamine diluted to 1000 mL. The samples were kept in the solution for 10 min at room temperature. The corrosivity at the various sites was evaluated according to weight loss and labeled with a corrosion category according to ISO 12944-2 [9]. A corrosion depth of 1.3–25 μm/y is category C2, 25–50 μm/y is C3, 50–80 μm/y is C4 and 80–200 μm/y is C5. Corrosion in the range 200–700 μm/y is category CX, but such a high corrosivity was not measured in this study. Each corrosivity measurement was assessed from three replicate samples.

The relative standard deviation for the measurements was 4% on average and ranging between 2% and 6% for the individual sets of three parallels. All measurements were performed during the 12 months in 2016–2017. Hence, variations from year to year are not accounted for.

The dominant wind direction is from the west on all the bridges. The Gjemnessund and Tjeldsund bridges are located behind larger islands and, therefore, somewhat shielded. Hardanger bridge is located deep inside a fjord and even more shielded. Freshwater from rivers reduce the salinity in the fjord to 3.1% at the bridge location, compared to 3.5% in the open sea, but this difference is assumed to be insignificant. Sotra bridge is quite exposed. Nessundet bridge is also somewhat exposed but crossing a strait in an inland lake where the only salt exposure is from winter deicing salt.

Annual precipitation varies significantly between the locations of the five bridges. The precipitation may have two opposite effects on corrosivity. Primarily, increasing precipitation will increase the time of wetness on the steel and thereby increase corrosion. However, in marine environments, precipitation may also decrease corrosion by washing off salt deposits. This effect is typically found under the bridgeway, where salt accumulates but never or rarely is washed away by rain [10]. Hence, in an inland environment, corrosivity will normally increase with annual precipitation, while in a coastal environment, the effect of precipitation may vary from place to place on the construction.

As Table 3 shows, the corrosivities measured at the level of the bridgeway are all within categories C2 and C3. On Nessundet bridge, the west side is more corrosive, probably because western winds are dominating. Under the bridgeway, the samples were shielded from rain, which resulted in a much lower corrosion rate than on the sides. Tjeldsund bridge is going east-west, so the difference between the directions are smaller, but the north side is slightly more exposed. On Gjemnessund, the corrosivity was highest under the bridgeway, where the samples were mounted on a vortex shedding blade. This is probably explained by the deposition of sea salt and shielding from rain that would wash the salt off the surface. The sailing clearance is 43 m, but salt deposits are nevertheless found under the bridgeway. In general, coastal bridges are more susceptible to coating degradation and corrosion on the underside of the bridgeway due to this effect. On Sotra bridge, C3 corrosivity was found on the pylon at the level of the bridgeway. No measurements were performed on the bridgeway, which may have shown higher corrosivity, as found on Gjemnessund. Hardanger bridge, which is the most shielded of the five bridges showed low corrosivity. As for Gjemnessund, the corrosivity may have been somewhat higher on the bridgeway, but due to the high sailing clearance and wind shielding, the corrosivity is assumed to be low even here. Figure 2 shows corrosivity as a function of the height along one pylon on three of the bridges, measured from 5 m above the sea level. On Gjemnessund and Hardanger, the corrosivity was generally low all the way. For both bridges, this was somewhat surprising since the pylons are located near the sea. Airflow patterns around the massive concrete pylons may partly explain the difference, but the lower annual precipitation at these two sites may also have contributed, decreasing the time of wetness on the samples. For the Sotra bridge, it is interesting to see how the corrosivity decreases as a function of height. Up to about 12 m, the corrosivity was category C5. Further up the corrosivity gradually decreased and reached almost category C2 from 45 m.

From the corrosivity measurements, it is evident that the height above sea level has a large influence on corrosivity. Annual precipitation also seems to be an important parameter. Wind conditions are expected to contribute, but wind measurements on the investigated bridges have not been performed and cannot be discussed in this study.

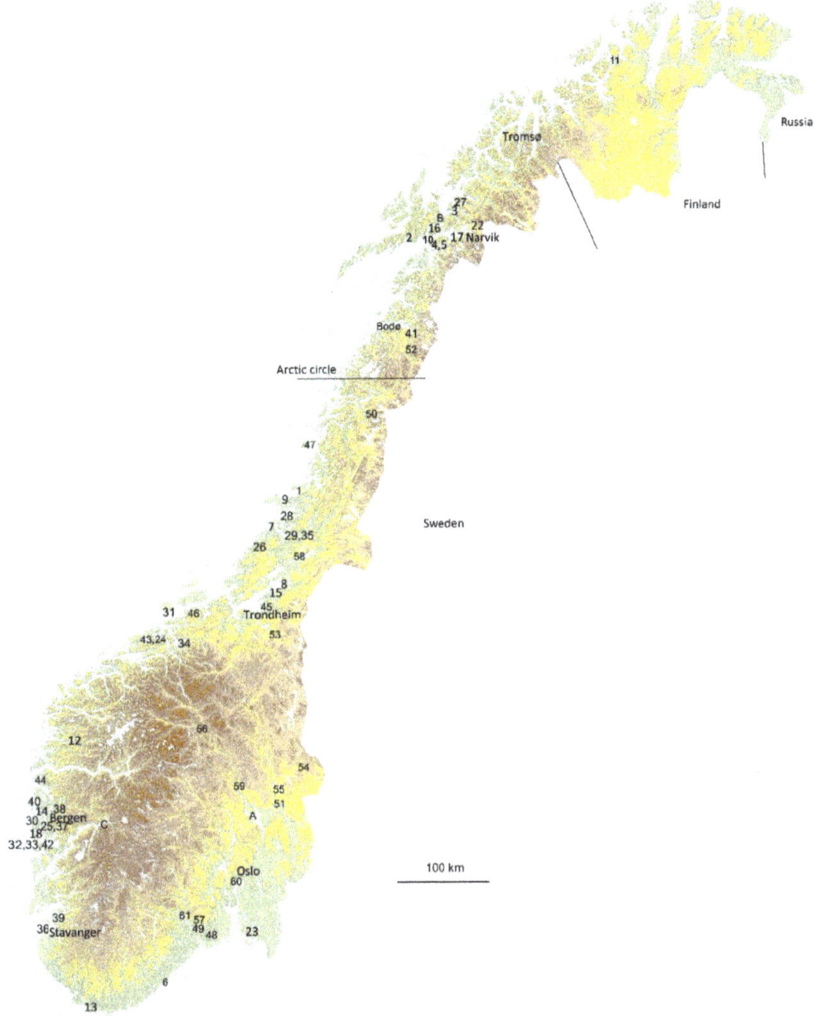

Figure 1. A map of Norway showing the location of the 61 bridges studied. Corrosivity was measured on bridges A, B, C, 14 and 43.

Panels of cold rolled mild steel, 150 mm × 100 mm × 2 mm in dimensions, were exposed for one year. Corrosivity was measured at the level of the bridgeway on all the bridges. For Gjemnessund, Sotra and Hardanger, corrosivity was also measured on one tower at various heights between 5 and 70 m above sea level. After retrieval, the samples were cleaned in a solution of 500 mL 37% HCl and 3.5 g hexamethylene-tetraamine diluted to 1000 mL. The samples were kept in the solution for 10 min at room temperature. The corrosivity at the various sites was evaluated according to weight loss and labeled with a corrosion category according to ISO 12944-2 [9]. A corrosion depth of 1.3–25 µm/y is category C2, 25–50 µm/y is C3, 50–80 µm/y is C4 and 80–200 µm/y is C5. Corrosion in the range 200–700 µm/y is category CX, but such a high corrosivity was not measured in this study. Each corrosivity measurement was assessed from three replicate samples.

The results from the corrosivity measurements were used to estimate corrosivity on all the 61 bridges in the study.

Table 1. The description of the bridges where corrosivity was measured. The location of the bridges is shown in Figure 1, labeled A, B, 43, 14 and C as given in the table.

Bridge	Climatic Conditions			Sailing Clearance (m)
	Geography	Temperature Avg. (°C)	Precipitation Avg. (mm/year)	
Nessundet (A)	Inland lake	3	700	10
Tjeldsund (B)	Shielded coast	3	1000	41
Gjemnessund (43)	Shielded coast	7	1300	43
Sotra (14)	Exposed coast	8	1900	50
Hardanger (C)	Shielded fjord	6	1100	55

2.2. Bridges and Coatings

Steel bridges longer than 100 m built between 1967 and 1995 were included in the study, 61 bridges in total. For C4 and C5 environments, shorter steel bridges were included to increase the number of bridges in these categories of corrosivity. The duplex coatings were specified from 1965, but assuming that it took some time to implement, only bridges from 1967 are included in this study. Bridges built after 1995 were not included due to the insufficient exposure time so far. The location of the 61 bridges is shown in Figure 1. For each bridge, corrosivity was estimated based on the corrosivity measurements on the five bridges described in Table 1. The estimates were based on the geographical location and sailing clearance. Average daily traffic was not assumed to affect the corrosivity significantly since the coated steel construction is located under the bridgeway on all the bridges in the study.

Two coating specifications were used in this period, as shown in Table 2. The only difference between the two specifications is that the zinc chromate in the two first alkyd coats was replaced with zinc phosphate in 1977.

Table 2. The coating specifications between 1965 and 1995.

Coat No.	1965		1977	
1	TSZ, pure Zn	100 μm	TSZ, pure Zn	100 μm
2	Phosphoric acid wash primer	–	Phosphoric acid wash primer	–
3	Alkyd with zinc chromate	50 μm	Alkyd with zinc phosphate	50 μm
4	Alkyd with zinc chromate	50 μm	Alkyd with zinc phosphate	50 μm
5	Alkyd	50 μm	Alkyd	50 μm
6	Alkyd	50 μm	Alkyd	50 μm

2.3. Coating Performance Data

Coating condition has been assessed based on pictures from routine bridge inspections performed by NPRA personnel. A thorough coating inspection is performed every 5 years and the results are documented in the NPRA bridge management database. The database also contains information about coating maintenance. In addition, information about the extent and scope of maintenance has been collected from the NPRA personnel responsible for the specific maintenance operations. The coating condition on the various bridges have been assessed according to four categories defined for this study:

- Good: The paint coating is in good condition and little or no degradation can be seen;
- Fair: There is some paint degradation, and zinc corrosion products (white) can be found locally;
- Poor: The steel has started to corrode and red rust is found;
- Repaired: Coating maintenance has been performed; in most cases, patch repair with a full topcoat.

The coating condition can vary across steel structures and typically overlapping joints, edges, welds and bolts are attacked by corrosion before the flat surfaces. Older truss construction bridges have more of these corrosion traps, while on modern box beam bridges these are, to a large extent, avoided. This has not been considered in the assessment. However, various coating application errors were found, which will be discussed.

2.4. Coating Performance Indicator

To enable a quantitative comparison of coating performance on the various bridges, a coating performance indicator (CPI) has been defined for this study:

$$CPI = L + L \cdot \frac{S}{C} \tag{2}$$

where L is the coating lifetime for a repaired coating or present coating age for not repaired coatings, C is the corrosivity category number (category 2–5) and S is the assessment of condition described above, but numerical. The coating condition has been assessed as either good, fair, poor or repaired, which have been given the values 3, 2, 1 and 0 respectively. For repaired coatings, CPI will then be the coating lifetime. For not repaired coatings, the CPI will be the current age of the coating, with an additional expected life proportional to current age and condition, and inversely proportional to corrosivity. Though CPI may be regarded as an expected lifetime of not repaired coatings, such a claim cannot be made without further investigation into coating performance and maintenance, which is beyond the scope of this work. Here, the CPI is only used for estimating an indicator for coating performance to enable comparison.

3. Results

3.1. Corrosivity

Corrosivity at the level of the bridgeway on the five bridges is shown in Table 3. Hardanger bridge and Gjemnessund bridge are modern box beam bridges, while the others are older truss constructions. Nessundet bridge is an arch bridge, while the rest are suspension bridges. The samples were placed at different locations on the bridges, as shown in the table.

Table 3. The description of the bridges where corrosivity was measured and the corrosivity found at the level of the bridgeway.

Bridge. Measurement Site	Climatic Conditions			Sailing Clearance (m)	Corrosivity	
	Geography	Temp. Avg. (°C)	Precipitation Avg. (mm/y)		μm/year	Category
Nessundet (A) Truss, west side Truss, middle Truss, east side	Inland lake	3	700	10	26 6 13	C3 C2 C2
Tjeldsund (B) Truss, south side Truss, middle Truss, north side	Shielded coast	3	1000	41	12 13 18	C2 C2 C2
Gjemnessund (43) Pylon, west side Bridgeway fence Under bridgeway	Shielded coast	7	1300	43	11 28 48	C2 C3 C3
Sotra (14) Pylon, west side	Exposed coast	8	1900	50	27	C3
Hardanger (C) Pylon, west side	Shielded fjord	6	1100	55	9	C2

The relative standard deviation for the measurements was 4% on average and ranging between 2% and 6% for the individual sets of three parallels. All measurements were performed during the 12 months in 2016–2017. Hence, variations from year to year are not accounted for.

The dominant wind direction is from the west on all the bridges. The Gjemnessund and Tjeldsund bridges are located behind larger islands and, therefore, somewhat shielded. Hardanger bridge is located deep inside a fjord and even more shielded. Freshwater from rivers reduce the salinity in the fjord to 3.1% at the bridge location, compared to 3.5% in the open sea, but this difference is assumed to be insignificant. Sotra bridge is quite exposed. Nessundet bridge is also somewhat exposed but crossing a strait in an inland lake where the only salt exposure is from winter deicing salt.

Annual precipitation varies significantly between the locations of the five bridges. The precipitation may have two opposite effects on corrosivity. Primarily, increasing precipitation will increase the time of wetness on the steel and thereby increase corrosion. However, in marine environments, precipitation may also decrease corrosion by washing off salt deposits. This effect is typically found under the bridgeway, where salt accumulates but never or rarely is washed away by rain [10]. Hence, in an inland environment, corrosivity will normally increase with annual precipitation, while in a coastal environment, the effect of precipitation may vary from place to place on the construction.

As Table 3 shows, the corrosivities measured at the level of the bridgeway are all within categories C2 and C3. On Nessundet bridge, the west side is more corrosive, probably because western winds are dominating. Under the bridgeway, the samples were shielded from rain, which resulted in a much lower corrosion rate than on the sides. Tjeldsund bridge is going east-west, so the difference between the directions are smaller, but the north side is slightly more exposed. On Gjemnessund, the corrosivity was highest under the bridgeway, where the samples were mounted on a vortex shedding blade. This is probably explained by the deposition of sea salt and shielding from rain that would wash the salt off the surface. The sailing clearance is 43 m, but salt deposits are nevertheless found under the bridgeway. In general, coastal bridges are more susceptible to coating degradation and corrosion on the underside of the bridgeway due to this effect. On Sotra bridge, C3 corrosivity was found on the pylon at the level of the bridgeway. No measurements were performed on the bridgeway, which may have shown higher corrosivity, as found on Gjemnessund. Hardanger bridge, which is the most shielded of the five bridges showed low corrosivity. As for Gjemnessund, the corrosivity may have been somewhat higher on the bridgeway, but due to the high sailing clearance and wind shielding, the corrosivity is assumed to be low even here. Figure 2 shows corrosivity as a function of the height along one pylon on three of the bridges, measured from 5 m above the sea level. On Gjemnessund and Hardanger, the corrosivity was generally low all the way. For both bridges, this was somewhat surprising since the pylons are located near the sea. Airflow patterns around the massive concrete pylons may partly explain the difference, but the lower annual precipitation at these two sites may also have contributed, decreasing the time of wetness on the samples. For the Sotra bridge, it is interesting to see how the corrosivity decreases as a function of height. Up to about 12 m, the corrosivity was category C5. Further up the corrosivity gradually decreased and reached almost category C2 from 45 m.

From the corrosivity measurements, it is evident that the height above sea level has a large influence on corrosivity. Annual precipitation also seems to be an important parameter. Wind conditions are expected to contribute, but wind measurements on the investigated bridges have not been performed and cannot be discussed in this study.

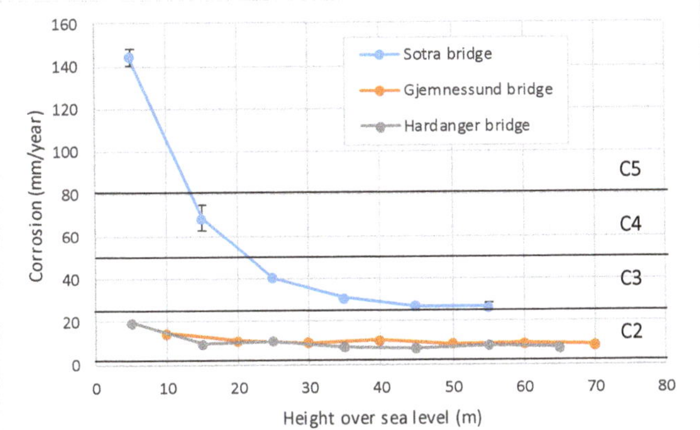

Figure 2. Corrosivity on three bridges as a function of height over sea level.

3.2. Coating Lifetimes

Tables 4 and 5 show the lifetime of duplex coatings on 23 steel bridges built between 1967 and 1977 and 38 bridges built between 1978 and 1995, respectively, 61 bridges in total. As shown in Table 2, the coating specification was changed during 1977 and the zinc chromate in the two first paint coats was replaced with zinc phosphate. Corrosivity is estimated for each bridge based on the geographical location and sailing clearance, relative to the five bridges where corrosivity was measured. Sailing clearance is given for all the coastal bridges. All the inland bridges are assumed to be in C2 environments. Winter salting of roads is assumed not to affect the corrosivity since the load bearing steel construction is located under the bridgeway for all the bridges in this study. There are few bridges in C2 environments in Table 4 since duplex coatings were not specified for inland bridges before 1977.

Table 4. The coating lifetime on coastal bridges built between 1967 and 1977. CPI is the coating performance indicator.

	Name	Corrosiv. (C)	Built	Clearance (m)	Condition (S)	Maintenance	Lifetime (years)	CPI
1	Madsøybrua	C5	1975	6	Repaired	2000	25	25
2	Erikstadstrøm		1975	3	Repaired	2017	42	42
3	Grov		1968	7	Repaired	1996	28	28
4	Kjerringstraum		1969	15	Repaired	1998	29	29
5	Sørstraumen		1969	3	Repaired	1998	29	29
6	Strømsund		1970	4	Repaired	2017	47	47
7	Stamnes	C4	1970	6	Repaired	2006	36	36
8	Verdal		1972	2	Repaired	1995	23	23
9	Kvalpsundet		1974	10	Repaired	1995	21	21
10	Slottvikstraumen		1974	6	Fair	–	>45	68
11	Kvalsundbrua		1977	26	Repaired	2001	24	24
12	Naustdal Bru		1970	2	Good	–	>49	90
13	Revøysund		1971	20	Good	–	>48	96
14	Sotrabrua		1971	50	Repaired	2011	40	40
15	Sundbrua		1971	3	Good	–	>48	96
16	Kjærfjorden	C3	1972	5	Repaired	2010	38	38
17	Skjomen		1972	35	Repaired	2000	28	28
18	Tofterøy		1975	20	Repaired	2011	36	36
19	Kjellingstraumen		1975	29	Good	–	>44	88
20	Randøy		1976	24	Fair	–	>43	72
21	Lokkarbrua		1977	30	Repaired	2005	28	28
22	Rombaksbrua	C2	1970	40	Repaired	2011	41	82
23	Rolvsøysund		1970	5	Good	–	49	123

Table 5. The coating lifetime on bridges built between 1978 and 1995. CPI is coating performance indicator.

	Name	Corrosiv. (C)	Built	Clearance (m)	Condition (S)	Maintenance	Lifetime (years)	CPI
24	Bergsøysundbrua	C5	1992	5	Repaired	2011	19	19
25	Nordhordlandsbrua		1994	5	Repaired	2017	23	23
26	Store Holmsund		1978	4	Repaired	2007	29	29
27	Åndervåg		1980	3	Repaired	2001	21	21
28	Nærøysund		1981	41	Repaired	2011	30	30
29	Ytterbystrømmen		1982	4	Good	–	>37	65
30	Torvsundet	C4	1984	5	Fair	–	>35	53
31	Kulisvabrua		1988	16	Fair	–	>31	47
32	Klubbasund		1989	20	Fair	–	>30	45
33	Djupasund		1990	20	Poor	–	>29	36
34	Bøfjordbrua		1992	2	Poor	–	>27	34
35	Høyknesbrua		1978	5	Fair	–	>41	68
36	Stavanger Bybru		1978	26	Poor	–	>41	55
37	Haglesundbrua		1982	50	Fair	–	>37	62
38	Eikanger I		1987	5	Good	–	>32	64
39	Helgøysund		1988	14	Fair	–	>31	52
40	Bukkholmstraum		1988	16	Fair	2012	24	52
41	Botn Bru	C3	1991	2	Good	–	>28	56
42	Brandasund		1991	24	Fair	–	>28	47
43	Gjemnessundbrua		1991	43	Fair	–	>28	47
44	Mjåsund		1993	30	Fair	–	>26	43
45	Sandfærhus		1995	5	Fair	–	>24	40
46	Dromnessundbrua		1995	16	Good	–	>24	48
47	Grimsøy Bru		1995	10	Good	–	>24	48
48	Hvåra		1978	–	Good	–	>41	103
49	Ulefossbrua		1978	–	Fair	–	>41	82
50	Korgen Bru		1978	–	Good	–	>41	103
51	Glåmbrua Elverum		1979	–	Good	–	>40	100
52	Pothus		1979	–	Good	–	>40	100
53	Moslett		1980	–	Good	–	>39	98
54	Jordet bru	C2	1981	–	Good	–	>38	95
55	Åmot bru		1981	–	Good	–	>38	95
56	Sundbru Ny		1981	–	Good	–	>38	95
57	Akkerhaugen		1981	–	Good	–	>38	95
58	Nødalsbrua		1982	–	Good	–	>37	93
59	Lillehammer		1984	–	Good	–	>35	88
60	Grinienga		1986	–	Good	–	>33	83
61	Nautesund		1986	–	Good	–	>33	83

A graphical summary of the coating performance on the bridges in Table 4 is given in Figure 3a. Of the 23 bridges, 16 have received coating maintenance, i.e., about 70%. Of these, eight are in C4 environments, five in C3 environments a one in a C2 environment. The repair of the C2 bridge was preventive maintenance due to topcoat flaking and not corrosion. Both the two bridges in the C5 environment have received maintenance. Of the remaining bridges, two are in a fair condition and five are in good condition.

The coating performance of bridges in Table 5 is shown graphically in Figure 3b for the 38 bridges built between 1978 and 1995. Of the 38 bridges, only five have received coating maintenance so far, i.e., about 13%. These five bridges are all exposed in C4 or C5 environments. Not surprising, the duplex coatings are performing very well in C2 environments. Contrary to Figure 3a, Figure 3b does show a correlation between corrosivity and coating performance.

The performance of the coatings in the C5 environment is similar for bridges coated with the 1965 specification and with the 1977 specification. The Madsøybrua and Erikstadstrøm bridges listed in Table 4 were repaired after 25 and 42 years, respectively. The Bergsøysundbrua and Nordhordlandsbrua bridges listed in Table 5 were repaired after only 19 and 23 years. A close inspection of the coatings reveals that application errors led to the failures on Bergsøysundbrua and Nordhordlandsbrua. Both bridges are floating bridges and may be regarded as mainly exposed in the marine splash zone. In the offshore industry, the corrosion rate in the marine splash zone is assumed to be 400 μm/y, i.e., corrosivity category CX [9]. Hence, to classify these bridges as category C5 may be too low.

The coatings were mainly degraded on the underside of the bridgeway, where marine salt deposits will never be washed off by rain. Hence, a concentrated and very aggressive brine is formed on the coating surface. For Bergsøysundbrua, the maintenance started too early. Even though the organic coating had partly failed, the zinc coating was still protecting the steel. For Nordhordlandsbrua, the degradation was more severe and the steel had started to corrode in some areas.

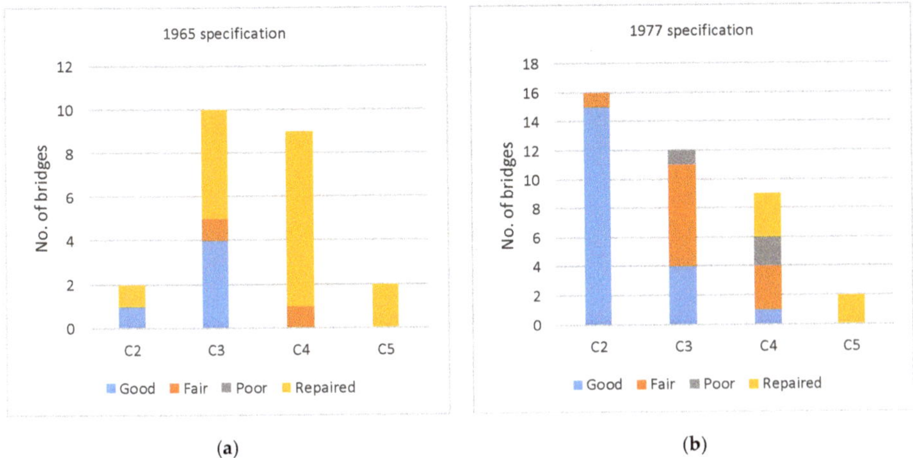

(**a**) (**b**)

Figure 3. The summary of coating condition for the two specifications as a function of corrosivity: (**a**) 1965 specification; (**b**) 1977 specification.

3.3. Coating Performance versus Corrosivity

Figure 4a shows the average age and age span of the bridges in the study, illustrating the difference in exposure time for the two coating specifications. Since only 21 of the 61 bridges has received coating maintenance so far, the average coating lifetimes as defined by time to first major coating maintenance (ISO 12944-1 [11]) cannot be calculated. Only the result for the C5 exposures represents actual lifetime since all bridges have had coating maintenance. In order to enable a quantitative comparison of coating performance, the coating performance indicator (CPI) was defined, as described above. The average CPI as a function of corrosivity is shown in Figure 4b, while the CPI for each bridge is given in Tables 4 and 5. The error bars show the standard deviation. Figure 4b shows that the coating performance increases with decreasing corrosivity for both coating specifications, as expected. There are only two bridges in both the C2 and C5 environments for the 1965 specification though, which gives a rather weak statistical base for making conclusions. Additionally, the standard deviation is rather high for several of the categories, but the trend in the results is evident.

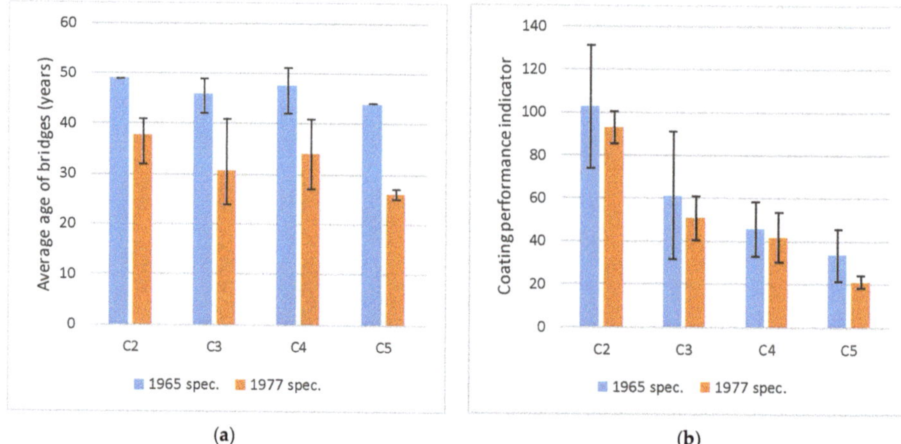

Figure 4. The coating age and performance: (**a**) Average age of the bridges. The bars show the highest and lowest age within each category of corrosivity; (**b**) Coating performance indicator as a function of corrosivity. The bars indicate standard deviation.

3.4. Coating Failure Mechanisms

Four different coating failure mechanisms have been identified in this study, of which two are specific to TSZ duplex coatings, and two are generally found in paint coatings. The most common failure mechanism was an insufficient barrier due to low paint film thickness. In many cases, this was found at steel edges, e.g., flanges on H-beams. This is typical for all paint coatings that are applied as a wet film. Due to surface tension, the paint will retract from sharp edges in order to reduce surface area, resulting in an inferior film thickness. Poor edge retention being a typical failure mode for paint has been subject to investigation, e.g., by Yun et al. [12].

Another reason for low film thickness that is specific to TSZ duplex coatings is the lack of an instrument that can measure paint film thickness over TSZ. The commonly used magnetic film thickness gauges will measure the total duplex film thickness, i.e., both TSZ and paint. The eddy current thickness gauges that will measure thickness over the first electrically conducting layer, i.e., the zinc coating, does not work well with TSZ due to its rough microstructure and porosity (see Figure 5). This makes large variations in conductivity and unstable measurements. The magnetic thickness gauge is, therefore, presently the only option. Ideally, each coating thickness should be measured as applied to ensure the coatings meet specifications.

The second failure mechanism is the formation of pinholes in the paint film [13]. When applying paint on TSZ, bubbles of trapped air and evaporating solvent are easily formed inside the film, probably due to the rough surface structure of the TSZ. As the bubbles grow, they attract paint from around and the wet film thickness grows locally. When the paint dries, the bubbles crack open, but if the opening is too narrow, it will not be filled by the subsequent coats. Thus, a pinhole is formed in the paint where the zinc will start to corrode at an early stage.

The third failure mechanism found was "spitting" during thermal spraying. If the arc spray gun is not properly adjusted, there will be incomplete melting of the zinc wire. These pieces of solid zinc will be sprayed along with the melted zinc but may protrude from the surface as tall peaks. When the paint is applied, the film thickness will be too low over these peaks and the paint will fail during exposure. A typical example is shown in Figure 6.

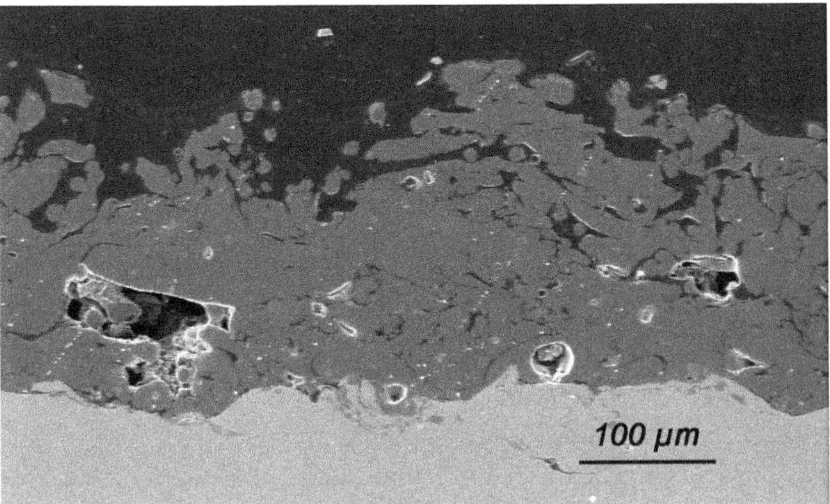

Figure 5. The cross-section of Thermal spray zinc (TSZ) coating. The very rough microstructure affects the conductivity of the coating, which results in a very large variation in paint film thickness measurements with Eddy current gauge.

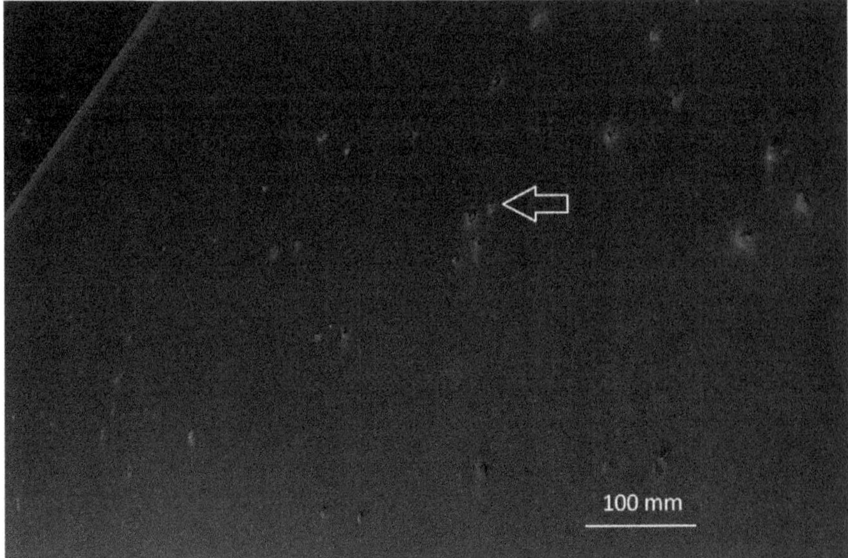

Figure 6. The corrosion of the TSZ due to "spitting" during thermal spraying. All the protrusions in the coating, one of them marked with an arrow, is due to spitting during thermal spraying. The paint film is too thin over the zinc protrusions and the zinc starts to corrode. Photo: NPRA.

The fourth failure mechanism is the hydrolysis of the alkyd paint in contact with concrete. The ester bonds in the alkyd binder are broken when exposed to an alkaline environment. This is also called saponification and is a well-known degradation mechanism for alkyds [14]. This was typically found on the upper flange of H-beams carrying a concrete bridgeway, see Figure 7.

Figure 7. The saponification of alkyd paint where the painted H-beam is in contact with the concrete (marked with an arrow) and subsequent corrosion of zinc and steel.

4. Discussion

4.1. TSZ Duplex Coating Failure Mechanisms

The coating failure mechanisms that were identified have different origins. Pinholes and spitting are typical application errors with TSZ duplex coatings. When coating a large bridge, such errors are likely to appear occasionally. By adjusting the application parameters immediately to prevent further formation of such errors, and light grinding with sandpaper before application of the next coat, the quality of the coating will improve significantly. Mainly pinholes and spitting were responsible for the degradation at Bergsøysundbrua and large amounts of pinholes were also found at Nordhordlandsbrua. In the very corrosive C5 environment, the TSZ started to corrode locally shortly after the bridge was built. The TSZ was protecting the steel, but the TSZ has a certain capacity for protection and when it is consumed, the steel starts to corrode.

Low film thickness results in coating failure due to the penetration of ions. Studies have shown that a protective coating should have a resistivity of more than 10^9 Ohm·cm^2 [15]. The film thickness required to reach this resistivity depends on the generic type of coating. The thickness specification given by the coating supplier will normally provide the required barrier. Low film thickness over edges due to low edge retention is a general problem with paint. TSZ duplex coatings will give longer lifetimes than pure paint coating systems since the TSZ coat is not susceptible to this. The TSZ film thickness will be the same over sharp edges as on flat surfaces. However, if the paint film degrades early, the TSZ will start to corrode, and depending on the corrosivity, it will eventually be consumed. Grinding and stripe-coating of edges are routinely performed to reduce this problem, but it will not be eliminated. On modern box beam bridges, the amount of edges is significantly reduced, compared to a truss bridge.

Low paint film thickness on flat surfaces was one of the main degradation mechanisms on Nordhordlandsbrua. The zinc started to corrode under the paint after a few years due to a combination of low film thickness and formation of a very aggressive brine under the bridgeway, where rain never could wash deposited sea salt away. Inspections and control during painting seem to have been insufficient. This problem may be negligible with modern epoxy coatings. The old alkyd paints were applied in thin coats specified at 50 μm each. High build epoxy mastic coatings typically must be applied in thicker coats in order to form a continuous film. Hence, the painter will see when enough paint has been applied, i.e., an inherent "smartness" in the paint.

The saponification problem was found on most bridges where a concrete bridgeway was resting on a coated H-beam. This problem was eliminated when the NPRA paint specification was changed from alkyds to epoxies. Epoxies are very tolerant to the alkaline conditions in concrete.

4.2. The Durability of TSZ Duplex Coatings

Coating maintenance constitutes the main operational cost for most steel bridges. Given the long lifetime of bridges, coating durability is a key factor for reducing life cycle costs. The TSZ duplex coatings investigated in this study provide very long lifetimes and, for most bridges, significantly longer than 25 years, which is defined as a "very high" lifetime in ISO 12944-1. Since only about 70% of the bridges with the 1965 coating specification and about 13% of the bridges with the 1977 coating specification have been repaired, and many of the bridges still have coatings in good or fair condition, even longer lifetimes will be achieved. For the less corrosive environments at least, it is likely that duplex coatings may last for the lifetime of the bridge.

There are several sources of variation that complicate the comparison of coating performances. Most important is the sample size. As stated earlier, the study includes few bridges in C5 environments, which constitute a weak statistical base for comparison. Geographical location plays a role. Bridges with the same sailing clearance may have different orientations or see different weather patterns leading to variable coating performance. Criteria for coating maintenance will affect the calculated performance indicator. Decisions about coating maintenance are made by different people and no firm criteria have been set by the NPRA. For example, the coating maintenance on Rombak bridge in 2011 was initiated as a preventive action. There was no corrosion on the bridge, but the topcoat had flaked off from parts of the bridge and the topcoat was renewed to prevent a more expensive maintenance operation in the future. The maintenance of the Nordhordland bridge, however, was initiated due to extensive coating degradation under the bridgeway and corrosion of the steel. In addition, non-technical factors affect the decision, like the available maintenance budget. Corrosion is a slow process and coating maintenance can often be postponed for many years without compromising the load carrying capacity of the bridge.

Both the 1965 and the 1977 coating specifications show a correlation between coating performance and corrosivity. The coatings have performed very well in mild conditions and failed earlier in corrosive conditions. Though the coating failures in C5 environments, at least partly, was due to coating application errors; this suggests that a stronger paint coating is required in corrosive environments. When the paint fails, the zinc coating will first act as a barrier, and later as a sacrificial anode when bare steel is exposed. Zinc is an active metal and even the 15% aluminum alloy that is used mainly today will have a certain corrosion rate. Hence, when the zinc is exposed, the remaining lifetime of the coating will be limited by its corrosion rate, which again is determined by the local corrosivity. The durability of a TSZ duplex coating in a corrosive environment is, therefore, very dependent on the performance of the paint coating.

5. Conclusions

Thermal spray zinc duplex coatings have provided long term corrosion protection to steel bridges in Norway. Considering the present coating age and condition, it is reasonable to assume that the duplex coating may last the entire 100-year bridge design life for many of the bridges in this investigation. For very corrosive environments, the two paint coating specifications appear to have been too weak.

Application quality has the strongest impact on coating life. Application errors like pinholes, spitting and low paint film thickness caused most coating failures found in this study. These errors decrease the protective properties of the paint film. The zinc coating provides active corrosion protection, but will have a definite lifetime, depending on the corrosivity. Hence, such application errors will reduce the coating lifetime more in corrosive environments than in mild environments.

Some of the coating degradation mechanisms found in this study will be significantly reduced or eliminated with modern bridge designs and modern paint systems, which probably will increase TSZ duplex coating durability in the future. The most important improvements being

- Box beam bridge designs with fewer edges, reducing the edge retention problem
- Using epoxy coatings with high tolerance to alkaline environments
- Coatings with an inherent "smartness", that tell the painter when enough paint is applied
- Awareness of the spitting and pinhole problems so that adequate measures can be made during the application

Author Contributions: Conceptualization, O.Ø.K., C.D. and H.M.; Investigation, O.Ø.K.; Data curation, O.Ø.K., C.D. and H.M.; Writing—original draft preparation, O.Ø.K.; Writing—review and editing, M.G.; Funding acquisition, C.D., H.M. and M.G.

Funding: This research was funded by the Norwegian Public Roads Administration, contract number 2014059758, and the International Zinc Association.

Acknowledgments: Funding from the Norwegian Public Roads Administration and the International Zinc Association is gratefully acknowledged.

Conflicts of Interest: The International Zinc Association promotes the use of zinc worldwide, including for corrosion protection. However, they do not sell any products or services, and were not involved in the data collection or interpretation in this study. The NPRA is a governmental body and responsible for all public road bridges in Norway but has no commercial interests in coatings. They made their entire bridge administration database available to the project and put no limitations on the data selection. The decision to publish this study was made by O.Ø. Knudsen, SINTEF.

References

1. *ISO 12944-5 Paints and Varnishes-Corrosion Protection of Steel Structures by Protective Paint Systems. Part 5: Protective Paint Systems*; The International Organization for Standardization: Geneva, Switzerland, 2018.
2. Klinge, R. Protection of norwegian steel bridges against corrosion (Korrosionsschutz von Norwegischen Stahlbrücken). *Stahlbau* **1999**, *68*, 382–391. [CrossRef]
3. Klinge, R. Altered specifications for the protection of norwegian steel bridges and offshore structures against corrosion. *Steel Constr. Des. Res.* **2009**, *2*, 109–118. [CrossRef]
4. Porter, F.C. *Zinc Handbook: Properties, Processing, and Use in Design*; CRC Press: Boca Raton, FL, USA, 1991.
5. Perkins, R.A. *Metallized Coatings for Corrosion Control of Naval Ship Structures and Components*; National Materials Advisory Board: Washington, DC, USA, 1983; No. NMAB-409.
6. AWS. *Corrosion Test of Flames-Sprayed Coated Steel. 19-Year Report*; American Welding Society: Miami, FL, USA, 1974; No. C2.14-74.
7. Eijnsbergen, J.F.H.v. *Duplex Systems. Hot-Dip Galvanizing Plus Painting*; Elsevier Science: Amsterdam, The Netherlands, 1994; p. 222.
8. *ISO 9226 Corrosion of Metals and Alloys—Corrosivity of Atmospheres—Determination of Corrosion Rate of Standard Specimens for the Evaluation of Corrosivity*; The International Organization for Standardization: Geneva, Switzerland, 2012.
9. *ISO 12944-2 Paints and Varnishes-Corrosion Protection of Steel Structures by Protective Paint Systems. Part 2: Classification of Environments*; The International Organization for Standardization: Geneva, Switzerland, 2017.
10. Kreislova, K.; Geiplova, H. Evaluation of corrosion protection of steel bridges. *Procedia Eng.* **2012**, *40*, 229–234. [CrossRef]
11. *ISO 12944-1 Paints and Varnishes-Corrosion Protection of Steel Structures by Protective Paint Systems. Part 1: General Introduction*; The International Organization for Standardization: Geneva, Switzerland, 2017.
12. Yun, J.T.; Kwon, T.K.; Kang, T.S.; Kim, K.L.; Kim, T.K.; Han, J.M. A critical study on edge retention of protective coatings for a ship hull. In Proceedings of the CORROSION 2005, Houston, TX, USA, 3–7 April 2005; National Association of Corrosion Engineers (NACE): Houston, TX, USA, 2005.
13. Hasselø, J.-A.; Djuve, G. Coating systems with long lifetime–paint on thermally sprayed zinc. In Proceedings of the CORROSION 2016, Vancouver, BC, Canada, 6–10 March 2016; National Association of Corrosion Engineers (NACE): Houston, TX, USA, 2016.

14. Knudsen, O.Ø.; Forsgren, A. *Corrosion Control through Organic Coatings*; Taylor & Francis: London, UK, 2017.
15. Bierwagen, G.P.; He, L.; Li, J.; Ellingson, L.; Tallman, D. Studies of a new accelerated evaluation method for coating corrosion resistance—Thermal cycling testing. *Prog. Org. Coat.* **2000**, *39*, 67–78. [CrossRef]

Review

Sacrificial Thermally Sprayed Aluminium Coatings for Marine Environments: A Review

Berenika Syrek-Gerstenkorn [1,2], Shiladitya Paul [3,4,*] and Alison J Davenport [1]

[1]   School of Metallurgy and Materials, University of Birmingham, Edgbaston, Birmingham B15 2TT, UK; BAS684@student.bham.ac.uk (B.S.-G.); a.davenport@bham.ac.uk (A.J.D.)
[2]   National Structural Integrity Research Centre, Cambridge CB21 6AL, UK
[3]   TWI, Cambridge CB21 6AL, UK
[4]   School of Engineering, University of Leicester, University Road, Leicester LE1 7RH, UK
*   Correspondence: Shiladitya.Paul@twi.co.uk

Received: 16 January 2020; Accepted: 4 March 2020; Published: 12 March 2020

Abstract: One of the corrosion mitigation methods that is used for the protection of steel operating in seawater environments involves the application of sacrificial metallic coatings (such as aluminium, zinc, and their alloys). This paper reviews current knowledge about thermally-sprayed (TS) and cold-sprayed (CS) Al coatings for the corrosion protection of steel. It also summarises the key findings of the substantial amount of work that has been devoted to understanding mechanisms and the parameters that control the performance of TS Al coatings, such as the spraying method and its parameters like coating thickness and the application of sealer. The paper includes suggestions for areas of further research that could lead to the development of more resilient and longer-lasting coatings, based on the results from both laboratory and field tests that have been published in the literature. It also highlights the need for conducting simulated laboratory tests at conditions of intended service and the importance of long-term testing.

Keywords: thermally-sprayed aluminium (TSA); anti-corrosion coatings; cold spray; cathodic protection; seawater; corrosion

---

1. Introduction

The corrosion protection of structures that operate in seawater environments is a particularly important issue. It has been estimated that the cost of corrosion is high, not only in terms of financial loss (3.4% of global GDP in accordance to the NACE report [1], but widely believed to be between 3% and 4%) but also in the number of lives lost due to catastrophic failures and environmental damage (when toxic products leak out from corroded equipment). The battle against corrosion can be supported in several ways: by appropriate material selection (e.g., corrosion-resistant alloys (CRAs)), by the application of coatings (organic or metallic), by cathodic protection (impressed current or sacrificial anodes), or by the use of corrosion inhibitors and corrosion allowance. The most common corrosion mitigation methods for the protection of offshore structures that are made of steel (usually mild or low-alloyed [2]) involve the application of cathodic protection and/or protective coatings, depending on the conditions to which the structure is subjected. The types of exposure for these materials can be divided into four main zones: atmospheric, splash/tidal, submerged, and buried. Parts of an offshore structure that are constantly immersed in seawater are usually protected by the application of cathodic protection, while parts that are exposed to the marine atmosphere are usually protected by protective coatings. In the splash/tidal zone however, the corrosion of steel is usually controlled by robust coatings and corrosion allowance (in the splash zone) as well as cathodic protection (for parts located below mean water level) [3]. This is the most severe and the most challenging area for corrosion protection due to alternating wetting and drying conditions, easy access to oxygen, atmospheric pollutants, UV radiation, and erosive actions combined with frequent contact with floating objects such as debris, boats or ice.

Protective coatings can be divided into: inorganic (for example ceramic (e.g., alumina) or carbon coatings), organic (such as epoxy and polyurethane), organic–inorganic hybrids (also known as smart or functionalised coatings), and metallic coatings [4]. Depending on the electrode potential with respect to the substrate metal (usually steel), metallic coatings can be formed either anodic (e.g., aluminium and zinc) or cathodic (e.g., nickel and copper). The primary function of protective coatings is their barrier properties. However, if the integrity of the coating is compromised and the substrate is exposed, the corrosion of either the substrate (in case of organic and cathodic coatings) or the coating (in case of sacrificial anodic coatings) will develop [5].

The aim of this paper was to review and summarise current knowledge about thermally-sprayed aluminium (TSA) coatings (formed by electric arc or wire flame) for the corrosion protection of steel operating in seawater environments and to suggest potential areas for further research. Sacrificial protection mechanisms that are provided by those coatings and the effect of various parameters (such as coating thickness and composition, spraying parameters, application of sealants and operating conditions) are discussed. A comparison with cold-sprayed aluminium coatings is also presented. Laboratory tests and in-service performance of the coatings are discussed and summarised.

2. Sacrificial Coatings

Metallic coatings (which are more active than steel) such as Al, Zn, and their alloys (Figure 1), supply cathodic protection to the substrate by working as a sacrificial anode. When Al (or Zn) and steel are coupled together and exposed to seawater, a galvanic couple is established between those two metals, and electrons flow from the coating (anode) towards the substrate (cathode), which is accompanied by the dissolution of the coating, as schematically shown in Figure 2. When steel is cathodically protected, calcareous matters deposit on its surface (due to local pH changes explained in Section 4.5), which decrease the cathodic protection (CP) demand.

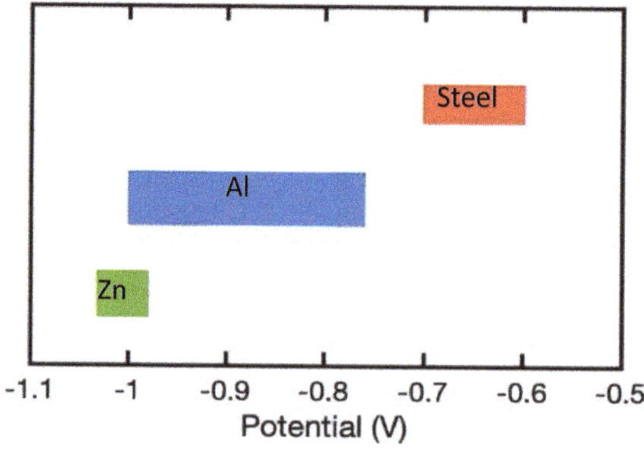

Figure 1. Galvanic series in seawater showing corrosion potential of steel, Al, Zn and their alloys. All values are against saturated calomel electrode (SCE).

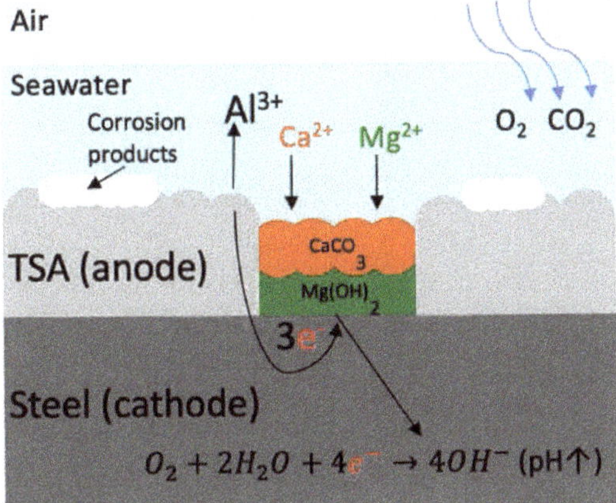

Figure 2. Sacrificial protection mechanisms provided by Al coatings

There are two key factors that determine the performance of sacrificial coatings:

- Low self-corrosion rate.
- Current output enabling the polarisation of the steel in the protective regime, usually considered to be between −0.8 and −1.1 V (Ag/AgCl/seawater) [6]. Values more negative than −1.1 V are generally not recommended to avoid excessive hydrogen generation.

There are many advantages of sacrificial metallic coatings:

- Barrier properties combined with cathodic protection.
- Newly applied coatings can be handled immediately (i.e., no drying time) [7].
- Spraying and repair can be performed on-site.
- More economical over the lifetime than organic coatings (total life cycle cost).
- Good resistance to mechanical damage.

3. Thermal Spraying

Though there are several thermal spray processes that can be used for the preparation of coatings, only two of them are commonly used for the deposition of sacrificial metallic coatings, namely electric arc and flame spray. Both involve propelling molten or semi-molten metal particles towards the substrate by a stream of air, thus creating a layer-by-layer deposition, until the required thickness of the coating is achieved. In contrast to other techniques, the two spraying processes above can be performed on-site and are economical.

A less commonly used technique for the deposition of metallic coatings is cold spray. This process, even though it is the costliest, has some advantages over the aforementioned techniques. Cold spray relies on the deposition of metallic powders in the solid-state, which means that no heat source such as flame or arc is needed, which can be critical when spraying materials that are highly sensitive to oxidation. The process can be performed in an area with a high risk of catching fire (e.g., an oil rig). There is no UV radiation during cold spraying, so no special eye protection measures are needed. Moreover, coatings that are prepared by cold spray tend to have a lower level of porosity [8], as can be seen in Figure 3.

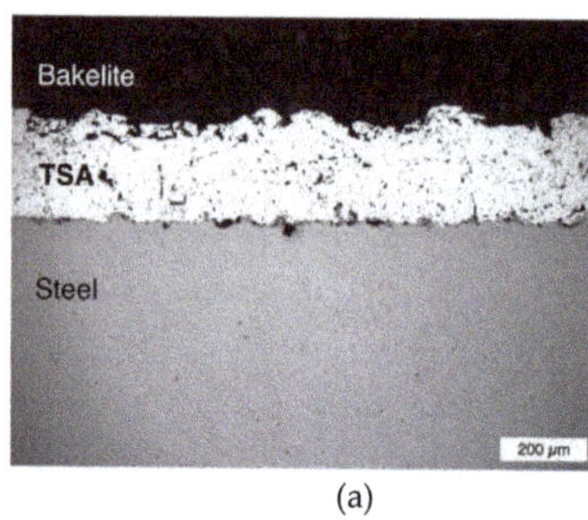

(a)

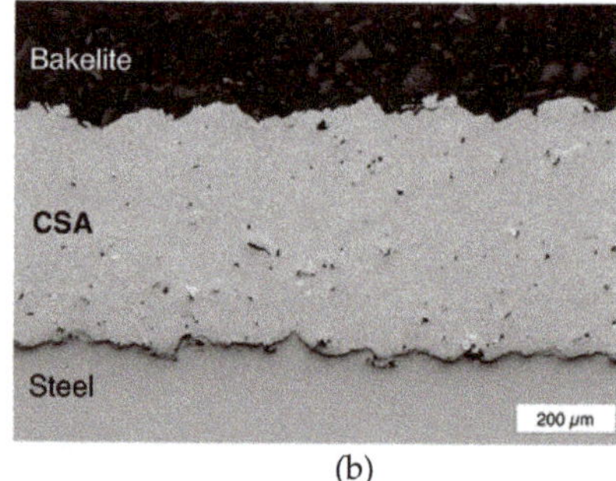

(b)

Figure 3. Cross-section of arc-sprayed Al coating (**a**) and cold-sprayed Al–12%Si coating (**b**) on steel.

3.1. Electric Arc Spraying

This technique, also known as twin wire arc or wire arc spray, is based on melting the tips of two metallic wires with an electric arc, which is established between them through the application of direct current (DC). Molten particles are then accelerated and propelled towards the substrate via an air stream. More information about the process can be found elsewhere: [9–11].

3.2. Wire Flame Spraying

In this technique, a metallic wire is melted by a flame, and the molten particles are accelerated towards the workpiece by pressurised air. More details can be found elsewhere: [12,13].

3.3. Cold Spraying

In contrast to thermal spraying processes, cold spray relies on the deposition of particles in the solid state, and these particles plastically deform upon impact [14]. Since the melting of the feedstock material is avoided, a coating with a negligible oxide content can be obtained. Moreover, cold-spray coatings tend to be denser (achievable porosity level below 1% [15]) than the coatings that are deposited by using low-velocity thermal spray processes, such as electric arc and wire flame spraying (typical porosity level between 5%–15% [16]), due to the lack of the splashing of molten droplets and the peening effect of solid particles, which close voids in the underlaying layer upon impact. Furthermore, since there is no melting of the feedstock material, the build-up of unfavourable tensile residual stresses arising from the thermal contraction of splats upon cooling is also reduced. This, combined with the peening effect, allows for the deposition of thicker coatings than the ones that are prepared by using arc or wire flame spraying [16]. A solid-state deposition also eliminates problems with the preferential evaporation of elements from feedstock material during spraying and maintains an unchanged phase composition of the coating as compared with the initial powder [16]. However, due to the low spray plume (typically <1 cm in diameter), the coating of large surface areas can be time-consuming. A detailed description of the process can be found elsewhere [17].

4. Laboratory Tests

Corrosion protection that is provided by thermally-sprayed aluminium coatings has been a subject of many investigations during the past three decades. Attempts have been made to understand the mechanisms and parameters that influence the performance of the coatings by using different methods. The most notable findings are reviewed below.

4.1. Effect of the Spraying Method and Its Parameters

A comparison between TSA coatings that are deposited by using electric arc (EA) and flame spraying (FS) was conducted by several researchers. Rodriguez et al. [18] observed differences in the oxide layers of coatings that were deposited by using arc and flame spraying through the use X-ray photoelectron spectroscopy (XPS). They also detected the different corrosion products that were formed on those coatings after a 4000-hour saline mist test: bayerite in the FS process and boehmite in the coating that was deposited by EA. Gartland and Eggen [19] concluded, based on their lab and field tests on TSA that was prepared by using both methods, that the spraying method had little influence on the performance of the coating, but the arc-sprayed coatings exhibited better adhesion (arc-sprayed Al: 9.0 ± 1 MPa, flame-sprayed Al: 3.5 ± 0.34 MPa). A higher adhesion of arc-sprayed coatings was also reported by others [20,21]. Moreover, coatings that are prepared by flame spraying tend to exhibit a higher level of porosity in comparison to the ones that are prepared by arc-spraying [9]. Porosity is an inevitable characteristic feature of thermally-sprayed coatings, but its level can be altered by selecting appropriate feedstock material, application method, and spraying parameters. To obtain a less porous coating, oxyfuel flame spraying and small diameter wire (1.6 and 2.3 mm), or low current (100–200 A) arc spraying should be used [22]. It has also been estimated that arc-spraying is more economical [9].

For cold-sprayed Al coatings, it has been recognised that a finer powder size provides better corrosion protection to the underlying steel due to its lower porosity level. Coatings that were prepared with 40–50 μm and 50–75 μm Al powders performed well, both under NaCl immersion and 11 years of atmospheric exposure [23]. A better performance of cold-sprayed Al coatings (attributed to their lower porosity level) than arc-sprayed coatings in natural seawater was observed by other researchers [8].

4.2. Effect of Coating Thickness

The effect of TSA coating thickness on corrosion performance was studied by Han et al. [24] by using electrochemical methods. They used stainless steel STS 304 steel substrates. Several experiments were performed, and their data were presented in the paper. However, based on the presented results, meaningful conclusions could not be drawn.

The results of the mechanical testing of arc-sprayed Al coatings (222–397 μm) on mild steel showed that the adhesion strength of the coatings increased with the coating thickness [25]. However, even the 222 μm coating showed a very good adhesion strength of 10.74 MPa. The minimal required adhesion strength of TS aluminium coatings varies across different standards from 4.5 to 10.3 MPa, as shown in Table 1.

Table 1. Adhesion strength specifications for thermally-sprayed aluminium coatings in key standards.

Standard	Min. Adhesion Strength [MPa]
ISO 2063	4.5
SSPC-CS 23.00/AWS C2.23M/NACE No. 12	6.89
NORSOK M-501	9
AWS C2.18-93	10.3

However, laboratory tests that were performed by Thomason et al. [26] on flame-sprayed Al coatings revealed that a high coating thickness (~400–450 μm) may lead to the development of blisters, especially in an aggressive splash zone environment. It was suggested that residual stresses are higher in thick coatings. A similar correlation between the coating thickness and the tendency to develop blisters was also reported in a 19-year study conducted by the American Welding Society [27].

In order to provide a sufficient level of protection for the whole service life of a structure, the coating must not be too thin. The corrosion rates of TSA in salt water environments that have been reported in the literature (Table 2)—although estimated by using different experimental methods, exposure periods and temperatures—suggest that the dissolution of the coating in seawater is probably less than 20 μm year. This indicates a minimum coating thickness of 200 μm for 10 years exposure. The AWS C.2 18–93 standard recommends thickness between 200 and 350 μm for seawater immersion.

Table 2. Corrosion rates of Al-coatings in simulated or natural seawater environments.

Coating Consumable	Spraying Method	Substrate	Electrolyte	Temperature [°C]	Duration	Method	Corrosion Rate [μm/year]	Ref.
Al 99.5%	Electric arc	22%Cr Duplex stainless steel	Artificial seawater	18 ± 2 80 ± 2	25 days	Linear polarisation resistance (LPR)	5–8 6–7	[28]
Al 99.5%	Electric arc	glass	Artificial seawater	25 50 100	22 days	LPR	0.2–1.5	[29]
Al 99.6% AlMg5%	Electric arc and flame	Steel	Natural seawater	6.5–11	11 months	LPR Polarisation curves	~3.3 (Al arc) ~2.4 (AlMg arc) 2.0–2.7 (Al flame) ~3.2 (AlMg flame)	[19]
Al 99.5%	Electric arc	Carbon Steel	Natural Seawater	10 ± 2	30 days	Polarisation resistance Polarisation curves	20	[30]
Al 99.0%	Electric arc	Mild steel	3.5% NaCl	Room	44 days	Tafel Electrochemical impedance spectroscopy (EIS)	Not given	[31]
AlMg5%	Electric arc	Carbon steel	Natural seawater	20 50 70 90 (internal temp)	280 days 280 days 60 days 280 days	Polarisation curves LPR	~1.0 ~2.2 ~4.0 ~4.8	[32]
Al 1100	Electric arc	Carbon steel	Natural seawater	10 50 70	(1)230 days (2)250 days	LPR (1) Polarisation curves (2)	7 and 5 6 and 5 16 and 8	[33]
Al 99.7%	Electric arc	Stainless steel (SS)	Natural seawater	Room	24 hours	Polarisation curves	Not given	[24]
Al 99.5%	Electric arc	Carbon steel	Artificial seawater	25 ± 1	32 days	LPR	Not given	[34]
Al 99.5%	Electric arc	Carbon steel	Artificial seawater	26 ± 1	90 days	LPR	<20	[35]
Al 99.99%	Electric arc	Mild steel	3.5% NaCl	26–28	264 h	Polarisation curves EIS	Not given	[36]
Al Al–Al$_2$O$_3$	Electric arc and Plasma spray	SS 316L	Artificial seawater	30	30 days	Polarisation curves EIS	Not given	[37]
Commercially available Al powder	Cold spray	Mild steel	3.5% NaCl	Room temperature	96 h	Polarisation curves EIS	Not given	[23]
Al–Al$_2$O$_3$	Cold spray	Mild carbon steel	0.01%NaCl 0.1%NaCl 1%NaCl	80	21 days	Mass change	~0.04 ~0.05 ~0.06	[38]
Al 99.5%	Electric arc	Carbon steel	Artificial seawater	5 ~101.5	50 days	LPR	5–7	[39]
Al 99.5%	Electric arc	SS404L	3.5% NaCl	Ambient temperature	0 h 500 h 1000 h	Tafel analysis	17.1 0.1 0.2	[40]

4.3. Effect of Coating Composition

The most commonly used alloys for the preparation of thermally-sprayed Al coatings for the corrosion protection of steel are Al 1050 (99.5% Al) and Al–5%Mg, as can be seen from Table 2. ISO 2063-1 and NORSOK M-501 also recommend those alloys. The required compositions of relevant wires provided in AWS C2.25 and ISO 14919 are shown in Table 3.

Table 3. Composition of wire feedstock reported in relevant codes and standards.

Standard	UNS	Common Name	Element (wt%)									
			Al	Cr	Cu	Fe	Mn	Si	Ti	Zn	Mg	Other
AWS C2.25	A91100	Al (1100)	99.00 min	-	0.05–0.20	0.95 (Fe and Si)	0.05	0.95 (Fe and Si)	-	0.10	-	-
	A91350	Al (1350)	99.50 min	0.01	0.05	0.40	0.01	0.10	0.02 (V+Ti)	-		GaB (0.03-0.05)
	A95356	Al–5Mg	Rem.	0.05–0.20	0.10	0.40	0.05–0.20	0.25	0.06–0.20	0.10	4.5–5.5	
	A71001	Al MMC	88 min	-	-	-	-	-	-	-		Al_2O_3 (8–12)
ISO 14919	Not provided	Al 99.5	99.5 min	-	≤0.02	≤0.40	≤0.02	≤0.25	≤0.02	≤0.07		≤0.03
		Al–5Mg	Rem.	0.05–0.2	≤0.10	≤0.40	0.05–0.20	≤0.30	0.06–0.20	≤0.10	4.5–5.6	≤0.15

Pure aluminium shows excellent corrosion resistance due to its passive nature. When exposed to an oxidising environment (e.g., air and water), a continuous and uniform natural oxide film (Al_2O_3) develops on the metal's surface, which works as an electrical insulator. It prevents the movement of electrons (produced during the anodic dissolution of the metal) from the metal to the oxide/solution interface, which results in the inhibition of cathodic reactions.

Mg, when added in small amounts, improves the mechanical properties of solid Al, but it does not decrease its corrosion performance. However, alloys containing 3% Mg or more are susceptible to the precipitation of intermetallic particles (Al_3Mg_2) at the grain boundaries when exposed to elevated temperatures for a longer period of time [41]. Those particles are anodic in relation to the Al matrix and corrode preferentially, leading to intergranular corrosion and stress corrosion cracking [42]. In the case of thermally-sprayed Al, Morakul et al. [43], who tested Al–2%Mg and Al–5%Mg in 3.5% NaCl at 25 °C, reported a lower corrosion resistance and a shorter fatigue life of the coating with a higher Mg concentration. At elevated temperatures, thermally-sprayed Al–5%Mg coatings with and without external CP were tested by Wilson et al. [32], who observed coating degradation (thickness reduction) at higher temperatures. They suggested the possible degradation mechanism as being chemical dissolution. Moreover, one of the samples that was exposed to 90 °C (internal temperature of the pipe) exhibited blistering.

In 2017, Quale et al. [30] tested a thermally-sprayed sacrificial coating comprised of Al, Zn and In, based on the assumption that the addition of Zn and In to aluminium prevents its passivation. The results of the tests revealed that the open circuit potential (OCP) of the freely corroding Al–Zn–In coating reached -1000 mV (Ag/AgCl) after 60 days in natural seawater and was 65 mV more negative than the Al (99.5%) coating. The difference was even more pronounced after 235 days when the coatings were coupled to carbon steel in a 10:1 ratio and the Al–Zn–In coating reached −972 mV, whereas the Al coating stabilised at −803 mV. It was concluded that the Al–Zn–In coating provides better protection than a conventional TSA coating due to its better CP efficiency. Moreover, the Al–Zn–In coating can be used in conjunction with conventional anodes with minimal current drain. It was also noticed that thermal spraying leads to a reduction of Zn content in the deposited coating.

In 2019, Adamiak et al. [44] tested arc-sprayed Al coatings with and without NiAl-buffered sub-coating on armour-grade steel. They observed that the use of the sub-coating increases adhesion and improves erosion wear resistance of the Al layer.

Al/Al$_2$O$_3$ Composites

Some attempts have been made to improve the corrosion and mechanical properties of Al cold-sprayed coatings by reinforcing Al with Al$_2$O$_3$ [38,45,46].

Based on the lab tests in a 3.5% NaCl solution, it was noticed that even though the reinforced coating exhibited a higher corrosion resistance during a short immersion (<200 h), its corrosion performance significantly worsened during longer immersion due to the severe corrosion of the aluminium matrix [46].

Huang et al. [47] tested an Al/Al$_2$O$_3$ flame-sprayed coating for marine applications. They observed a formation of an alumina skeleton inside the coatings, which gives rise to increased mechanical and barrier properties for the penetration of Cl$^-$ ions. Moreover, they reported the dissolution of Al splats which were not connected to the Al$_2$O$_3$ frame.

A better performance of the composite coating was also reported by Abdoli et al. [37] after a 30-day immersion in artificial seawater (ASTMD1141) (with and without the addition of *Escherichia coli* bacteria) for 30 days.

4.4. Effect of Sealing

As shown by Lee et al. [36], corrosion products that form on Al-coated steel enhance the barrier properties of the coating by blocking the pores and cracks inside the coating. However, during the initial immersion, when corrosion deposits are not present, the dissolution of the Al coating can be relatively high. Moreover, if the coated structure is stored in a humid environment before being placed in service, rust staining may occur [7].

To prolong the lifetime of TSA coatings and prevent the development of discoloured areas, suitable sealant systems can be applied. Sealants are designed to penetrate and fill the surface-connected porosity, suppressing the diffusion of corrosive molecules from the environment through the coating.

Organic sealants are comprised of epoxies, phenolics, furans, polymethacrylates, silicones, polyesters, polyurethanes, and polyvinyl esters. For TSA, aluminium-filled vinyl and silicone have been used [48]. In accordance with the NORSOK M-501 standard [49], for low-temperature operations (below 120 °C), two-component epoxy should be used, whereas for high-temperature applications (above 120 °C), aluminium silicone should be used. Information on organic materials that are compatible with TSA is provided in ISO 1244-5. ISO 2063 and AWS C2.23, mention that the thickness of sealants should not exceed 40 µm. López-Ortega et al. [2,50] conducted a series of experiments in which arc-sprayed aluminium with an organic topcoat (epoxydic paint [51]) on high strength, low alloy steel R4 grade was studied. Based on weathering aging tests in different climatic cabinets, as well as immersion tests and tribocorrosion tests in artificial seawater, it was concluded that this duplex system exhibits good corrosion and tribocorrosion properties. The same group [51] also studied a functionalised topcoat system containing 25% wt of SiO$_2$ and a 1.5 wt% Cu$_2$O (to obtain superhydrophobicity and antibacterial characteristics) on a TSA coating that was modified by plasma electrolytic oxidation (PEO). The use of PEO on TSA was previously reported to be a promising technique for improved corrosion and wear resistance [52,53]. The effect of an epoxy sealant applied on an arc-sprayed Al coating by using a cathode electrophoresis method was investigated by Pang et al. [54] and compared with sealing by using boiling water. After immersion in 3.5% NaCl at 40 °C for seven days, it was observed that the thickness of epoxy-sealed TSA was unchanged but the thickness of the TSA that was sealed by using boiling water decreased from 100 to 40 µm.

Other inorganic sealing methods that have been investigated to enhance the behaviour of thermally-sprayed coatings have involved the thermal diffusion of Zn [55], the use of phosphate-containing salts that chemically react with Al [56], calcium nitrate [57], hydrothermal treatment in boiling deionised (DI) water [58], and the sol–gel method [59]. The use of glass powders to seal porosity in an arc-sprayed aluminium coating was also investigated [60].

The reduced porosity level of an arc-sprayed aluminium coating was also achieved by Wenming et al., who used a CO_2 laser to re-melt the coating [61]. They also noticed that the re-melting of the coating changed the way the coating adhered to the substrate from mechanical to metallurgical bonding.

Though the use of sealants can significantly improve the performance of TSA, it should be noted that there have been some examples where sealed coatings developed blisters (Hutton tension leg platform [62]) or failed prematurely (Heidrun platform [26]). The degradation mechanisms of painted TSA was studied first by Knudsen [63] and later by Sumon et al. [64]. Knudsen attributed the accelerated degradation of TSA to the development of an acidic environment underneath the paint. His work focused on the scenarios where TSA/organic system was either in electrical contact with bare steel or contained a defect. Sumon et al. tested several organic coatings on TSA (with and without a scribe) and bare steel by using salt spray exposure. They observed organic coating delamination on TSA-coated samples, not only on scribed samples but also on intact ones due to the diffusion of water and chloride ions through the epoxy layer. The proposed organic coating disbonding mechanisms included the development of voluminous corrosion products lifting the organic layer, anodic undermining, and cathodic disbonding.

4.5. Effect of Damage

As mentioned in previous sections, when an aluminium coating gets damaged and the steel is in direct contact with an electrolyte, a galvanic couple is established between the coating and the substrate due to the potential difference between those two metals. Since Al exhibits more active potential than steel in seawater, it undergoes dissolution accordance to Equation (1):

$$Al \rightarrow Al^{3+} + 3e^-, \tag{1}$$

Electrons that are produced during the above reaction are consumed at the steel (cathode) according to the following reactions:

$$O_2 + 2H_2O + 4e^- \rightarrow 4OH^-, \tag{2}$$

$$2H_2O + 2e^- \rightarrow H_2 + 2OH^-, \tag{3}$$

The production of OH^- ions during cathodic reactions causes an increase in the pH of the solution in the vicinity of the cathode, which triggers the precipitation of calcareous matters. These deposits, depending on the temperature, provide higher or lower barrier properties to the underlying steel.

In a paper published by Thomason, it was suggested that a holiday (a discontinuity: defect, pinhole, crack etc.) in a coating as large as 50% could be successfully protected by a TSA coating for a few years and for 20 years in the case of a 6% holiday in a 200 µm flame-sprayed coating [65].

Modelling work done by using the final element method (FEM) that was conducted by Huang et al. showed that a cold-sprayed AlZn coating can only provide protection to the substrate when the damage in the coating is not larger than 1 mm in width [66]. However, it should be pointed out that, in this study, the effects of calcareous deposits and corrosion product formations were not included.

4.6. Effect of Temperature

The behaviour of TSA (Al–5%Mg) coatings (with and without CP) in natural seawater at different temperatures was studied by Wilson et al. [32]. In their experiments, carbon steel pipes with TSA coatings were internally heated to 20, 50, 70 and 90 °C and exposed to seawater (8–10 °C). It was observed that the corrosion rate increased with increasing temperature. Moreover, high initial corrosion rates that were recorded at all temperatures decreased with time below 10 µm/year after approximately 40–50 days. Based on the analysis of the thickness loss of the coatings, it was suggested that TSA (in conjunction with CP) might undergo chemical dissolution at elevated temperatures [32]. It was also noted that temperature influences the type and amount of calcareous deposits that were formed on the coatings. Deposits that were formed at higher temperatures contained a higher amount of Mg, whereas at lower temperatures, the deposits mainly consisted of $CaCO_3$. Experiments that were conducted by other researchers [67] on TSA (AA1050) that contained defects (4%) in boiling synthetic seawater (ASTM 1141) for up to 5000 h also revealed the formation of brucite as well as β-alumina and mixed hydrated oxides of Mg and Al. No $CaCO_3$ was detected in the holiday (defect) region.

The precipitation of calcareous deposits during the corrosion of TSA coatings with a 4% defect was also studied in synthetic seawater at 30 and 60 °C by Ce and Paul [68]. The authors conducted pH measurements during the corrosion of samples that were exposed to seawater for up to 350 h and observed that the local pH of the TSA coating increased with time at 60 °C (to the value between 7 and 8.5), whereas the opposite behaviour occurred at 30 °C (decreased to pH 4.76). A higher pH was linked to the passivation phenomena. Moreover, different types of products were detected at those temperatures: $Al(OH)_3$ at 30 °C and κ-Al_2O_3 at 60 °C. Some Fe-containing corrosion products were observed on the sample exposed at 30 °C, and both aragonite and brucite were detected at both temperatures in the holiday region.

The effect of temperature on the precipitation of calcareous scales on cathodically protected steel was studied by Lin and Dexter [69]. They noticed that less calcareous deposits formed at temperatures <~10 °C and that at lower temperatures, the predominant phase was calcite ($CaCO_3$), whereas at higher temperatures, aragonite ($CaCO_3$) dominated. The difference was attributed to the inhibitive properties of Mg ions, which inhibit both nucleation and growth of calcite but only the nucleation of aragonite. Studies performed by Yang et al. [70] on cathodically protected steel in artificial seawater at 20 °C revealed the presence of double-layered deposits consisting of an Mg-rich inner layer and a calcium-rich outer layer. The development of calcareous layers on exposed steel is important in the context of the cathodic protection that is provided by TSA coatings because the deposits impede oxygen diffusion to the steel surface and reduce the rate of dissolution of aluminium.

The performance of cathodically polarised (−1, −1.1 and −1.2 V Ag/AgCl) TSA coatings that were exposed to seawater and mud at elevated temperatures was studied by Knudsen et al. [33,71]. They observed an increase in the corrosion rate of TSA with temperature. They also pointed out that the excessive polarisation to −1.2 V of TSA that was exposed to mud led to the very rapid dissolution of the coating.

Significantly less attention has been given to the performance of TSA coatings at lower temperatures. However, two papers have been published by Paul on the performance on TSA coatings (with defects) at 5 °C, the first one at ambient pressure [39] and the second one [72] under high pressure (50 MPa) to simulate deep sea conditions. Both tests revealed the formation of calcareous deposits in the holiday region, which consisted of a thin layer of $Mg(OH)_2$ and a thick layer of $CaCO_3$. In both cases, TSA provided full protection to the underlying steel.

Tests that were performed by Dexter [73] on aluminium metal in seawater revealed an ennoblement of the potential with temperature decrease. A similar trend should be expected for sprayed aluminium. If the potential of a coating is less negative at lower temperatures, its ability to provide cathodic protection (when coating gets damaged) is probably lower too. No papers focusing on the performance of TSA operating at temperatures below 5 °C have been found in the literature.

4.7. Effect of Oxygen

Eighteen-day lab-based experiments [74] that were conducted in aerated and deaerated artificial seawater at 25 °C on TSA with 3% holiday showed that under deaerated conditions, the corrosion rate of TSA coatings is lower and calcareous deposits that are formed consist mainly of brucite ($Mg(OH)_2$).

The effect of oxygen on the galvanic corrosion of aluminium–steel couple in an NaCl solution was studied by Pryor and Keir [75]. They noticed that at pH lower than 4 and higher than 10, the galvanic current flow and mass change of Al were independent on the dissolved oxygen concentration, which was attributed to the main cathodic reaction being the reduction of water leading to the evolution of the hydrogen gas. Within the 4–10 pH range, however, dissolved oxygen had a strong effect on the galvanic corrosion. The saturation of the solution with oxygen resulted in an increase in galvanic current, whereas deaeration caused a decrease of the current to nearly zero.

4.8. Effect of the Solution Chemistry

It is well known that aluminium owes its good corrosion resistance in neutral aqueous environments to its passivating nature. Aluminium oxide, which is formed naturally on the surface of the bare metal, works as an electrical insulator. It inhibits the movement of electrons between the metal and the electrolyte, which results in the inhibition of cathodic reactions. However, for effective galvanic protection, Al needs to be able to corrode at a certain rate to successfully protect the steel substrate. It has been well documented in the literature that chloride ions are capable of breaking down the protective oxide layer, which leads to a localized form of corrosion [76–78]. Even though one would usually try to avoid any form of corrosion, in the case of sacrificial coatings, the presence of chloride ions can be beneficial. It has been observed that for atmospheric exposures, there is a critical threshold value of 100 mg Cl^- m^{-2} day^{-1}, below which Al and Al-based coatings are ineffective [79]. Similar behaviour has also been observed under full immersion laboratory tests that were performed in 0.6 M NaCl and 0.6 M Na_2SO_4 [80]. The presence of Cl^- ions was necessary for effective sacrificial performance, but no recommendations regarding minimum concentration were given.

The effect of magnesium and calcium ions on the performance on TSA coatings containing defects was recently studied by Echaniz et al. [35]. The 90-day immersion tests that were performed in different solutions revealed differences between calcareous deposits that were formed on the exposed steel. When samples were exposed to solutions that contained either magnesium or calcium ions, a single layer of either brucite or calcite, respectively, was formed. However, when both ions were present in the solution, a bi-layer consisting of aragonite on top of brucite was detected. Moreover, electrochemical measurements revealed more noble potential of the sample that was exposed to artificial seawater than to the solutions containing only a combination of the following salts: $MgCl_2$, $CaCl_2$, and $NaHCO_3$.The reason for this could be due to the presence of SO_4^{2-} ions in artificial seawater. Those ions were detected in Al corrosion products by Syrek-Gerstenkorn et al. on TSA coatings after immersion in artificial seawater for 32 days [34]. Other researchers who studied the effect of SO_4^{2-} on the corrosion of aluminium in NaCl solution observed that the addition of those ions inhibited the initiation of pitting corrosion, but it enhances the growth of pre-existing pits [81]. The inhibitive effect of SO_4^{2-} was also reported by others [82].

5. Long-Term Field Studies

TSA coatings have proven to be an effective corrosion mitigation control for steel during several long-term field studies that were performed in the past three decades. The most notable ones are summarised in the Table 4 below.

Table 4. Long-term field studies involving testing of thermally-sprayed aluminium (TSA) coatings.

Name/ Organisation	Exposure Time	Exposure Type	Coating Material	Coating Thickness	Spraying Method	Sealed	Holiday/ Artificial Defect	Findings	Ref.
BISRA	12 years	Atmospheric (rural, coastal industrial) Seawater immersion	Zn, Al, Cd, Pb, Sb	3–4 mils	Thermal Spraying (type not specified), electroplating, and hot dipping.	No	No	Life of the coating roughly proportional to the coating thickness regardless of the application process.	[83,84]
LaQue Centre	12 years extended to 34 years	Atmospheric (marine and industrial)	Powders of various compositions of Zn, Al, Zn and Al Zn and Al and Mg Mn, Al and Mg	0.08–0.2 mm	Powder flame spraying	No	No	Al-based coatings performed better than Zn-based ones in industrial atmospheres but worse in marine atmospheres after 12 years exposure. There were more resistant coatings with high Al content than Zn after longer (34 years) exposure in marine atmosphere.	[83,85]
AWS	19 years	Atmospheric and immersion	Zn, Al	80–460 µm	Wire flame spraying	Wash primer, clear vinyl, Al-pigmented vinyl, Chlorinated rubber	No	80–150 um TSA coatings with and without sealing protect steel in seawater and in atmosphere. The coating that gave the best performance was a thin TSA. It exhibited the lowest level of pitting and blistering.	[27]
JACC	18 years	Atmospheric, immersion	Zn, Al, Zn–Al	175 ± 25 µm (flame and arc-sprayed coating) 400 ± 25 µm only for Al (arc-sprayed)	Wire flame and electric arc spraying	Epoxy	No	TSA not effective in splash zone when there is a damage in the coating.	[83]
PATINA project	3.5 years	Atmospheric	Al, Zn, Zn–Al	150 µm	Flame spraying (type not specified)	No	Yes	TSA only effective in atmosphere with high chloride content.	[79]
the Corrosion Advice Bureau of BISRA	10 years	Atmospheric, immersion	Al, Zn, Zn and Al	50,100,150 µm	Wire and powder spraying (exact method not specified)	Variety of paints and topcoats	Yes	Painted TSA can provide 10 years life without maintenance under atmospheric exposures. When damaged, rusting of the substrate occurs under all conditions. Unpainted Zn coatings failed under immersion conditions, whereas unpainted Al coatings provided protection.	[86]
US Army, Buzzard's Bay	20 years	Atmospheric, splash, tidal, immersion in natural seawater	Various organic and metallic coatings	150 µm (TSA without sealant) 80–90 µm sealed	Wire flame spraying	Wash coati primer and Al vinyl	Yes	TSA with sealant was the best performing coating among 24 systems tested Unsealed Al coating failed in atmospheric and splash zones (red rust present).	[62,87]
US Army, La Costa Island	21 years, results reported after 10 years	Atmospheric, splash, tidal, immersion in natural seawater	Various organic and metallic coatings	150 µm	Wire flame spraying	Vinyl topcoat	Yes	The most severe corrosion in the splash zone.	[87,88]

6. Laboratory Corrosion Testing Methods

In order to predict the long-term performance of anti-corrosive coatings, long-term field testing is the most reliable approach; however, these are expensive and require long-term access to marine test sites. Short laboratory testing can give some indication and insight into the mechanisms that dictate the protectiveness of the system. The most common lab-based testing methods involve salt spray tests and electrochemical measurements, such as monitoring of the open circuit potential (OCP) and electrochemical impedance spectroscopy (EIS), during immersion tests. In order to estimate the rate of corrosion of the coatings, several methods have been employed by different researchers, such as linear polarisation resistance (LPR), polarisation curves, and mass change, as shown in Table 2.

Unfortunately, all of the above methods have some limitations. For example, the LPR method is not applicable when a non-conductive sealant or paint is used. Moreover, this technique should be used when the system undergoes general corrosion, whereas Al is known to suffer from localised corrosion in aqueous solutions containing Cl^- ions.

Additionally, it should be mentioned that if the samples contain defects, the use of polarisation curves is not accurate, because anodic and cathodic polarization curves reflect reactions that occur on both metals simultaneously. Therefore, the Tafel slopes (which are needed to calculate corrosion rates) that are obtained from those curves are not accurate.

EIS is a powerful technique that is especially useful for studying coatings. However, the selection of appropriate equivalent electrical circuit that explains the mechanism occurring in the system can be very difficult, especially in case of porous coatings, such as TSA. Nonetheless, some researchers have done EIS modelling to understand the corrosion behaviour of sacrificial coatings [8,23,31,37,89–91].

The common industrial practices of the monitoring of corrosion based on salt spray testing and visual examination has its limitations, particularly for long-term prediction of performance. It is common laboratory practice to use 5 wt% or 3.5 wt% NaCl solutions for salt spray tests, even though natural seawater contains other ions, such as Mg^{2+}, Ca^{2+}, and SO_4^{2-}, which play important roles in corrosion processes.

The estimation of corrosion rates based on the mass change of the samples is not applicable to thermally-sprayed coatings due to their porous nature. Corrosion products form on the coating, fill the pores, and are not possible to remove. Therefore, the mass of the corroded samples is changed due to both the dissolution of the coating and the corrosion products that get trapped inside it.

7. In-Service Performance of Al Coatings

One of the earliest applications of TSA coatings for the corrosion protection of an offshore structure was the Hutton tension leg platform (TLP) which was installed in 1984 in the North Sea [92]. TSA was applied on risers and tethers and flare boom. The risers were connected to the TLP hull (protected by an impressed current CP) and a subsea template (connected to sacrificial anodes). In 1986, one of the tethers was removed for inspection, and one of the risers was removed in the following year. The inspections revealed the presence of blisters on the coating that was applied on tethers (with vinyl sealer), and no signs of the degradation of the coating were observed on the risers (with silicone sealer). Apart from the blisters, TSA was in very good condition even after 12 years in service [26]. The possible reasons for the development of the blisters were related to the vinyl sealer, which either reacted with TSA or did not penetrate the coating.

In the Norwegian offshore sectors, TSA (mainly Al–5%Mg) has been used for the protection of several offshore structures (e.g., Shell Draugen field, Sleipner riser platform, Troll Gas, Heidrun field, Heimdal platform). By 1997, approximately 400,000 m^2 of steel was protected by using TSA coatings. The most common areas that were protected by the TSA coatings include flare booms, crane booms, pipes, under cellar decks, vessels under thermal insulation, and burner booms [93]. An excellent performance of TSA with aluminium silicone paint on the Heimdal platform was reported after 10 years in service. However, a premature degradation of TSA was observed on the Sleipner platform (after seven years in service) as well as on Jotun B. In both cases, TSA coatings were applied as duplex systems consisting of sprayed aluminium and a thick overcoat [94]. A similar problem occurred on a gas platform near East Timor in Australia, where TSA failed after one year in service [95].

Another example where TSA did not provide adequate protection was the Heindrun TLP installed in 1995 in the Norwegian North Sea. After less than four years in service, a serious damage of the TSA coating (supported by CP from sacrificial anodes) with silicone sealer was observed on two oil export risers and one gas export riser operating in the splash zone. Further lab-based studies that were conducted to identify possible reasons for the TSA failure on the Heindrun risers revealed that the tendency of blistering increases with increases of coating thickness and the rise of pipe temperature. It was also observed that blistering can be mitigated by using a silicone sealer [26].

A very good performance of sealed TSA coatings that were applied on risers connected to sacrificial anodes was reported on the Jolliet tension leg platform operating in the Gulf of Mexico for 13 years [26].

TSA with organic topcoats was also used for the protection of bridges, such as Nidelv bridge in Trondheim, where the duplex system provided protection to the bridge for 30 years. However, when applied to Tromsø bridge, duplex coating failed after a year or less [93]. Some of the suggested possible factors that might have promoted the development of many blisters in the splash zone region included the insufficient quality of blast-cleaning and insufficient coating thickness [96].

Recently, TSA has been tested for the protection of offshore wind turbines as a part of the Cost Reduction for Offshore Wind Now (Crown) project [97]. TSA with epoxy paint as a sealant is also being used for the protection of offshore wind turbines operating in the Baltic Sea as a part of the Arkona project [98].

8. Summary and Future Outlook

The literature review revealed that TSA has proven its capability to successfully protect steel from corrosion in harsh marine environments when applied properly. The main conclusions that can be drawn from the substantial amount of work that has been devoted to understanding mechanisms and parameters controlling the performance of TSA are as follows:

1. It appears that the most economical spraying technique that is capable of producing coatings with low level of porosity and good adhesion is the electric arc spraying method.

2. Coating thickness is an important parameter that influences performance. The optimal thickness has been reported to be between 150–375 μm. A lower thickness can result in an insufficient amount of aluminium to provide long-term protection to the substrate. An excessive coating thickness may lead to high residual coating stress leading to its premature failure.

3. The application of sealants can prolong the lifetime of the coating. Moreover, it has been suggested that sealing can prevent blistering. However, a low viscosity sealant that flows easily inside the pores without staying on top of the TSA should be used. The application of a thick organic layer on TSA may lead to an accelerated dissolution of the coating. The suggested failure mechanisms include the development of an acidic environment underneath the organic layer, the development of voluminous corrosion products rising the organic coating, anodic undermining, and cathodic disbondment.

4. It is important to differentiate between the self-corrosion of the coatings and how protective they are for steel when the coating is damaged or connected to bare steel. If large area of steel is exposed, the corrosion of TSA can be significantly higher due to the anodic nature of the coating.

5. Simulated laboratory tests should be conducted under intended service conditions. The testing of the coatings in NaCl solutions is not representative of in-service conditions and should be avoided. Seawater temperature and ions present in seawater play important roles in the corrosion of Al and the precipitation of calcareous deposits on cathodically polarised steel. Calcareous minerals are capable of impeding the diffusion of oxygen to the steel surface and reducing the dissolution rate of aluminium. Hence, constituents of these minerals such as Ca^{2+} and Mg^{2+} ions should be included in the test electrolyte.

6. TSA has mostly been used in offshore structures with some form of CP system in place. Recently, TSA has been applied on offshore wind turbines as a stand-alone corrosion control method. However, no standards that cover the suitability of using TSA as a primary CP system exist.

Despite well-documented successful performance history of TSA, some gaps in knowledge still remain. Areas that have not been fully understood or addressed yet include:

- The determination of minimal salinity of water needed for the TSA to work effectively, especially if TSA contains defects.
- The performance of damaged TSA coatings in cold seawater, especially when not in conjunction with external CP.
- The suitability of using TSA coatings in splash zones.
- The suitability of using TSA coatings for the protection of high strength steels where hydrogen embrittlement is a concern.

Though cold-sprayed coatings have many advantages, such as low porosity and oxide levels, they have not been used for corrosion protection of steel structures in-service yet due to the higher costs (CAPEX AND OPEX) that are associated with cold spray systems. Only some limited laboratory data exist. To evaluate the corrosion performance of the coatings, long-term testing is needed.

Author Contributions: Conceptualization, B.S.-G.; writing—original draft preparation, B.S.-G.; writing—review A.J.D.; Review and editing—S.P.; supervision, A.J.D. and S.P. All authors have read and agreed to the published version of the manuscript.

Funding: This publication was made possible by the sponsorship and support of Lloyd's Register Foundation, a charitable foundation helping to protect life and property by supporting engineering-related education, public engagement and the application of research. The work was enabled through, and undertaken at, the National Structural Integrity Research Centre (NSIRC), a postgraduate engineering facility for industry-led research into structural integrity established and managed by TWI through a network of both national and international Universities. Authors gratefully acknowledge financial support from the Centre for Doctoral Training in Innovative Metal Processing (IMPaCT) funded by the UK Engineering and Physical Sciences Research Council (EPSRC), grant reference EP/L016206/1.

Acknowledgments: The authors would like to thank P. McNutt for the micrograph of cold-sprayed coating.

Conflicts of Interest: The authors declare no conflict of interest.

References

1. Koch, G.; Varney, J.; Thompson, N.; Moghissi, O.; Gould, M.; Payer, J. International measures of prevention, application, and economics of corrosion technologies study. *NACE Int.* **2016**, 2–6.
2. López-Ortega, A.; Bayón, R.; Arana, J.L. Evaluation of protective coatings for high-corrosivity category atmospheres in offshore applications. *Materials* **2019**, *12*, 1325. [CrossRef]
3. Det Norske Veritas. *Recommended Practice DNVGL-RP-0416 Corrosion Protection for Wind Turbines*; DET NORSK VERITAS: Oslo, Norway, 2016.
4. Harb, S.V.; Trentin, A.; Torrico, R.F.O.; Pulcinelli, S.H.; Santilli, C.V.; Hammer, P. Organic-inorganic hybrid coatings for corrosion protection of metallic surfaces. In *New Technologies in Protective Coatings*; Giudice, C., Canosa, G., Eds.; IntechOpen: London, UK, 2017; pp. 19–52. [CrossRef]
5. Paul, S. Corrosion control for marine- and land-based infrastructure applications. In *ASM Handbook—Volume 5A: Thermal Spray Technology*; Tucker, R.C., Ed.; ASM International: Materials Park, OH, USA, 2013; Volume 5, pp. 248–252.
6. Det Norske Veritas. *Recommended Practice DNVGL-RP-B401 Cathodic Protection Design*; DET NORSK VERITAS: Oslo, Norway, 2017.
7. Cunningham, T.; Avery, R. Sealer coatings for thermal-sprayed aluminum in the offshore industry. *Mater. Perform.* **2000**, *39*, 46–48.
8. Zhu, Q.J.; Wang, K.; Wang, X.H.; Hou, B.R. Electrochemical impedance spectroscopy analysis of cold sprayed and arc sprayed aluminium coatings serviced in marine environment. *Surf. Eng.* **2012**, *28*, 300–305. [CrossRef]

9. Malek, M.H.A.; Saad, N.H.; Abas, S.K.; Shah, N.M. Thermal arc spray overview. *IOP Conf. Ser. Mater. Sci. Eng.* **2013**, *46*, 012028. [CrossRef]

10. Fournier, J.; Miousse, D.; Legoux, J.-G. Wire-arc sprayed nickel based coatings for hydrogen evolution reaction in alkaline solutions. *Int. J. Hydrog. Energy* **1995**, *24*, 519–528. [CrossRef]

11. Tejero-Martin, D.; Rezvani Rad, M.; McDonald, A.; Hussain, T. Beyond traditional coatings: A review on thermal-sprayed functional and smart coatings. *J. Therm. Spray Technol.* **2019**, *28*, 598–644. [CrossRef]

12. Amin, S.; Panchal, H. A review on thermal spray coating processes. *Int. J. Curr. Trends Eng. Res.* **2016**, *2*, 556–563.

13. Davis, J.R. Introduction to thermal spray processing. In *Handbook of Thermal Spray Technology*; ASM International: Materials Park, OH, USA, 2004.

14. Sabard, A.; Hussain, T. Bonding mechanisms in cold spray deposition of gas atomised and solution heat-treated Al 6061 powder by EBSD. *arXiv* **2018**, arXiv:1811.08694.

15. Champagne, V.; Helfritch, D. Critical assessment 11: Structural repairs by cold spray. *Mater. Sci. Technol.* **2015**, *31*, 627–634. [CrossRef]

16. Champagne, V.K. (Ed.) *The Cold Spray Materials Deposition Process: Fundamentals and Applications*; Woodhead Publishing Limited: Cambridge, UK, 2007; ISBN 9781845691813.

17. Pathak, S.; Saha, G.C. Development of sustainable cold spray coatings and 3D additive manufacturing components for repair/manufacturing applications: A critical review. *Coatings* **2017**, *7*, 122. [CrossRef]

18. Pombo, R.R.M.H.; Paredes, R.S.C.; Wido, S.H.; Calixto, A. Comparison of aluminum coatings deposited by flame spray and by electric arc spray. *Surf. Coat. Technol.* **2007**, *202*, 172–179.

19. Gartland, P.O.; Eggen, T.G. Cathodic and anodic properties of thermally sprayed Al and Zn-based coatings in seawater Paper No. 367. In *Proceedings of the Corrosion 90*; NACE International: Houston, TX, USA, 1990.

20. Bardal, E. The effect of surface preparation on the adhesion of arc and flame-sprayed aluminum and zinc coatings to mild steel. In Proceedings of the 7th International Metal Spraying Conference, London, UK, 10–14 September 1973.

21. Lieberman, E.S.; Clayton, C.R.; Herman, H. *Thermally Sprayed Active Metal Coatings for Corrosion Protection in Marine Environments*; Report; Defense Technical Information Center: Fort Belvoir, VA, USA, 1984.

22. *American Welding Society Guide for the Protection of Steel with Thermal Sprayed Coatings of Aluminium and Zinc and their Alloys and Composites*; AWS C2.18: Florida, FL, USA, 1993.

23. Li, N.; Li, W.Y.; Yang, X.W.; Alexopoulos, N.D.; Niu, P.L. Effect of powder size on the long-term corrosion performance of pure aluminium coatings on mild steel by cold spraying. *Mater. Corros.* **2017**, *68*, 546–551. [CrossRef]

24. Han, M.S.; Woo, Y.B.; Ko, S.C.; Jeong, Y.J.; Jang, S.K.; Kim, S.J. Effects of thickness of Al thermal spray coating for STS 304. *Trans. Nonferrous Met. Soc. China (Engl. Ed.)* **2009**, *19*, 925–929. [CrossRef]

25. Malek, M.H.A.; Saad, N.H.; Abas, S.K.; Roselina, N.R.N.; Shah, N.M. Performance and microstructure analysis of 99.5% aluminium coating by thermal arc spray technique. *Procedia Eng.* **2013**, *68*, 558–565. [CrossRef]

26. Thomason, W.H.; Olsen, S.; Haugen, T.; Fischer, K. Deterioration of thermal sprayed aluminum coatings on hot risers due to thermal cycling Paper No. 04021. In Proceedings of the Corrosion 2004, New Orleans, LA, USA, 28 March–1 April 2004; NACE International: New Orleans, LA, USA, 2004.

27. American Welding Society. *Corrosion Testing of Flame-Sprayed Coated Steel—19 Year Report C2.14-74*, 1974.

28. Paul, S.; Lee, C.M.; Harvey, M.D.F. Improved coatings for extended design life of 22%Cr duplex stainless steel in marine environments. In *Proceedings of the Thermal Spray 2012: Proceedings from the International Thermal Spray Conference and Exposition*; ASM International: Houston, TX, USA, 2012; pp. 544–549.

29. Paul, S.; Harvey, M.D.F.; Ho, Q.Y.; Yunus, K.; Fisher, A.C. Corrosion testing of thermally sprayed aluminum. In *Proceedings of the Thermal Spray 2015: Proceedings from the International Thermal Spray Conference*; ASM International: Long Beach, CA, USA, 2015; pp. 964–970.

30. Quale, G.; Årtun, L.; Iannuzzi, M.; Johnsen, R. Cathodic protection by distributed sacrificial anodes—A new cost- effective solution to prevent corrosion of subsea structures Paper No. 8941. In *Proceedings of the Corrosion 2017*; NACE International: New Orleans, LA, USA, 2017.

31. Abedi Esfahani, E.; Salimijazi, H.; Golozar, M.A.; Mostaghimi, J.; Pershin, L. Study of corrosion behaviour of Arc sprayed aluminum coating on mild steel. *J. Therm. Spray Technol.* **2012**, *21*, 1195–1202. [CrossRef]

32. Wilson, H.; Johnsen, R.; Rodriguez, C.T.; Hesjevik, S.M. Properties of TSA in natural seawater at ambient and elevated temperature. *Mater. Corros.* **2019**, *7*, 293–306. [CrossRef]

33. Knudsen, O.Ø.; Van Bokhorst, J.; Clapp, G.; Duncan, G. Technical note: Corrosion of cathodically polarized thermally sprayed aluminum in subsea mud at high temperature. *Corrosion* **2016**, *72*, 560–568.

34. Syrek-Gerstenkorn, B.; Paul, S.; Davenport, A.J. Use of thermally sprayed aluminium (TSA) coatings to protect offshore structures in submerged and splash zones. *Surf. Coat. Technol.* **2019**, *374*, 124–133. [CrossRef]

35. Echaniz, R.G.; Paul, S.; Thornton, R. Effect of seawater constituents on the performance of thermal spray aluminum in marine environments. *Mater. Corros.* **2019**, *70*, 996–1004. [CrossRef]

36. Lee, H.S.; Singh, J.K.; Park, J.H. Pore blocking characteristics of corrosion products formed on Aluminium coating produced by arc thermal metal spray process in 3.5 wt.% NaCl solution. *Constr. Build. Mater.* **2016**, *113*, 905–916. [CrossRef]

37. Abdoli, L.; Huang, J.; Li, H. Electrochemical corrosion behaviors of aluminum-based marine coatings in the presence of Escherichia coli bacterial biofilm. *Mater. Chem. Phys.* **2016**, *173*, 62–69. [CrossRef]

38. Bai, X.; Tang, J.; Gong, J.; Lü, X. Corrosion performance of Al–Al$_2$O$_3$ cold sprayed coatings on mild carbon steel pipe under thermal insulation. *Chin. J. Chem. Eng.* **2017**, *25*, 533–539. [CrossRef]

39. Paul, S. Corrosion performance of damaged thermally sprayed aluminiun in synthetic seawater at different temperatures. *Therm. Spray Bull.* **2015**, *67*, 139–146.

40. Yung, T.Y.; Chen, T.C.; Tsai, K.C.; Lu, W.F.; Huang, J.Y.; Liu, T.Y. Thermal spray coatings of Al, ZnAl and Inconel 625 alloys on SS304L for anti-saline corrosion. *Coatings* **2019**, *9*, 32. [CrossRef]

41. Ghali, E. *Corrosion Resistance of Aluminium and Magnesium Alloys: Understadning, Performance and Testing*; John Wiley & Sons: Hoboken, NJ, USA, 2010.

42. Yan, J.; Heckman, N.M.; Velasco, L.; Hodge, A.M. Improve sensitization and corrosion resistance of an Al-Mg alloy by optimization of grain boundaries. *Sci. Rep.* **2016**, *6*, 1–10. [CrossRef] [PubMed]

43. Morakul, S.; Otsuka, Y.; Miyashita, Y.; Mutoh, Y. Effect of Mg concentration on interfacial strength and corrosion fatigue behaviour of thermal-sprayed Al-Mg coating layers. *Eng. Fail. Anal.* **2018**, *88*, 13–24. [CrossRef]

44. Adamiak, M.; Czupr Nski, A.; Kopy, A.; Monica, Z.; Olender, M.; Gwiazda, A. The properties of arc-sprayed aluminum coatings on armor-grade steel. *Metals* **2018**, *8*, 142. [CrossRef]

45. Irissou, E.; Legoux, J.G.; Arsenault, B.; Moreau, C. Investigation of Al-Al$_2$O$_3$ cold spray coating formation and properties. *J. Therm. Spray Technol.* **2007**, *16*, 661–668. [CrossRef]

46. Silva, F.S.D.; Bedoya, J.; Dosta, S.; Cinca, N.; Cano, I.G.; Guilemany, J.M.; Benedetti, A.V. Corrosion characteristics of cold gas spray coatings of reinforced aluminum deposited onto carbon steel. *Corros. Sci.* **2017**, *114*, 57–71. [CrossRef]

47. Huang, J.; Liu, Y.; Yuan, J.; Li, H. Al/Al$_2$O$_3$ composite coating deposited by flame spraying for marine applications: Alumina skeleton enhances anti-corrosion and wear performances. *J. Therm. Spray Technol.* **2014**, *23*, 676–683. [CrossRef]

48. Fauchais, P.; Vardelle, A. Thermal sprayed coatings used against corrosion and corrosive wear. In *Advanced Spray Applications*; Jazi, H.S., Ed.; IntechOpen: London, UK, 2012; pp. 3–39. ISBN 978-953-51-0349-3.

49. *NORSOK Standard M-501 Surface Preparation and Protective Coating*, 2004.

50. López-Ortega, A.; Bayón, R.; Arana, J.L. Evaluation of protective coatings for offshore applications. Corrosion and tribocorrosion behaviour in synthetic seawater. *Surf. Coat. Technol.* **2018**, *349*, 1083–1097. [CrossRef]

51. López-Ortega, A.; Areitioaurtena, O.; Alves, S.A.; Goitandia, A.M.; Elexpe, I.; Arana, J.L.; Bayón, R. Development of a superhydrophobic and bactericide organic topcoat to be applied on thermally sprayed aluminum coatings in offshore submerged components. *Prog. Org. Coat.* **2019**, *137*, 105376. [CrossRef]

52. Gu, W.; Shen, D.; Wang, Y.; Chen, G.; Feng, W.; Zhang, G.; Fan, S.; Liu, C.; Yang, S. Deposition of duplex Al2O3 /aluminum coatings on steel using a combined technique of arc spraying and plasma electrolytic oxidation. *Appl. Surf. Sci.* **2006**, *252*, 2927–2932. [CrossRef]

53. López-Ortega, A.; Arana, J.L.; Rodríguez, E.; Bayón, R. Corrosion, wear and tribocorrosion performance of a thermally sprayed aluminum coating modified by plasma electrolytic oxidation technique for offshore submerged components protection. *Corros. Sci.* **2018**, *143*, 258–280. [CrossRef]

54. Pang, X.; Wang, R.; Wei, Q.; Zhou, J. Effect of epoxy resin sealing on corrosion resistance of arc spraying aluminium coating using cathode electrophoresis method. *Mater. Res. Express* **2018**, *5*, 016527. [CrossRef]

55. Wang, Y.; Zhang, T.; Zhao, W.; Tang, X. Sealing Treatment of Aluminium Coating on S235 Steel with Thermal Diffusion of Zinc. *J. Therm. Spray Technol.* **2015**, *24*, 1052–1059. [CrossRef]

56. Lee, H.-S.; Singh, J.K.; Ismail, M.A. An effective and novel pore sealing agent to enhance the corrosion resistance performance of Al coating in artificial ocean water. *Sci. Rep.* **2017**, *7*, 41935. [CrossRef] [PubMed]

57. Lee, H.; Kumar, A.; Mandal, S.; Singh, J.K.; Aslam, F.; Alyousef, R.; Albduljabbar, H. Effect of sodium phosphate and calcium nitrate sealing treatment on microstructure and corrosion resistance of wire arc sprayed aluminum coatings. *Coatings* **2020**, *10*, 33. [CrossRef]

58. Liu, L.M.; Wang, Z.; Song, G. Study on corrosion resistance properties of hydrothermal sealed arc sprayed aluminium coating. *Surf. Eng.* **2010**, *26*, 399–406. [CrossRef]

59. Armada, S.; Tilset, B.G.; Pilz, M.; Liltvedt, R.; Bratland, H.; Espallargas, N. Sealing HVOF thermally sprayed WC-CoCr coatings by sol-gel methods. *J. Therm. Spray Technol.* **2011**, *20*, 918–926. [CrossRef]

60. Wang, R.; Zhou, J. Effect of glass powder sealings on the corrosion resistance of arc sprayed Al coating. *Mater. Res. Express* **2019**, *6*, 086566. [CrossRef]

61. Wenming, L.; Tianyuan, S.; Dejun, K. Effects of laser remelting on surface-interface morphologies, bonding modes and corrosion performances of arc-sprayed Al coating. *Anti Corros. Methods Mater.* **2017**, *64*, 43–51. [CrossRef]

62. Fischer, K.P.; Thomason, W.H.; Rosbrook, T.; Murali, J. Performance history of thermal-sprayed aluminum coatings in offshore service. *Mater. Perform.* **1995**, *34*, 27–35.

63. Knudsen, O.Ø.; Rogne, T.; Røssland, T. Rapid degradation of painted TSA Paper No. 04023. In *Proceedings of the Corrosion 2004*; NACE International: New Orleans, LA, USA, 2004.

64. Sumon, T.A.; Lyon, S.B.; Scantlebury, J.D. Failure of aluminium metal spray/organic duplex coating systems on structural steel. *Corros. Eng. Sci. Technol.* **2013**, *48*, 552–557. [CrossRef]

65. Thomason, W.H. Offshore corrosion protection with thermal-sprayed aluminum. In Proceedings of the Offshore Technology Conference, Houston, TX, USA, 6–9 May 1985.

66. Huang, G.; Lou, X.; Wang, H.; Li, X.; Xing, L. Investigation on the cathodic protection effect of low pressure cold sprayed AlZn coating in seawater via numerical simulation. *Coatings* **2017**, *7*, 93. [CrossRef]

67. Ce, N.; Paul, S. Thermally Sprayed Aluminium Coatings for the Protection of Subsea Risers and Pipelines Carrying Hot Fluids. *Coatings* **2016**, *6*, 58. [CrossRef]

68. Ce, N.; Paul, S. The effect of temperature and local pH on calcareous deposit formation in damaged thermal spray aluminum (TSA) coatings and its implication on corrosion mitigation of offshore steel structures. *Coatings* **2017**, *7*, 52. [CrossRef]

69. Lin, S.-H.; Dexter, S.C. Effects of temperature and magnesium ions on calcareous deposition. *Corrosion* **1988**, *44*, 615–622. [CrossRef]

70. Yang, Y.; Scantlebury, J.D.; Koroleva, E.V. A Study of calcareous deposits on cathodically protected mild steel in artificial seawater. *Metals* **2015**, *5*, 439–456. [CrossRef]

71. Knudsen, O.Ø.; Van Bokhorst, J.; Clapp, G.; Duncan, G. Corrosion of cathodically polarized TSA in subsea mud at high temperature Paper No. 4196. In Proceedings of the Corrosion 2004, San Antonio, TX, USA, 9–13 March 2014; NACE International: San Antonio, TX, USA, 2014.

72. Paul, S. Protection of deep sea steel structures using thermally sprayed aluminium Paper No. 9009. In Proceedings of the Corrosion 2017, New Orleans, LA, USA, 26–30 March 2017; NACE International: New Orleans, LA, USA, 2017.

73. Dexter, S.C. Effect of variations in sea water upon the corrosion of aluminum. *Corrosion* **1980**, *36*, 423–432. [CrossRef]

74. Paul, S. Behavior of damaged thermally sprayed aluminum (TSA) in aerated and dearated seawater Paper No. 12766. In Proceedings of the Corrosion 2019, Nashville, TN, USA, 24–28 March 2019; NACE International: Nashville, TN, USA, 2019.

75. Pryor, M.J.; Keir, D.S. Galvanic corrosion: I. current flow and polarization characteristics of the aluminium-steel and zinc-steel couples in sodium chloride solution. *J. Electrochem. Soc.* **1957**, *104*, 269–275. [CrossRef]

76. Frankel, G.S. Pitting corrosion of metals. A review of the critical factors. *J. Electrochem. Soc.* **1998**, *145*, 2186–2198. [CrossRef]

77. Szklarska-Smialowska, Z. Pitting corrosion of aluminum. *Corros. Sci.* **1998**, *41*, 1743–1767. [CrossRef]

78. Natishan, P.M.; O'Grady, W.E. Chloride ion interactions with oxide-covered aluminum leading to pitting corrosion: A review. *J. Electrochem. Soc.* **2014**, *161*, C421–C432. [CrossRef]

79. Panossian, Z.; Mariaca, L.; Morcillo, M.; Flores, S.; Rocha, J.; Peña, J.J.; Herrera, F.; Corvo, F.; Sanchez, M.; Rincon, O.T.; et al. Steel cathodic protection afforded by zinc, aluminium and zinc/aluminium alloy coatings in the atmosphere. *Surf. Coat. Technol.* **2005**, *190*, 244–248. [CrossRef]

80. Rios, G. Effect of chlorides on the electrochemical behaviour of thermally sprayed aluminium protective coatings. Ph.D. Thesis, University of Manchester, Manchester, UK, 2012.

81. Na, K.H.; Pyun, S. Il Effects of SO42-, S2O 32- and HSO4- Ion additives on the pitting corrosion of pure aluminium in 1 M NaCl solution at 40–70 °C. *J. Solid State Electrochem.* **2005**, *9*, 639–645. [CrossRef]

82. Wu, T.I.; Wu, J.K. Effect of sulfate ions on corrosion inhibition of AA 7075 aluminum alloy in sodium chloride solutions. *Corrosion* **1995**, *51*, 185–190. [CrossRef]

83. Kuroda, S.; Kawakita, J.; Takemoto, M. An 18-year exposure test of thermal-sprayed Zn, Al, and Zn-Al coatings in marine environment. *Corrosion* **2006**, *62*, 635–647. [CrossRef]

84. Mansford, R.E. Sprayed aluminium and zinc in corrosive environments. *Corros. Technol.* **1956**, 314–316. [CrossRef]

85. Hoar, T.P.; Radovici, O. Zinc-aluminium sprayed coatings. *Trans. IMF* **1964**, *42*, 211–222. [CrossRef]

86. Watkins, K.O. Painting of metal-sprayed structural steelwork: Report on the condition of specimens after 10 years' exposure. *Br. Corros. J.* **1974**, *9*, 204–210. [CrossRef]

87. Kumar, A.; Van Blaricum, V.; Beitelman, A.; Boy, J. Twenty year field study of the performance of coatings in seawater. In *Corrosion Testing in Natural Waters*; Young, W., Kain, R., Eds.; ASTM International: West Conshohocken, PA, USA, 1997; Volume Second, pp. 74–90.

88. Bukowski, J.; Kumar, A. *Coatings and Cathodic Protection of Piling on Seawater: Results of 10-Year Exposure at LaCosta Island, Fl*; Tech. Rep. M-321; Constr. Eng. Lab, US Army: Champaign, IL, USA, 1982.

89. Jiang, Q.; Miao, Q.; Liang, W.; Ying, F.; Tong, F.; Xu, Y.; Ren, B.; Yao, Z.; Zhang, P. Corrosion behaviour of arc sprayed Al–Zn–Si–RE coatings on mild steel in 3.5wt% NaCl solution. *Electrochim. Acta* **2013**, *115*, 644–656. [CrossRef]

90. Lee, H.S.; Kwon, S.J.; Singh, J.K.; Ismail, M.A. Influence of Zn and Mg alloying on the corrosion resistance properties of Al coating applied by arc thermal spray process in simulated weather solution. *Acta Metall. Sin. (Engl. Lett.)* **2018**, *31*, 591–603. [CrossRef]

91. Lee, H.S.; Singh, J.K.; Ismail, M.A.; Bhattacharya, C.; Seikh, A.H.; Alharthi, N.; Hussain, R.R. Corrosion mechanism and kinetics of Al-Zn coating deposited by arc thermal spraying process in saline solution at prolong exposure periods. *Sci. Rep.* **2019**, *9*, 1–17. [CrossRef]

92. Tiong, D.K.-K.; Pit, H. Experiences on "thermal spray aluminium (TSA)" coating on offshore structures Conference paper No. 04022. In Proceedings of the Corrosion 2004; NACE International: New Orleans, LA, USA, 2004.

93. Doble, O.; Pryde, G.; Oil, K. Use of thermally sprayed aluminium in the Norwegian Offshore Industry. *Prot. Coat. Eur.* **1997**, *2*, 1–10.

94. Knudsen, O.Ø. *Coating Systems for LONG lifetime: Thermally Sprayed Duplex Systems SINTEF Report A14189*; SINTEF: Trondheim, Norway, 2010.

95. Mandeno, W.L. Thermal metal spray: Successes, failures and lessons learned. In Proceedings of the Proceedings of Australasian Corrosion Association Corrosion & Prevention Conference, Melbourne, Australia, 11–14 November 2012.

96. Klinge, R. Altered specifications for the protection of Norwegian steel bridges and offshore structures against corrosion. *Steel Constr.* **2009**, *2*, 109–118. [CrossRef]

97. TWI Ltd. CROWN Project Commended for Offshore Win Corrosion Work. Available online: https://www.twi-global.com/media-and-events/press-releases/2018-02-collaborative-crown-project-commended-for-offshore-wind-corrosion-work (accessed on 7 January 2020).

98. Matthiesen, H. Arkona Offshore Wind Project. Available online: Ttps://www.norwep.com/content/download/32957/239720/version/1/file/Mathiesen+2017-12-04_Norwegian+Offshore+Delegation_Arkona+Offshore+Wind+Project.pdf (accessed on 7 January 2020).

Article

Discoloration Resistance of Electrolytic Copper Foil Following 1,2,3-Benzotriazole Surface Treatment with Sodium Molybdate

Dong-Jun Shin [1,2,†], **Yu-Kyoung Kim** [3,†], **Jeong-Mo Yoon** [1] **and Il-Song Park** [1,*]

[1] Division of Advanced Materials Engineering, Research Center for Advanced Materials Development and Institute of Biodegradable Materials, Chonbuk National University, Jeonju 561-756, Korea; dongjun519@naver.com (D.-J.S.); yoonjm@jbnu.ac.kr (J.-M.Y.)

[2] Graduate School of Engineering, Tohoku University, Sendai 980-8579, Japan

[3] Department of Dental Biomaterials, Institute of Biodegradable Materials, BK21 plus Program, School of Dentistry, Chonbuk National University, Jeonju 561-756, Korea; yk0830@naver.com

* Correspondence: ilsong@jbnu.ac.kr; Tel.: +82-63-270-2294

† These authors contributed equally to this study.

Received: 26 October 2018; Accepted: 20 November 2018; Published: 26 November 2018

Abstract: The copper which an important component in the electronics industry, can suffer from discoloration and corrosion. The electrolytic copper foil was treated by 1,2,3-benzo-triazole (BTA) for an environmentally friendly non-chromate surface treatment. It was designed to prevent discoloration and improve corrosion resistance, consisted of BTA and inorganic sodium molybdate (Na_2MoO_4). Also the ratio of the constituent compounds and the deposition time were varied. Electrochemical corrosion of the Cu-BTA was evaluated using potentiodynamic polarization. Discoloration was analyzed after humidity and heat resistance conditioning. Surface characteristics were evaluated using scanning electron microscopy (SEM) and X-ray photoelectron spectroscopy (XPS). Increasing corrosion potential and decreasing current density were observed with increasing Na_2MoO_4 content. A denser protective coating formed as the deposition time increased. Although chromate treatment under severe humidity (80% humidity, 80 °C, 100 h) provided the highest humidity resistance, surface treatment with Na_2MoO_4 had better heat discoloration inhibition under severe heat-resistant conditions (180 °C, 10 min). When BTA reacts with Cu to form the Cu-BTA-type insoluble protective film, Na_2MoO_4 accelerates the film formation without being itself adsorbed onto the film. Therefore, the addition of Na_2MoO_4 increased anticorrosive efficiency through direct/indirect action.

Keywords: electrolytic copper foil; 1,2,3-benzotriazole; inorganic sodium molybdate (Na_2MoO_4); electrochemical corrosion; discoloration; insoluble protective film

---

1. Introduction

The copper is an important component in the electronics industry owing to its superior heat conductivity, electrical conductivity and workability. Copper foils with thicknesses of 5–100 µm are used for wiring the electronic circuits of printer and battery electrodes [1]. Recently, the materials used in electronic components are becoming thinner, simpler, more complex and highly functional with the rapid development of the electronic information industry. Since the rolled copper foil is susceptible to cyclic strain hardening and fatigue [2,3], thinner electrolytic copper foil is increasingly being used. Both copper foils are usually manufactured using a sulfuric acid-copper sulfate electrolytic solution. Electrolytic copper foil has the advantages of faster production speed, lower cost, constant strength and greater flexibility of the thin film [4]. However, electrolytic copper foil produced using plating baths is easily oxidized and suffers from surface discoloration in humid environments [5–7].

This discoloration and corrosion results in reduced adhesive strength and a failure of the electrical conductivity when applied to a PCB (Printed Circuit Board) substrate. Therefore, post-treatment (e.g., chromate immersion) is used during the production of electrolytic copper foil to provide resistance to discoloration and corrosion resistance [8,9] and to improve adhesion [10]. Chromate treatment contains hexavalent chromium and reducing trivalent chromium, both of which are designated as environmentally hazardous elements in the restriction of hazardous substances (EU RoHS). Their usage and surface-treatment content are subject to strict international restrictions and controls [11]. Therefore, research related to the development of new rust-preventive materials and processing technologies is needed.

In previous studies, materials for enhancing the rust-prevention effect of copper or copper alloys have been heterocyclic compounds, containing large numbers of N and S atoms, that form a physical barrier on the copper surface [12], or a-based compounds of benzotriazole containing large amounts of C, H and N atoms [13,14]. For example, benzotriazole (BTA) has been reported to have excellent antirust qualities when it forms an insoluble compound in the form of Cu-BTA on the copper surface. This physical protective film inhibits the diffusion of permeable ions such as P, S and Cl [13,15]. However, alone, it cannot match the tarnish-inhibitive properties of chromate and many studies have used a combination of organic and inorganic compounds with BTA to improve the rust-prevention effect of the surface treatment [16–19]. In particular, the addition of potassium-sorbate [18] or sodium-dodecylsulfate [19] has been reported to increase the adsorption rate of BTA onto the copper surface and to improve corrosion resistance by forming a thicker Cu-BTA film. Molybdate (MoO_4) is known to provide corrosion resistance when mixed with organic materials owing to its non-oxidative property [20–22]. It has been applied as a corrosion inhibitor for copper-coupled steel [23]. The addition of sodium molybdate (Na_2MoO_4) to low-concentration solutions is often used in combination with other inhibitors because it accelerates corrosion resistance due to oxidation [24]. Ramesh [22] reported the rust-inhibition effect for copper by mixing a triazole compound with molybdate but relatively few studies have considered the complex anti-rusting effect of BTA and molybdate [16,25].

In this study, copper foil surface treatment was performed by adding Na_2MoO_4 to BTA. The results showed that this pretreatment improved corrosion and discoloration resistance and an optimum pretreatment procedure was developed by varying the concentrations of additives and deposition time.

2. Materials and Methods

Untreated copper foil (Iljin Copper Foil Co., Ltd., Iksan, Korea), with a thickness of 12 μm, was cut into 30 mm × 50 mm pieces and subjected to surface treatment. Commercialized chromate copper (Iljin Copper Foil Co., Ltd., Korea) foil was used for comparison [26]. The copper foil was pickled in a 20 vol.% H_2SO_4 aqueous solution to remove the oxide film formed by contact with the atmosphere. The electrolyte was prepared as 5 and 50 mM BTA (1,2,3-benzotriazole, $C_6H_5N_3$) without Na_2MoO_4 and 1 and 5 mM of Na_2MoO_4 was added in 5 and 50 mM BTA (1,2,3-benzotriazole, $C_6H_5N_3$), respectively. The copper foils was immersed for 3 or 5 s in one of the two surface-treatment solutions. The concentrations of BTA and Na_2MoO_4 and the surface-treatment time conditions are summarized in Table 1.

To investigate the electrochemical properties, potentiodynamic polarization tests were performed with EG&G PAR (273A, Princeton Applied Research Corporation, Princeton, NY, USA). A test sample was used as a working electrode and Pt was connected to a counter electrode. Ag/AgCl (saturated KCl) was used as a reference electrode at 3 mV/s in 3.5 wt.% NaCl solution. Potential corrosion and current density were measured by Tafel extrapolation to evaluate the electrochemical corrosion characteristics. The test was performed 5 times and the mean and standard deviation were obtained.

Table 1. Experimental conditions.

Sample Code		BTA (mM)	Sodium-Molybdate (mM)	Surface Treatment Time (s)
Comparison Groups	Untreated	–	–	–
	Chromate	–	–	–
Modified Groups	B5	5	–	
	B5M1	5	1	
	B5M5	5	5	3/5
	B50	50	–	
	B50M1	50	1	
	B50M5	50	5	

Evaluation of discoloration resistance of the surface-treated specimens was carried out under severe high-temperature (heat resistance) and humidity (moisture resistance) environments. To evaluate discoloration at high temperatures, the samples were exposed to a constant temperature of 180 °C for 10 min. Discoloration in humid environments was evaluated by exposing the specimen to 80% humidity at 80 °C for 100 h. To prevent local corrosion, each sample was held at least 30 mm from the ground and the intersample distance was maintained at 50 mm to control interferences between the samples. The coloration of the electrolytic copper foil was analyzed using a colorimeter (Color i5, X-rite, Grand Rapids, MI, USA). A D65 light source was chosen for analyzed according to the guidelines of the Commission Internationale de I'Eclairage (CIE) and the values of color coordinates L^*, a^* and b^* were measured with a viewing angle of $2°$ using the SCE (Specular Component Excluded) method. The average mean value of L^*, a^* and b^* was obtained by measuring the center of the sample 10 times to calculate a ΔE^* value. The color coordinates were derived using the equation:

$$\Delta E^* = [(\Delta L^*)^2 + (\Delta a^*)^2 + (\Delta b^*)^2] \tag{1}$$

where $\Delta L^* = L_1^* - L_2^*$, $\Delta a^* = a_1^* - a_2^*$ and $\Delta b^* = b_1^* - b_2^*$. L_1^*, a_1^* and b_1^* represent values before harsh condition testing of the surface-treated sample and L_2^*, a_2^* and b_2^* represent values after harsh condition testing. Discoloration resistance data were analyzed for statistical significance using a one-way ANOVA-test which a p value of less than 0.05 was considered significant. After treatments and tarnish testing, surface morphology and color were observed using a stereomicroscope (DE/EZ4, Leica, Milton Keynes, UK). The thickness of the film was measured seven times and the average and standard deviation were obtained by excluding the minimum value and the maximum value.

The copper foils treated for 5 s were solidified in liquid nitrogen and then cut to observe a cross-section using FE-SEM (field emission scanning electron microscopy, SU-70, Hitachi, Tokyo, Japan).

Surface component analysis of Na_2MoO_4 concentrations was performed using X-ray photoelectron spectroscopy (XRP, Kα Model, Thermo VG Scientific, Waltham, MA, USA).

3. Results and Discussion

3.1. Characterization of Electro-Chemical Corrosion

Potentiodynamic polarization is an electrochemical method that is mainly used to measure the corrosion resistance of a material with a coating layer. The corrosion currents (I_{corr}) of the samples were determined at the intersection of the extrapolated cathodic and anodic Tafel lines using the linear polarization method. Figure 1 and Table 2 show the results of the potentiodynamic polarization of copper foil treated at 20 °C for 3 s under the addition of Na_2MoO_4. The BTA treatment groups showed higher corrosion potential and lower corrosion current than those of untreated and chromate groups. In terms of concentration polarization in hydrogen reduction, the reduction rate is dominated by the diffusion of hydrogen ions toward the metal surface [27]. The limiting diffusion current density is affected by the diffusion coefficient, the concentration of reactive ions in the solution and the thickness of the diffusion layer [28]. In this study, the diffusion coefficient and the concentration of reactive ions in solution are not different, so the change in cathodic current densities will be determined by the thickness of the diffusion layer. Thin and dense chromate coatings of several nanometers in thickness will have relatively low cathodic polarization currents [29]. The BTA coating is less dense than the chromate treatment but it has a thick and stable coating of 60 nm or more, increasing the corrosion potential and the fitting potential at the higher potential.

Table 2. Corrosion potential and current densities from potentiodynamic polarization.

Sample Code	E_v (V)	Standard Deviation	I_{corr} (A/cm^2)	Standard Deviation (10^{-6})
Untreated	−0.278	±0.012	1.152×10^{-5}	±1.984
Chromate	−0.227	±0.008	1.906×10^{-6}	±0.331
B5(3 s)	−0.152	±0.007	2.392×10^{-6}	±2.616
B5M1(3 s)	−0.115	±0.007	6.880×10^{-7}	±0.149
B5M5(3 s)	−0.096	±0.005	7.249×10^{-7}	±0.271
B50(3 s)	−0.158	±0.007	5.336×10^{-7}	±0.099
B50M1(3 s)	−0.120	±0.006	6.469×10^{-7}	±0.151
B50M5(3 s)	−0.104	±0.011	7.223×10^{-7}	±0.143
B5(5 s)	−0.155	±0.004	3.708×10^{-7}	±0.090
B5M1(5 s)	−0.098	±0.007	1.623×10^{-6}	±0.551
B5M5(5 s)	−0.081	±0.005	3.749×10^{-7}	±0.115
B50(5 s)	−0.101	±0.005	8.213×10^{-7}	±0.245
B50M1(5 s)	−0.074	±0.009	6.505×10^{-7}	±0.203
B50M5(5 s)	−0.070	±0.036	4.357×10^{-7}	±0.135

Figure 1b shows experiments with higher BTA (50 mM) after immersion for 3 s. It was observed no differences in the corrosion current density and corrosion potential as compared with those observed for 5 mM BTA (Figure 1a). Lower corrosion current density increased the surface-insoluble coating and corrosion resistance. Although increasing the amount of Na$_2$MoO$_4$ had a positive effect on corrosion resistance, the corrosion potentials and current densities of 5 and 50 mM BTA concentrations did not show significant differences. There is a threshold value of BTA ion concentration for the reaction to form an insoluble Cu-BTA film during the reaction of BTA and copper. It was observed that the copper removal rate gradually increased with higher BTA concentrations, when BTA addition was lower than 2 mM, a uniform and compact Cu-BTA passivating film could not be effectively built up [30]. Finsgar [31] considered BTA concentrations of 0.1–10 mM and found that rust-inhibition efficiency decreased with the addition of more than 10 mM BTA during co-electrospinning potential in 3% NaCl solution.

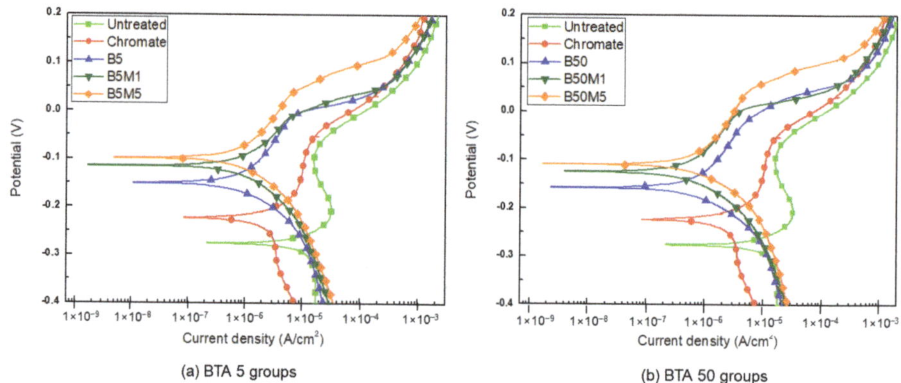

(a) BTA 5 groups   (b) BTA 50 groups

Figure 1. Potentiodynamic polarization curves of surface treatment immersed for 3 s in (**a**) 5 mM and (**b**) 50 mM 1,2,3-benzotriazole (BTA) combined with sodium molybdate (Na$_2$MoO$_4$).

Figures 1 and 2 and Table 2 show potentiodynamic polarization as a function of surface-treatment time. The corrosion resistances of 5 mM BTA concentration did not change with surface-treatment time

(3 s or 5 s; Figures 1a and 2a). However, the corrosion potentials increased by the addition of Na_2MoO_4. For a treatment time of 5 s, the 50 mM BTA groups had a higher potential voltage than did the 5 mM BTA groups. Therefore, the increase in the corrosion potential was proportional to the immersion time and to the addition of high amounts of BTA and Na_2MoO_4. However, the addition of more than 1 mM Na_2MoO_4 resulted in reduced potential voltage (Figure 2b). Increasing the amount of Na_2MoO_4 (i.e., over 5 mM) did not affect the rate of corrosion, because Na_2MoO_4 at 0.01 M concentration in the electrolyte represents a critical quantity for stable passivation [32]. The presence of molybdate results in the passivation of copper, which enhances the adsorption of BTA [25]. Moshier et al. [33] showed that both the Mo +4 and Mo +6 states is limited by solubility and the molybdate had been reduced during the formation of the passive film. However, a certain amount of molybdenum concentration does not increase the film. In other words, just 1 mM of molybdate anions in the solution is sufficient to quickly absorb the surface and form a thin MoO_2 film and that film growth is controlled by the diffusion of anions. In this study, the critical amount of Na_2MoO_4 of copper foil was identified to improve corrosion resistance. For Q235 steel, Zhou et al. [34] found that the optimum ratio of Na_2MoO_4 to BTA is 4:1.

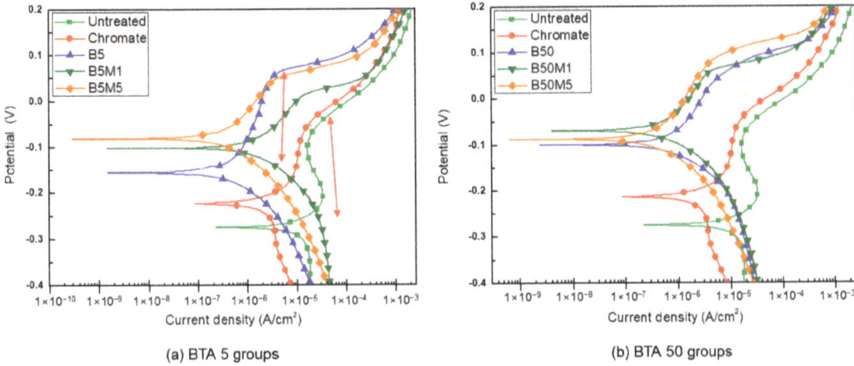

Figure 2. Potentiodynamic polarization curves of surface treatment immersed for 5 s in (**a**) 5 mM and (**b**) 50 mM 1,2,3-benzotriazole (BTA) combined with sodium molybdate (Na_2MoO_4)

A passive film was clearly observed after immersion for 5 s for all BTA-treated groups and $CuMoO_4$ and BTA-Cu synthesized from BTA enhanced the stability of the passive film. In the part of anodic current curve, the breakdown potential was shown with higher anodic current. Passivity means that a thin surface film is formed in the anode polarization to have corrosion resistance. If the cathodic reduction reaction current density is greater than the critical anodic current density, the passive state is stable. In other words, it is a section showing stable current density with increasing potential due to the protective film [35]. Since the thickness of the passive film obtained by anodic polarization was limited to a few nanometers [36], corrosion of the fitting occurred on the surface while the potential increases and the current density raised again as the passive film was destroyed.

After the copper foil was immersed in the surface-treatment solution, the reaction of the insoluble Cu-BTA film started on the surface and it reached an equilibrium reaction at a certain thickness. In other words, 5 s provided sufficient time for the formation of the Cu-BTA film and for the reaction to reach equilibrium. Longer treatment time (i.e., 5 s) provided increased corrosion potential owing to the thicker Cu-BTA film. In other words, setting optimum conditions for both Na_2MoO_4 content and treatment time effectively increased the corrosion potential and reduced the corrosion rate.

3.2. Anti-Discoloration Characteristic

Figure 3a shows the color difference ΔE^* measured after samples underwent the wet condition discoloration test. The chromate group exhibited excellent moisture resistance and has the lowest color difference ΔE^*. Color difference tended to decrease with increasing deposition time and the addition of Na_2MoO_4. An improvement in moisture resistance was not observed with increasing BTA addition. Although electrochemical corrosion resistance in the Na_2MoO_4 + BTA treated groups was higher than in the chromate group, humidity resistance was lower.

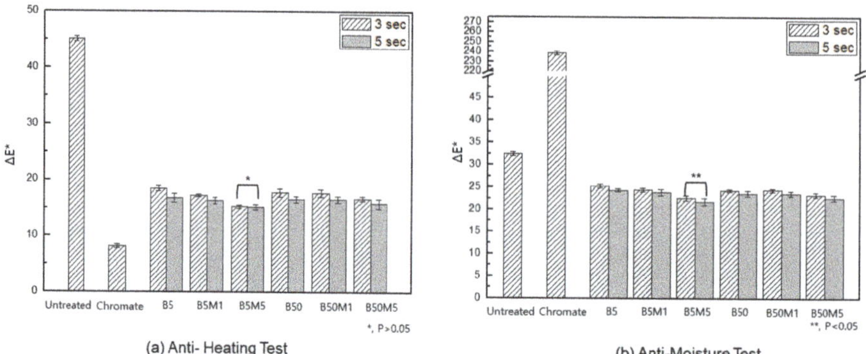

Figure 3. Color difference (ΔE^*) data from discoloration tests under (**a**) high-humidity (80% humidity at 80 °C for 100 h) and (**b**) high-temperature (180 °C for 10 min) conditions.

Chromate samples maintained the same pink surface color after 100 h of moisture resistance evaluation (Figure 4). Chromate with a yellow to pink color has a thickness of just a few microns and the total Cr in the film is just 1.3–1.6 mg·dm^{-2} for gloss [37]. The untreated group and the BTA surface-treatment groups formed local corrosion in the form of black CuCl and copper oxide (II) CuO spots on the copper surface [38]. These spots formed most frequently in the untreated sample and the amounts and sizes of the spots decreased with the addition of Na_2MoO_4.

Figure 3b shows the color difference ΔE^* measured after heat resistance discoloration tests under severe conditions. The chromate sample showed the highest value of discoloration (a ΔE^* value of more than 10 times that of the other groups and a color change from pink to green-blue in Figure 4), followed by the untreated group. The BTA modified groups showed the best heat resistance characteristics. Among the surface-treatment groups using BTA, heat discoloration resistance increased with increasing Na_2MoO_4. This result was similar to that of the electrochemical analysis, which showed relatively better corrosion potential when Na_2MoO_4 was added. During high-temperature testing, the Cr component of a copper surface reacts with O, N and other airborne elements. Owing to the formation of binary compounds (Cr_2O_3 and CrN) with an oxidation state of +3, the chemical bonding of Cr is altered, which may cause a change in the color of the surface [39]. Furthermore, when chromate treatment was carried out at 70 °C or higher, corrosion resistance reduced, and, its corrosion resistance was badly deteriorated at 80 °C. Therefore, when the temperature exceeds 70 °C, fine uniformity occurs in the chromate film causing corrosion and color change [40].

There was no difference in the BTA concentrations for the heat and moisture resistance tests and the results were proportional to the amount of Na_2MoO_4. In terms of moisture resistance, BTA treatment and Na_2MoO_4 addition were lower than those required for the chromate group but higher than those for the untreated group. In particular, thermal resistance was greater than that of the chromate group and the untreated group. Therefore, treatment of BTA with Na_2MoO_4 is effective for surface treatment that requires environmentally friendly heat resistance.

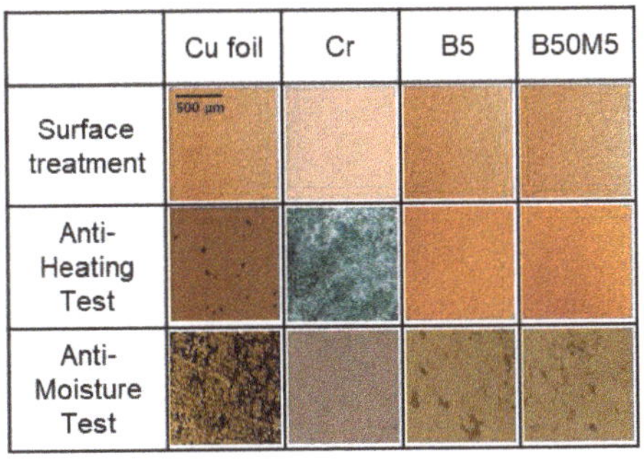

	Cu foil	Cr	B5	B50M5
Surface treatment				
Anti-Heating Test				
Anti-Moisture Test				

Figure 4. Optical images of Cu foil surfaces after surface treatment and harsh-environment (high humidity and high temperature) tests.

3.3. Surface Morphology Analysis

Morphology analysis based on cross-sections observed by FE-SEM (Figure 5) showed that films prepared by adding Na_2MoO_4 were more densely attached to the Cu substrate and a uniform considering the standard deviation of the thickness data. This was because Na_2MoO_4 accelerated the formation of Cu-BTA, even at relatively low concentrations (1 mM of Na_2MoO_4).

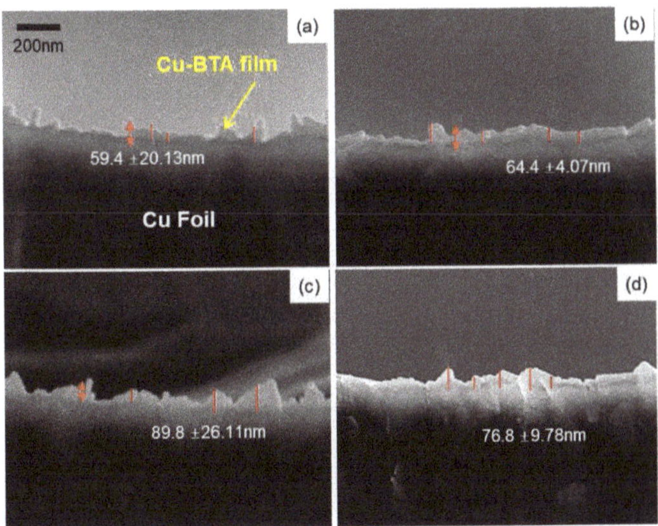

Figure 5. Field emission scanning electron microscopy (FE-SEM) images of sample cross sections: (**a**) B5, (**b**) B5M1, (**c**) B50 and (**d**) B50M1 immersed for 5 s.

In XPS analysis (Figure 6), all of the groups (irrespective of Na_2MoO_4 content) contained C 1s, N 1s, O 1s, Cu 3d, Cu 3p 1/2, Cu 3s and Cu 2p 1/2 peaks. The C, N, O elements were components of the BTA and the peaks related to Cu-BTA formation [25]. When 1 mM Na_2MoO_4 was added, Mo-related peaks were not detected in the surface components (binding energy = 228 eV: Mo 3d 5/2, 231 eV: Mo 3d 3/2, Figure 6(b1)), suggesting that Na_2MoO_4 was involved in the formation of the Cu-BTA film but was not itself part of the chemical bonding [25]. By the way, after treatment with 5 mM Na_2MoO_4, Mo was present in the film and directly involved in corrosion (binding energy = 228 eV: Mo 3d 5/2, 231 eV: Mo 3d 3/2).

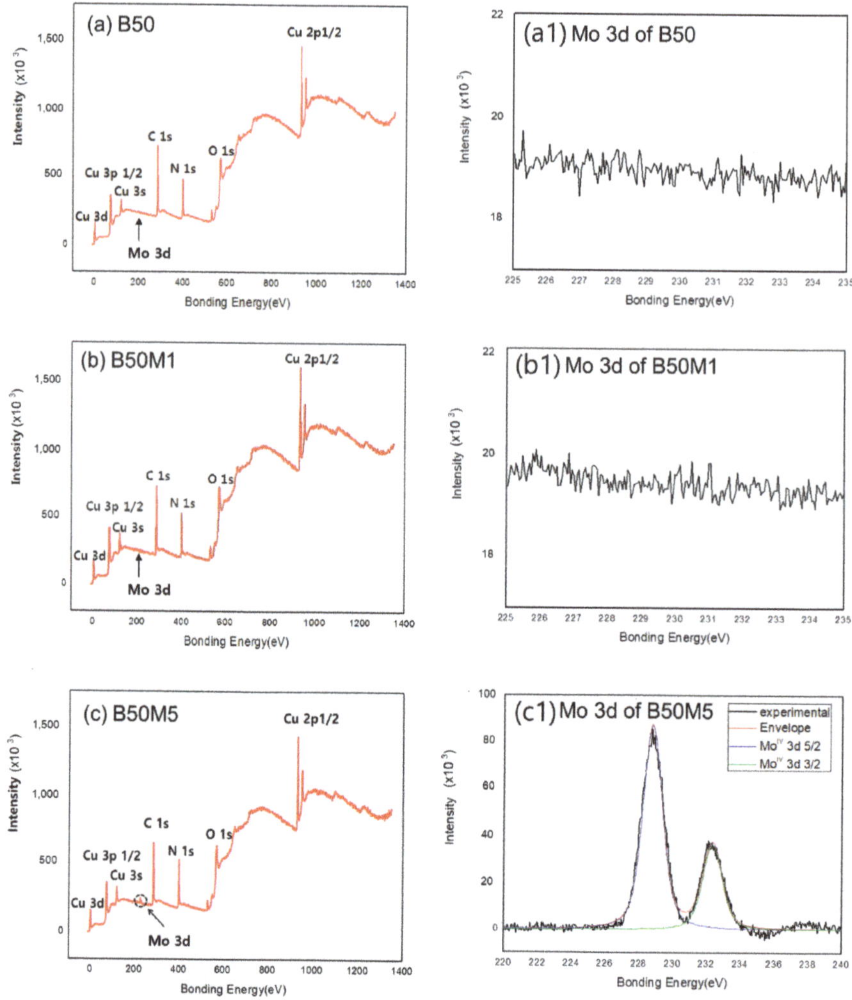

Figure 6. X-ray photoelectron spectroscopy (XPS) patterns of the (**a**) B50, (**b**) B50M1 and (**c**) B50M5 groups immersed for 5 s and XPS spectra of Mo 3*d* (same sample as survey) from each group.

After MoO_4 is reduced to $MoO(OH)_2$, it oxidizes and adheres to intermetallic compound particles [41]. The anion MoO_4^{2-} is preferentially adsorbed on the metal surface to inhibit the entry of chloride ions, thus effectively inhibiting corrosion. In Figure 6(c1), a new Mo 3*d* peaks were observed around the binding energy of 228 eV and 231 eV [42], which can be attributed to Mo in Na_2MoO_4 [43]. This would strengthen the inhibition of BTA on Cu. Na_2MoO_4 itself did not join the inhibition film but promoted the adsorption of BTA at the Cu_2O bottom layer, which is beneficial for the formation of Cu(I) BTA films.

4. Conclusions

This study investigated the influence of additive Na_2MoO_4 and surface-treatment time on the use of BTA as a surface treatment to improve the corrosion resistance and discoloration resistance of the electrolytic copper foil. The corrosion resistances of 5 mM BTA and 50 mM BTA groups were not effective following 3 s treatment, however the corrosion resistance also increased owing to the formation of a thicker Cu-BTA film as the surface-treatment time increased. In contrast, the addition of Na_2MoO_4 greatly improved the corrosion resistance and the ridging property even after a short treatment time. The increase in the concentration of Na_2MoO_4 was promoted by the adsorption of Cu ions onto the surface of the Cu layer, which is beneficial for the formation of Cu(I) BTA films. It was preferentially adsorbed on the Cu surface where it inhibited the entry of the chloride ions that cause corrosion. Therefore, the surface treatment using Na_2MoO_4 and BTA, which control the concentration and the treatment time, has better discoloration characteristics than the chromate treatment in high temperature (heat resistance) environment, so it is expected to be used in the surface treatment requiring environment-friendly heat resistance.

Author Contributions: Conceptualization, D.-J.S. and I.-S.P.; Methodology, Y.-K.K.; Formal Analysis, D.-J.S. and Y.-K.K.; Writing-Original Draft Preparation D.-J.S. and Y.-K.K.; Writing-Review & Editing, Y.-K.K. and J.-M.Y.; Visualization, I.-S.P.; Supervision, J.-M.Y.; Project Administration, J.-M.Y.; Funding Acquisition, I.-S.P.

Funding: This research was supported by "Research Base Construction Fund Support Program" funded by Chonbuk National University in 2018.

Acknowledgments: This study has reconstructed the data of Dong-Jun Shin's master thesis.

Conflicts of Interest: The authors declare no conflict of interest.

References

1. Weiss, B.; Gröger, V.; Khatibi, G.; Kotas, A.; Zimprich, P.; Stickler, R.; Zagar, B. Characterization of mechanical and thermal properties of thin Cu foils and wires. *Sens. Actuators A Phys.* **2002**, *99*, 172–182. [CrossRef]

2. Merchant, H.D.; Minor, M.G.; Liu, Y.L. Mechanical fatigue of thin copper foil. *J. Electron. Mater.* **1999**, *28*, 998–1007. [CrossRef]

3. De Angelis, R.; Knorr, D.; Merchant, H. Through-thickness characterization of copper electrodeposit. *J. Electron. Mater.* **1995**, *24*, 927. [CrossRef]

4. Merchant, H.D.; Liu, W.C.; Giannuzzi, L.A.; Morris, J.G. Grain structure of thin electrodeposited and rolled copper foils. *Mater. Charact.* **2004**, *53*, 335–360. [CrossRef]

5. FitzGerald, K.P.; Nairn, J.; Skennerton, G.; Atrens, A. Atmospheric corrosion of copper and the colour, structure and composition of natural patinas on copper. *Corros. Sci.* **2006**, *48*, 2480–2509. [CrossRef]

6. Rickett, B.I.; Payer, J.H. Composition of copper tarnish products formed in moist air with trace levels of pollutant gas: Sulfur dioxide and sulfur dioxide/nitrogen dioxide. *J. Electrochem. Soc.* **1995**, *142*, 3713–3722. [CrossRef]

7. Watanabe, M.; Tomita, M.; Ichino, T. Analysis of tarnish films on copper exposed in hot spring area. *J. Electrochem. Soc.* **2002**, *149*, B97–B102. [CrossRef]

8. Kajiwara, T.; Tanii, Y.; Hashimoto, K. Method of Making Thin Copper Foil for Printed Wiring Board. U.S. Patent 5,114,543 A, 19 May 1992.

9. Yamanishi, K.; Oshima, H.; Sakaguchi, K. Copper Foil for Printed Circuits and Process for Producing the Same. U.S. Patent 5,366,814 A, 22 November 1994.

10. Morisaki, S.; Mase, K. Copper Foil Having Bond Strength. U.S. Patent 4,010,005 A, 1 March 1977.

11. Hua, L.; Chan, Y.C.; Wu, Y.P.; Wu, B.Y. The determination of hexavalent chromium (Cr^{6+}) in electronic and electrical components and products to comply with RoHS regulations. *J. Hazard. Mater.* **2009**, *163*, 1360–1368. [CrossRef]

12. Zucchi, F.; Trabanelli, G.; Monticelli, C. The inhibition of copper corrosion in 0.1 M NaCl under heat exchange conditions. *Corros. Sci.* **1996**, *38*, 147–154. [CrossRef]

13. Finšgar, M.; Milošev, I. Inhibition of copper corrosion by 1,2,3-benzotriazole: A review. *Corros. Sci.* **2010**, *52*, 2737–2749. [CrossRef]

14. Ishida, H.; Johnson, R. The inhibition of copper corrosion by azole compounds in humid environments. *Corros. Sci.* **1986**, *26*, 657–667. [CrossRef]

15. Xu, Z.; Lau, S.; Bohn, P.W. The role of benzotriazole in corrosion inhibition: Formation of an oriented monolayer on Cu_2O. *Surf. Sci.* **1993**, *296*, 57–66. [CrossRef]

16. Shams El Din, A.M.; Mohammed, R.A.; Haggag, H.H. Corrosion inhibition by molybdate/polymaliate mixtures. *Desalination* **1997**, *114*, 85–95. [CrossRef]

17. Ramesh, S.; Rajeswari, S.; Maruthamuthu, S. Corrosion inhibition of copper by new triazole phosphonate derivatives. *Appl. Surf. Sci.* **2004**, *229*, 214–225. [CrossRef]

18. Gelman, D.; Starosvetsky, D.; Ein-Eli, Y. Copper corrosion mitigation by binary inhibitor compositions of potassium sorbate and benzotriazole. *Corros. Sci.* **2014**, *82*, 271–279. [CrossRef]

19. Villamil, R.; Corio, P.; Rubim, J.; Agostinho, S. Sodium dodecylsulfate–benzotriazole synergistic effect as an inhibitor of processes on copper | chloridric acid interfaces. *J. Electroanal. Chem.* **2002**, *535*, 75–83. [CrossRef]

20. Saremi, M.; Dehghanian, C.; Sabet, M.M. The effect of molybdate concentration and hydrodynamic effect on mild steel corrosion inhibition in simulated cooling water. *Corros. Sci.* **2006**, *48*, 1404–1412. [CrossRef]

21. Vukasovich, M.S.; Farr, J.P.G. Molybdate in corrosion inhibition—A review. *Polyhedron* **1986**, *5*, 551–559. [CrossRef]

22. Ramesh, S.; Rajeswari, S. Evaluation of inhibitors and biocide on the corrosion control of copper in neutral aqueous environment. *Corros. Sci.* **2005**, *47*, 151–169. [CrossRef]

23. Mustafa, C.; Shahinoor Islam Dulal, S. Molybdate and nitrite as corrosion inhibitors for copper-coupled steel in simulated cooling water. *Corrosion* **1996**, *52*, 16–22. [CrossRef]

24. Deyab, M.A.; Abd El-Rehim, S.S. Inhibitory effect of tungstate, molybdate and nitrite ions on the carbon steel pitting corrosion in alkaline formation water containing Cl^- ion. *Electrochim. Acta* **2007**, *53*, 1754–1760. [CrossRef]

25. Zhang, D.-Q.; Joo, H.G.; Lee, K.Y. Investigation of molybdate–benzotriazole surface treatment against copper tarnishing. *Surf. Interface Anal.* **2009**, *41*, 164–169. [CrossRef]

26. Yang, J.-S.; Lim, S.-L.; Kim, S.-B.; Kim, K.-J. Electrolyte Solution for Manufacturing Electrolytic Copper Foil and Electrolytic Copper Foil Manufacturing Method Using the Same. U.S. Patent 20040104117A1, 3 June 2004.

27. Stansbury, E.E.; Buchanan, R.A. Kinetics of Single Half-Cell Reactions. In *Fundamentals of Electrochemical Corrosion*; Stansbury, E.E., Buchanan, R.A., Eds.; ASM International: Novelty, OH, USA, 2000.

28. Bard, A.J.; Faulkner, L.R.; Leddy, J.; Zoski, C.G. *Electrochemical methods: Fundamentals and Applications*; Wiley: Hoboken, NJ, USA, 1980.

29. Domínguez-Crespo, M.A.; Onofre-Bustamante, E.; Torres-Huerta, A.M.; Rodríguez-Gómez, F.J. Morphology and corrosion performance of chromate conversion coatings on different substrates. *J. Mex. Chem. Soc.* **2008**, *52*, 241–248.

30. Jiang, L.; He, Y.; Niu, X.; Li, Y.; Luo, J. Synergetic effect of benzotriazole and non-ionic surfactant on copper chemical mechanical polishing in KIO_4-based slurries. *Thin Solid Films* **2014**, *558*, 272–278. [CrossRef]

31. Finšgar, M.; Lesar, A.; Kokalj, A.; Milošev, I. A comparative electrochemical and quantum chemical calculation study of BTAH and BTAOH as copper corrosion inhibitors in near neutral chloride solution. *Electrochim. Acta* **2008**, *53*, 8287–8297. [CrossRef]

32. Breslin, C.; Treacy, G.; Carroll, W. Studies on the passivation of aluminium in chromate and molybdate solutions. *Corros. Sci.* **1994**, *36*, 1143–1154. [CrossRef]

33. Moshier, W.; Davis, G. Interaction of molybdate anions with the passive film on aluminum. *Corrosion* **1990**, *46*, 43–50. [CrossRef]

34. Zhou, Y.; Zuo, Y.; Lin, B. The compounded inhibition of sodium molybdate and benzotriazole on pitting corrosion of Q235 steel in NaCl + $NaHCO_3$ solution. *Mater. Chem. Phys.* **2017**, *192*, 86–93. [CrossRef]

35. Jones, D.A. *Principles and Prevention of Corrosion*, 2nd ed.; Prentice Hall: Upper Saddle River, NY, USA, 1996; pp. 168–198.

36. Chandrasekaran, M. Metals for Biomedical Devices. In *Forging of Metals and Alloys for Biomedical Applications*; Niinomi, M., Ed.; Woodhead Publishing: Cambridge, UK, 2010; pp. 235–250.

37. Kim, S.W.; Lee, C.T. Environment-friendly trivalent chromate treatment for Zn electroplating. *J. Korean Ind. Eng. Chem.* **2006**, *17*, 433.

38. Youda, R.; Nishihara, H.; Aramaki, K. SERS and impedance study of the equilibrium between complex formation and adsorption of benzotriazole and 4-hydroxybenzotriazole on a copper electrode in sulphate solutions. *Electrochim. Acta* **1990**, *35*, 1011–1017. [CrossRef]

39. Dorward, R.C.; Hasse, K.R. Incubation effects in precracked stress corrosion specimens from Al-Zn-Mg-Cu alloy 7075. *Corros. Sci.* **1979**, *19*, 131–140. [CrossRef]

40. Laget, V.; Jeffcoate, C.; Isaacs, H.; Buchheit, R. Dehydration-induced loss of corrosion protection properties in chromate conversion coatings on aluminum alloy 2024-T3. *J. Electrochem. Soc.* **2003**, *150*, B425–B432. [CrossRef]
41. Lopez-Garrity, O.; Frankel, G. Corrosion inhibition of aluminum alloy 2024-T3 by sodium molybdate. *J. Electrochem. Soc.* **2014**, *161*, C95–C106. [CrossRef]
42. Zhang, B.; Wu, J.; Peng, D.; Li, X.; Huang, Y. In-Situ Scanning Micro-Electrochemical Characterization of Corrosion Inhibitors on Copper. Available online: https://dr.ntu.edu.sg/handle/10220/40846 (accessed on 26 October 2018).
43. Mert, B.D.; Mert, M.E.; Kardaş, G.; Yazıcı, B. Experimental and theoretical investigation of 3-amino-1,2, 4-triazole-5-thiol as a corrosion inhibitor for carbon steel in HCl medium. *Corros. Sci.* **2011**, *53*, 4265–4272. [CrossRef]

MDPI

St. Alban-Anlage 66

4052 Basel

Switzerland

Tel. +41 61 683 77 34

Fax +41 61 302 89 18

www.mdpi.com

Coatings Editorial Office

E-mail: coatings@mdpi.com

www.mdpi.com/journal/coatings